Sinkholes and Subsidence

Karst and Cavernous Rocks in Engineering and Construction

Tony Waltham
Fred Bell and
Martin Culshaw

Sinkholes and Subsidence

Karst and Cavernous Rocks in Engineering and Construction

Springer

Published in association with
Praxis Publishing
Chichester, UK

Dr Tony Waltham
Lecturer in engineering geology
Civil Engineering Division
Nottingham Trent University
Nottingham
UK

Professor Fred. G. Bell
Visiting Research Associate
British Geological Survey
Keyworth
Nottinghamshire
UK

Martin G. Culshaw
Manager of the Urban Geoscience and
Geological Hazards Programme
British Geological Survey
Keyworth
Nottinghamshire
UK

SPRINGER–PRAXIS BOOKS IN GEOPHYSICAL SCIENCES
SUBJECT *ADVISORY EDITOR*: Dr Philippe Blondel, C.Geol., F.G.S., Ph.D., M.Sc., Senior Scientist, Department of Physics, University of Bath, Bath, UK

ISBN 978-3-540-20725-2 Springer-Verlag Berlin Heidelberg New York

Springer is part of Springer-Science + Business Media (springeronline.com)

Bibliographic information published by Die Deutsche Bibliothek

Die Deutsche Bibliothek lists this publication in the Deutsche Nationalbibliografie; detailed bibliographic data are available from the Internet at http://dnb.ddb.de

Library of Congress Control Number: 2004112901

Cover design: Jim Wilkie
Project Management: Originator Publishing Services, Gt Yarmouth, Norfolk, UK

Printed on acid-free paper

Contents

Preface

It is the authors' and publisher's intention that this book provides an accessible information source for civil engineers who encounter one of the most serious types of difficult ground conditions – on cavernous karst. Essentially it is written by geologists for engineers. It aims to explain and improve the understanding of ground cavities, subsurface processes, sinkhole collapses and ground subsidence that together constitute a significant geohazard on terrains of limestone and certain other rocks. It is the authors' belief that, once the processes of karst terrains are fully appreciated, the modern generation of construction engineers will be well able to design structures and buildings that will stand safely on this difficult ground. Sadly there is a welter of misunderstanding and misconception about "holes in the ground" that needs to be rectified, and hopefully this book does just that.

The main chapters first review ground conditions and processes, and then move on to the thorny problems of assessment, before reviewing appropriate engineering practices. They were written with extensive cooperation between the three authors. The case studies are all by invited specialists who have contributed material from their own experiences on karst. Each contributor is credited in his own case study heading, and the authors are delighted to be able to thank all of them for their welcome efforts that have created a most valuable section of the book.

Illustrations in the text are all credited to their appropriate sources, except that the photographs from Tony Waltham's Geophotos picture library are credited as TW for purposes of repetitive brevity. The authors are grateful to the colleagues and companies who have permitted their own photographs to be used in these pages. Copyright is retained by all the photographers, except Martin Culshaw and Anthony Cooper whose images are copyrighted by the British Geological Survey (NERC). The book is published with the permission of the Executive Director of the British Geological Survey (NERC).

The authors are grateful to their many friends who contributed to the preparation of this book, especially Lu Zhengxin (Nottingham University) who investigated cave roof stability through his numerical modelling, Cath Poulton (British Geological Survey) for assistance in literature search and Hilary Lambert for providing the small, central cover photograph. We also thank Trevor Ford (Leicester University), Ian Jefferson (Birmingham University), Dave Reddish (Nottingham University) and Dave McCann (British Geological Survey) for their valuable comments during peer review of various chapters. Finally, we thank Clive Horwood at Praxis Publishing and Neil Shuttlewood at Originator for seeing the book successfully into print.

Tony Waltham, Fred Bell and Martin Culshaw
Nottingham, 2004

Postscript

Just as this text had been completed, its significance was demonstrated by an event in Tampa, Florida. In April 2004, an elevated section of an extension to the Lee Roy Selmon Expressway collapsed while under construction. A single massive column, that supported the entire 3-lane width of the road between spans of 30 m, dropped abruptly by 5 m. The column extended down into a caisson 1.8 m in diameter, which was founded in apparently strong limestone 19 m below the surface; rockhead was 11 m deep. The column carried a dead load of 11 MN, increased at the time of failure by 3 MN from a massive temporary truss used in the construction process. The caisson appears to have punched through the roof of a cave just below its base. Prior to construction, each column had been proven by a borehole that reached just 3 m below the caisson base. This was simply inadequate for such heavily loaded columns in limestone that is well known for its mature karst features. Repairs are now budgeted for $11M, whereas prior probing to 5 m or even to 7 m beneath every caisson toe would have added only a few thousand dollars to the total project cost. This failure was avoidable.

Press comments that "A sinkhole as deep as this is undetectable", "It was just a bizarre event", "A small problem with the soil, something 1–6 m across, is easy to miss; if you try to find every one, you could not afford the project", and "It was an act of God" showed a complete lack of understanding of the karst. The authors hope that this book will improve understanding in the future.

Contributors

Fred G. Bell, Consultant geologist, Nottinghamshire, U.K.
fred.g.bell@btinternet.com

Martin G. Culshaw, British Geological Survey, U.K.
mgc@bgs.ac.uk

Tony (A.C.) Waltham, Nottingham Trent University, U.K.
tony@geophotos.co.uk

Roger Brucker, Consultant geomorphologist, Ohio, U.S.A.
brucker@graphtronics.net

David Buttrick, VGI Consult, South Africa
intrac@mweb.co.za

Fred Calitz, Africa Geo-Environmental Services, South Africa
fcalitz@ages-group.com

C. Warren Campbell, Western Kentucky University, Alabama, U.S.A.
warren.campbell@wku.edu

Anthony H. Cooper, British Geological Survey, U.K.
ahc@bgs.ac.uk

John Deceuster, Faculté Polytechnique de Mons, Belgium
john.deceuster@fpms.ac.be

Percy H. Dougherty, Kutztown University, Pennsylvania, U.S.A.
dougherty@kutztown.edu

Clive Edmonds, Peter Brett Associates, U.K.
cedmonds@pba.co.uk

Kenneth S. Johnson, Oklahoma Geological Survey, U.S.A.
ksjohnson@ou.edu

Patricia Kambesis, Western Kentucky University, U.S.A.
pnkambesis@juno.com

Olivier Kaufmann, Faculté Polytechnique de Mons, Belgium
olivier.kaufamann@fpms.ac.be

Martin Knez, Karst Research Institute, Slovenia
knez@zrc-sazu.si

Mingtang Lei, Institute of Karst Geology, China
sinkhole@mail.karst.ac.cn

Junlin Liang, Institute of Karst Geology, China
sinkhole@mail.karst.ac.cn

Peter McDowell, Portsmouth University, U.K.
pmcdowell@waitrose.com

James T. Neal, Sharlot Hall Museum, Arizona, U.S.A.
hjneal@commspeed.net

Hyeong-Dong Park, Seoul National University, Korea
hpark@snu.ac.kr

Bernardas Paukštys, Public Establishment Water House, Lithuania
bernardas@iti.lt

Tadej Slabe, Karst Research Institute, Slovenia
slabe@zrc-sazu.si

Peter Styles, Keele University, U.K.
p.styles@esci.keele.ac.uk

Figures

Cover photographs

Front main	Polygonal karst on the Sinkhole Plain, Kentucky
Front central	Collapse sinkhole at Dishman Lane, Bowling Green, Kentucky
Front right	Re-activated collapse sinkhole over gypsum at Ripon, U.K.
Back left	Large collapse sinkhole containing lake (Modro Jezero), Croatia
Back right	New subsidence sinkhole in Sivas gypsum karst, Turkey

Tables

Boxes

Glossary of sinkhole terminology

Blue hole	Deep water-filled sinkhole in a coastal karst or the floor of a shallow sea.
Breakdown	Blocks of rock fallen from the walls or roof of a cave.
Breccia pipe	Column of breakdown debris above a collapsed cave chamber.
Buried sinkhole or **doline**	Sinkhole filled with loose sediment beneath a soil cover.
Caprock sinkhole or **doline**	Sinkhole in insoluble rock formed by collapse into underlying cavernous rock.
Cave	Natural hole in the ground, large enough for human entry.
Cavern	Cave or cave chamber usually of large dimensions.
Cenote	Steep-sided collapse sinkhole floored by a lake whose surface is at the regional water table.
Closed depression	Karst hollow with internal drainage, including sinkhole, doline, uvala, polje and cockpit.
Cockpit	Large stellate doline between the conical hills of cone karst.
Collapse chamber	Cave chamber modified by wall and roof collapse.
Collapse sinkhole or **doline**	Sinkhole formed by collapse of rock into a cave passage or chamber.

Cutter	Soil-filled fissure in rockhead of a soluble rock (*used in U.S.A.*)
Daya or **daia**	Wide and shallow, flat-floored depression in a desert limestone plateau.
Dissolution sinkhole or **doline**	Same as solution sinkhole or doline.
Doline	Closed depression in karst, often known as a sinkhole.
Doline karst	Karst terrain where the dominant landforms are solution dolines.
Dropout sinkhole or **doline**	Subsidence sinkhole that forms rapidly in a soil cover.
Filled sinkhole	Same as buried sinkhole.
Fissure	Cavity opened by dissolution along a rock discontinuity, but smaller than a cave passage.
Grike or **gryke**	Dissolution fissure within the bare rock of a limestone pavement (*used in Great Britain*).
Karren	Small dissolutional runnels etched into bare limestone surfaces.
Karst	Landscape created on soluble rock with efficient underground drainage.
Pinnacled rockhead	Extremely irregular rockhead with soil-filled fissures and buried sinkholes between remnant rock pinnacles.
Pipe	Cylindrical or conical mass of clay and sand that fills a solution sinkhole, shaft or cave.
Polje	Closed depression with wide alluviated floor, much larger than a doline.
Polygonal karst	Terrain composed entirely of internally drained dolines or sinkholes between a polygonal net of low ridges.
Ponor	Sink, normally into an open cave passage in the floor of a polje (*mainly used in eastern Europe*).
Pothole	Solutional shaft or mainly vertical cave system.
Pseudokarst	Terrain with caves and/or karst landforms not formed by dissolution of rock.

Ravelling	Breakdown and disassociation of soil that falls away from the roof and walls of a ground cavity.
Regolith	Soil cover that overlies bedrock (*mainly used in U.S.A.*).
Rockhead	Buried interface between the underlying bedrock and its soil cover.
Shaft	Vertical or steeply inclined cave passage.
Shakehole	Small suffosion sinkhole in till overlying limestone (*mainly used in England's Pennines*).
Sink	Point where a stream or river disappears underground.
Sinkhole	Small closed depression in karst, also known as a doline.
Solution sinkhole or **doline**	Sinkhole formed by dissolutional lowering of the rock surface in and around zones of drainage into a cavernous rock.
Stoping	Progressive collapse of roof rock that causes a cavern to migrate upwards.
Subsidence sinkhole or **doline**	Sinkhole created where soil is washed down into underlying cavernous rock.
Suffosion	Removal of fines by down-washing through unconsolidated sediment.
Suffosion sinkhole or **doline**	Subsidence sinkhole that develops slowly in a soil cover.
Swallet	Same as sink (*mainly used in southern England*).
Swallow hole	Same as sink.
Tiankeng	Extremely large collapse sinkhole.
Tumour sinkhole	Collapse sinkhole formed by undermining, where no large chamber ever existed.
Turlough	Karst depression that is seasonally flooded, larger than a sinkhole.
Uvala	Closed depression with multiple sink points (*now little used*).

1

Rocks, dissolution and karst

Karst refers to a distinctive terrain that evolves through dissolution of the bedrock and development of efficient underground drainage. It is therefore associated primarily with limestone, but also forms on other carbonates and other soluble rocks. The special landforms of karst include sinkholes, dry valleys, pavements, cave systems and associated springs. Karst terrain possesses not only topographic features peculiar to itself but also unique hydrogeological characteristics. The landforms of karst vary enormously in character, shape and size, and combine to create a terrain that may represent extremely difficult ground conditions for construction and engineering. Collapse of rock over caves formed by dissolution is fundamental to the evolution of karst terrains, but is the least important of karst hazards in civil engineering (Chapter 3).

Karstic rock masses may be overlain by allogenic sediments or by residual soils that represent the insoluble material left after dissolution. Residual soils are mainly sandy silty clays, and their wide range of plasticity reflects their clay content; chert fragments are a common component. Dissolutional weathering imparts a loose packing to residual soil, which may or may not be consolidated or indurated subsequently. In addition, the rockhead beneath the soil cover is typically abrupt rather than gradational, and may be highly irregular with pinnacles of rock protruding upwards into the soil. Although these soils may possess their own engineering problems, the most widespread hazard in karst terrains occurs where soil is washed into underlying bedrock openings so that a void migrates upwards by progressive collapse to form a subsidence or dropout sinkhole (Figure 1.1). Though these sinkholes form entirely within the soil cover, they are a major component of most karst terrains (Chapter 4). They also constitute the greatest hazard for construction engineers, especially as their development is so frequently induced by engineering activity (Chapter 8).

Cavities, collapse and ground subsidence may also develop as pseudokarst in insoluble rocks. Tubes and caves in some basaltic lava flows, along with

Figure 1.1. A new subsidence sinkhole in the garden of a house in Kentucky.
Photo: Art Palmer.

piping failures in clastic soils, are the most widespread forms of pseudokarst (Chapter 6).

Limestone is one of the world's most widespread sedimentary rocks, and karst is developed to some degree in almost every country of the world. Areas of limestone outcrop have potential for the development of karst and therefore the development of sinkholes (Figure 1.2). Available data varies greatly in quality, so the figured map varies in its detail, with high quality in Europe, Iran, U.S.A. and some other countries, but very generalised in Russia and Saudi Arabia for example. Furthermore, large areas marked on this map, including much of northern Canada, Siberia, Kazakhstan, Egypt and northern Australia have few sinkholes, because the karst is either buried or is poorly developed in desert or arctic environments. The potential impact on engineering is represented by the fact that in the U.S.A. 40% of land east of the Mississippi is underlain by some form of karstic rock – where hazardous bedrock cavities and surface sinkholes can therefore occur. The other most notable regions where large areas of karst lie beneath densely populated land are in southern China and the countries that once constituted Yugoslavia.

1.1 LIMESTONE LITHOLOGIES

Carbonate rocks are defined as those containing more than 50% by weight of carbonate minerals, although the proportion commonly exceeds 90%. The two

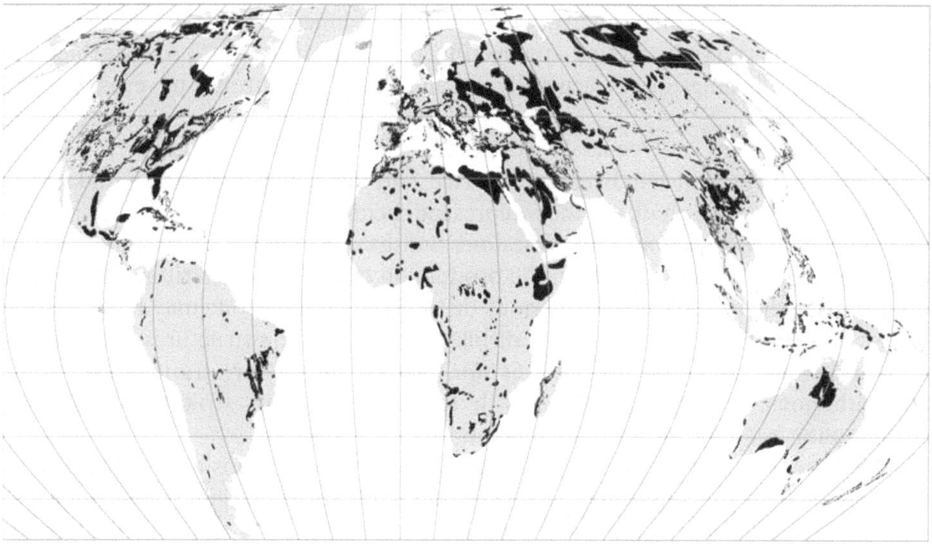

Figure 1.2. The worldwide distribution of karst, on an equal-area projection. The black areas are the main areas of limestone outcrop with potential development of karst with sinkholes. Major areas of pseudokarst on basalt and quartzite are also marked in dark grey, as in Iceland and Venezuela respectively, but small areas are barely distinguishable at this scale.
From Ford and Williams (2005).

major carbonate minerals are calcite ($CaCO_3$) and dolomite ($MgCa(CO_3)_2$), while aragonite (also $CaCO_3$), siderite ($FeCO_3$) and magnesite ($MgCO_3$) are rare in sedimentary rocks. Any non-carbonate fraction in these rocks is generally any of the clay minerals or cryptocrystalline silica (as chert or flint), which is left to form the residual soils during dissolutional weathering. The rocks known as dolomites contain high proportions of dolomite mineral, but are very similar to limestones with respect to their karst and their sinkhole hazards (Section 1.5).

The mechanical properties of old, well-lithified limestones cover a range of unconfined compressive strengths (UCS) of 30–100 MPa for the intact rock (Waltham and Fookes, 2003). Most caves, sinkholes and karst form in the stronger rocks with UCS > 60 MPa, unit weight 2.6 kN/m^3 and primary porosity $< 2\%$. Groundwater flow and dissolutional development are focused on fractures to create discrete conduits. These properties are largely dictated by tectonic history, but age of the limestone is irrelevant; major karst terrains lie on limestones that are Proterozoic in Brazil, Ordovician in Pennsylvania, Carboniferous in Great Britain, Permian in China, Mesozoic in Europe and Tertiary in Malaysia. Limestones of moderate strength, with UCS around 30 MPa, unit weight around 2.3 kN/m^3 and primary porosity $> 10\%$, have a larger proportion of diffuse groundwater flow, so their fissures are not so rapidly enlarged by dissolution. These include the Jurassic limestones of England's Cotswold Hills, where the extensive karst is significantly less cavernous and with fewer sinkholes than in the neighbouring Carboniferous limestone. Chalk and some other weaker varieties of limestone also support sinkhole karst, but have very distinctive engineering properties (Section 1.6). Marbles are metamorphosed limestones, generally strong and also prone to erosion into cavernous karst.

Limestones are polygenetic. Some are of mechanical origin representing carbonate detritus that has been transported and deposited or has accumulated *in situ*; these include the chalks. Others represent chemical or biochemical precipitates, or organic deposits such as coral limestone that have formed in place. Biological or biochemical processes dominate in the production of carbonate detritus, that mostly originates as shell debris. Allochthonous or transported limestone may have a fabric similar to that of sandstone and may display current bedding structures. Autochthonous limestones, that have formed *in situ*, possess only poor stratification. Some autochthonous limestones show growth bedding, the most striking of which is stromatolitic bedding, as in some algal and reef limestones.

Most limestones were formed as shallow-water marine deposits in environments that include tidal and supratidal flats, shelf and bank areas, marginal reefs and back-reef lagoons. They largely consist of varying proportions of complex derived grains (allochems), coarsely crystalline calcite (sparite) that may constitute a cement binder and microcrystalline calcium carbonate (micrite) that commonly forms the matrix (Folk, 1962). They may be described as calcilutites (carbonate mudstones) or calcarenites (of sand grain size), or by other terms based on their texture (Dunham, 1962). Ooliths are roughly spherical grains, 0.2–2.0 mm in diameter with concentric or radial structure, that formed by accretion on lagoon floors and may constitute the majority of the rock in oolitic limestones (also known as oolites). Most other carbonates are minor as rock units. Pelagic oozes and turbidites of deep-water marine basins leave little geological record. Those from evaporitic basins can be more extensive, but are mostly dolomitic (Section 1.5). Caliche is a carbonate duricrust that may be widespread in arid regions. Tufa is a soft porous carbonate precipitated by algal and bacterial action in springs and streams in limestone terrains. Travertine is a dense, banded deposit, similar to tufa in that it is deposited in flowing streams, but generally in response to changes in water chemistry or downstream of hydrothermal sources (the terms travertine, tufa and sinter are almost synonymous in different parts of the world). Stalagmite is the variety of travertine deposited in caves. Wind-blown carbonate sand may form on beaches and as dunes on coral islands, and is commonly capped by a partially cemented duricrust.

The mechanical behaviour of all carbonate sediments is influenced by grain size and those post-depositional changes that bring about induration, and thereby increase density and strength. Induration of limestones commonly starts during the early stages of deposition, by cementation that occurs where individual grains are in contact. Consequently, cementation is not solely dependent on consolidation due to increasing overburden pressure. Because induration can take place at the same time as deposition, carbonate sediments can sustain high overburden pressures, and can therefore retain high porosities at considerable depths. A bed of cemented grains may overlie one that is poorly cemented. Eventually, high overburden pressures, creep and recrystallization produce crystalline limestone with very low porosity.

1.2 LIMESTONE AND DISSOLUTION

The rate of dissolution of a rock generally depends on the solubility and specific solution rate constant of the constituent mineral, the degree of saturation of the solvent, the area presented to the solvent and the motion of the solvent (which may keep it undersaturated). The solubility of limestone in pure water is extremely low, and is comparable to that of silica. The key factor is therefore carbon dioxide, as aqueous solutions of this gas with limestone produce the bicarbonate radical, which is greatly soluble. The reaction may be simplified to the well-known equation:

$$CaCO_3 + CO_2 + H_2O = Ca^{2+} + 2HCO_3^-$$

This is however a gross over-simplification of the very complex dissolution processes and kinetics, which are described at length in reviews of various depth and complexity (Dreybrodt, 2000; White, 1988; Ford and Williams, 1989; Dreybrodt and Eisenlohr, 2000).

Direct rainfall has only a small carbon dioxide content, in equilibrium with about 0.03% CO_2 in the open atmosphere. Most carbon dioxide is biogenic, derived from organic activity within soils, and natural waters in equilibrium with the soil atmosphere (which contains 1–10% CO_2) have the highest chemical aggressivity towards carbonate rocks. Major concentrations of carbon dioxide are released when organic material decays, and are taken into solution by rainwater that percolates through the soil. This infiltration drainage is the prime source of chemically aggressive groundwater in limestone karst. Its aggressiveness may be enhanced by humic acids that are formed by the decay of humus in the soil; these can create groundwater with pH values of 4.5–5.0 or in some extreme cases even less than 4.0. Nitric and nitrous acids are also formed by organic decay and bacterial action in soils, but play only a minor part in karst dissolution.

Dissolutional aggressiveness of water to calcium carbonate relates to the content of dissolved carbon dioxide, which is a direct function of the partial pressures of the gas in its equilibrium atmosphere; hence the importance of soil-derived waters. It decreases slightly with increasing temperature (Table 1.1), but this is generally irrelevant, because natural waters are rarely saturated with the gas. The key factor is the availability of soil carbon dioxide, which is largely a function of biological activity and therefore increases at higher temperatures in terrains at low latitudes or altitudes. This more than offsets any direct effect of temperature on solubility. Limestone karst is better developed in humid temperate and tropical regions, due to both the greater availability of carbon dioxide in the soil and also the higher amounts of rainfall (Table 1.1). In arctic regions the amount of dissolved calcium carbonate in groundwaters issuing from karst springs varies from 0 to 170 mg/l, compared with 30 to 440 mg/l in humid temperate regions (Drew, 1985).

Table 1.1. The influence of climate on maturity in karst terrains, expressed by the mean rates of surface lowering, excluding dissolutional erosion that takes place underground. All figures are generalised to represent data derived from various sources (notably Smith, 1972; Balázs, 1973; Bögli, 1980; Ford and Williams, 1989).

	CO_2 content	Mean air temperature	$CaCO_3$ in solution (ppm)	Rainfall (mm/a)	Denudation (mm/ka)
Mean atmosphere	0.03%	10°C	70 in rainfall		
Mean atmosphere	0.03%	30°C	50 in rainfall		
Arctic karst	Soil air : 0.1%	0°C	50 in streams	100	10
Temperate karst	Soil air : 1%	10°C	200 in streams	1,500	30
Hot wet tropical karst	Soil air : 10%	30°C	300 in streams	2,500	60
Hot very wet tropical karst	Soil air : 10%	30°C	300 in streams	5,000	120

Additional factors that influence dissolutional aggressivity towards limestones include organic complexes, the common ion effect and mixing corrosion (Ford and Williams, 1989). Where dissolution continues, its rate slackens and it eventually ceases when saturation is reached. Maximum dissolution is therefore at sites where aggressive water first meets the limestone, and rockhead beneath a soil cover is therefore a focus of chemical activity. This accounts for pinnacled rockheads, sinkholes that narrow downwards into small fissures and the greater degree of cavitation that characterises the shallow zone of bedrock known as the epikarst (Williams, 1985). Continued dissolution at depth within a karst is due to the rapid transmission of unsaturated waters along open conduits. It is also due to the impact of mixing corrosion, whereby renewed dissolution is possible after two saturated waters mix and become unsaturated due to the non-linear equilibrium between dissolved carbon dioxide and the carbonate in solution (Bögli, 1964).

It is significant that dissolution of limestone is a very slow process. Mean rates of surface lowering on limestone range from <0.005 mm/y in arctic terrains with little soil, to >0.1 mm/y in wet equatorial regions beneath thick soil and plant cover, though exceptional rates of >5 mm/y have been recorded at specific sites on tropical islands (Trudgill, 1976). Fissure widening within limestone starts very slowly with laminar flow in narrow fissures, but then increases greatly when turbulent flow is initiated – in fissures about 5–10 mm wide under normal hydraulic gradients (White, 1988). The long-term pattern is therefore markedly non-linear, and it may take thousands of years to open a fissure to a few millimetres wide (Figure 1.3). A fissure may then grow into a large cave as dissolutional wall retreat takes place at around 0.1 mm/y (Gascoyne *et al.*, 1983; Mylroie and Carew, 1987; White, 1988). These rates are still so low that significant new cavities cannot be created in limestone within the lifetime of a built structure, and active dissolution of strong limestone is irrelevant to engineering, except in respect of reservoir leakage.

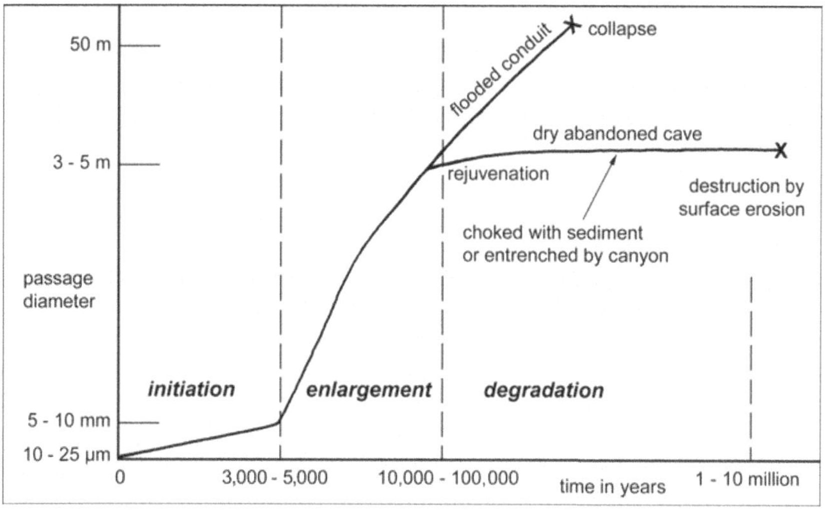

Figure 1.3. Dissolutional evolution from a narrow fracture to a wide fissure and then to an open cave, and its eventual destruction by collapse or loss through surface lowering. After White (1988).

Dissolution may however be accelerated by man-made changes in the groundwater conditions, or by changes in the character of the surface water that drains onto limestone. Importantly, accelerated dissolution by turbulent flows may be induced in smaller fissures by the substantial hydraulic gradients that engineering works may impose on soluble rocks. Experimental work on Jurassic Portland Limestone from England showed that the solution rate constant increased appreciably as flows became turbulent in fissures only about 2.5 mm wide under a hydraulic gradient of 0.2 (James and Kirkpatrick, 1980); seepage rates then increase rapidly and runaway situations can develop. Numerical modelling of very steep hydraulic gradients underneath dams indicated that the breakthrough to turbulent flow occurs in limestone once the fissure exit width reaches only 1 mm, at which point there is a dramatic increase in leakage rate (Dreybrodt *et al.*, 2002).

1.2.1 Cavernous ground

The long-term effect of dissolution in strong limestones (and dolomites) is to increase cavernous porosity by the expansion of integrated systems of cave conduits. Though large caves or heavily fissured zones can create very difficult conditions for engineering works, perception of the geohazard must be kept in proportion, as even the most mature karst is formed mainly of solid limestone. A massive limestone, with fissures 0.1 m wide spaced every 10 m on each of three rectilinear fracture systems, and with conduits 0.6 m in diameter along each fracture intersection, has a mean cavernous (secondary) porosity of 4.6%; this is distinct from any intergranular (primary) porosity of the intact rock. In such ground, the chance of a vertical borehole intersecting a void 0.6 m across is 12%. These figures give some measure of the scale of the collapse hazard in karstic limestone. However, they are representative of a most highly developed karst terrain, and typical values in most karst, especially outside the tropical zones, will be only a small fraction of these figures. Numerous caves were found on road construction projects in Slovenia (Case study #3) and southern China (Case study #5), but at both sites significant caves were exposed at rates of only about six per kilometre of road length – and these were in some of the world's most cavernous karst regions.

Though open voids typically constitute no more than a few percent of karstic rock masses, they can have a much greater impact on surface ground conditions where a thick soil mantles the cavernous bedrock. Vertical conduits on fracture intersections are potential sites for sinkhole development by soil suffosion into the bedrock. Following the example above, bedrock conduits on a 10-m grid beneath a soil mantle 5 m thick, may develop suffosion sinkholes with side slopes of 30° that are therefore 14 m across – and these combine to render 100% of the ground liable to potential subsidence. New sinkholes in thick soil mantles constitute the most widespread karst geohazard.

A tunnel excavated through a kilometre of mature karst typically meets only a few narrow fissures; it is rare that one encounters a large open cave. Inspections of tunnels, mines, road cuttings and quarry faces in karstic limestone show that they are largely in solid rock (Figure 1.4). Most rock faces expose karstic fissures that can be

Figure 1.4. Three dissolutionally opened fissures, one with a sediment-filled cave on it, exposed in a 30-m length of cut face on the new motorway from Postojna to Trieste across the highly cavernous limestone of the Kras in Slovenia.
TW.

readily sealed by modest amounts of masonry or concrete placed during excavation; these are small enough to create no major problems, but they do demand proper treatment and they cannot be ignored. In many limestone karsts, the ground may be 96–98% sound rock with a safe bearing capacity of 2–4 MPa, but the other 2–4% may be open voids with a bearing capacity of zero. Strong karstic limestone is unique in its enormous variation of ground conditions, where safe ground is randomly interspersed with relatively small areas that present significant hazards of catastrophic collapse. Highly variable ground is also a feature of karst on gypsum, which has cavernous porosities comparable to those of limestones, but its lower mechanical strength reduces its bearing capacity on sound rock.

1.3 KARST LANDFORMS DEVELOPMENT BY DISSOLUTION

All limestones and dolomites contain fractures that are susceptible to opening by dissolution. Bedding planes, commonly defined by thin partings of shale, are almost ubiquitous, and very thinly bedded carbonates are more prone to collapse before caves reach large dimensions. The dominant joint sets are commonly perpendicular to the bedding. Ultimately these become the open grikes in bare limestone pavements, the soil-filled fissures in pinnacled rockheads and the networks that increase karst permeability. Cave passages and chambers form along both joints and bedding planes, especially at their intersections, and also on faults that may provide optimum dissolution targets. Complex combinations of geological and hydrological influences determine which fissures out of many are enlarged to the point that they become potential geohazards. Impossible quandaries for construction engineers on karst are to decide just where voids exist within the ground or which areas of soil cover will next slump or collapse into unseen fissures.

Sinkholes are characteristic features of karst terrain. They are classified on a basis of origin into solution, collapse, caprock, dropout, suffosion and buried sinkholes (Chapter 2). They all constitute some form of significant geohazard in karst terrains (Figure 1.5). Solution, caprock and collapse sinkholes are initiated on the rocks' inherent weaknesses, and are integral components of the cave systems that evolve beneath any karst terrain (Chapter 3). In contrast, subsidence, dropout and suffosion sinkholes develop where soil is washed down into any one of the dissolutionally opened fissures exposed at rockhead (Chapter 4). They form not by dissolution of the rock but by ravelling of the overlying soils so that a void migrates up through the soil and ultimately creates a sinkhole in the surface. A sinkhole may be 30 m across where soil has been washed into a fissure less than a metre wide. The fissure may have taken 10,000 years to reach that size by slow dissolution, but the sinkhole may develop within hours as soil collapses and suffoses through the rockhead fissure and into an open cave system below.

Most caves are formed in competent limestones. They originate along bedding planes and tectonic fractures, which are enlarged into networks of open fissures, while favourable flow paths are enlarged selectively into caves (Palmer, 1991). There is an optimum zone of cave development just below the water table, where

Figure 1.5. A major sinkhole collapse under an urban highway in Bowling Green, a city that stands entirely on the Sinkhole Plain of Kentucky; the fog is due to warm air rising from the breached cave early on a cold winter morning.
Photo: Don Davis, KEEP.

efficient flow paths, dissolution capability and mixing corrosion are all concentrated. Zones of cave development are therefore inherited from levels just below successive water table positions in karsts with normal histories of evolution and rejuvenation. Cave levels are also created by enhanced dissolution at the saltwater/freshwater interface within coastal aquifers. All caves are parts of integrated cave systems that include interconnected fissures, conduits and locally enlarged voids. Sizes of individual caves range up to huge caverns, the largest being found in the humid tropics; Sarawak Chamber, in the Mulu caves of Sarawak, Malaysia, is over 300 m wide and 700 m long.

Many caves passages are abandoned as empty dry tunnels when their groundwater is captured by preferred routes. Alternatively they may be partially or wholly filled with clastic sediments with or without calcite deposits. If a cave is not abandoned or removed by a lowering ground surface, it may grow to a size where collapse becomes increasingly important. Roof collapse and void migration are features of every karst system, but their process rates are so low that they rarely influence engineering works except when the caves are at critical locations with respect to structural loading (Chapter 7). As a geohazard in karst, rock collapse is always subordinate to the processes of soil transport and the development of subsidence sinkholes.

1.4 CLIMATIC INFLUENCE ON LIMESTONE KARST

Suites of karst landforms in limestones and related carbonate rocks evolve through progressive erosion of the land surface while underground erosion is simultaneously enlarging cave conduits so that ever larger proportions of the drainage can pass underground. Both surface and underground erosion is by dissolution of the

Figure 1.6. Rounded pinnacles created by subsoil dissolution eroding deep fissures to form a large form of rundkarren beneath a thick clay soil at Shilin, southern China, now exposed in excavations on a construction site.
TW.

carbonate, at rates dependent on the flow and chemical aggressivity of the water. Both these factors are dependent on climate (Smith and Atkinson, 1976). The wet tropics support the most prolific vegetation with the highest contents of CO_2 in their soil air, and consequently their waters have the highest typical values for saturation with respect to calcium carbonate (Table 1.1). Water flows are largely a consequence of rainfall input (though they can be increased locally by supplies of allogenic water draining off adjacent outcrops of non-karstic rocks).

Highly aggressive, CO_2-rich, waters that percolate down through organic soils have maximum dissolution impact where they first meet limestone – at rockhead. Soils on karst are typically a permeable mixture of wind-blown loessic material and dissolution residue (terra rossa), with plant debris accounting for a significant proportion in all except the colder and drier climates. Subsoil dissolution is therefore a major process, and accounts for the very irregular rockhead typical of karst. Large, rounded rundkarren are the normal subsoil features in temperate climates (Bögli, 1960), with residual blocks and domes of bedrock between deep, dissolutionally-opened fissures (Figure 1.6). In the wet tropical regions, dissolutional fretting is more intense, so joint-guided fissures are deeper and the intervening residuals become pinnacles that may be 5–20 m tall, entirely beneath the soil profile. Some clay-rich glacial tills are so impermeable that they seal the rockhead from any dissolution effects, but most alluvial or glacial drift is permeable enough that rockhead relief is created by subsoil dissolution.

The wet tropics also have the most abundant rainfalls, and the total flows of water over a limestone terrain have the greatest control on denudation rates – which are fair indicators of the maturity, extent and scale of the caves, sinkholes and other karst landforms that may be anticipated (Table 1.1). Within the tropics there is a direct correlation between annual rainfalls and mean denudation rates. The combination of high rainfalls and dense plant covers in the islands and mainland of South-East Asia create optimum conditions for limestone dissolution. These areas therefore have the most spectacular karst, the largest caves, the largest sinkholes and the greatest engineering geohazards. In contrast, plant cover is constrained in the high latitudes and altitudes, as well as in the arid deserts (Jennings, 1983) – which consequently have poorly developed karst, generally with a smaller scale of sinkhole development.

An added factor in the maturity of the tropical karsts is their continued evolution throughout the climatic variations of the Pleistocene. Few karst landforms can be traced back more than a few million years, and within that time most limestone terrains in the higher latitudes have had long periods of glacial or periglacial conditions when dissolutional activity was curtailed by the shortage of biogenic carbon dioxide. Tropical karsts, however, evolved unabated. Paleoclimates also had their influence in the development of karst features during the Pleistocene pluvial stages in regions that are now deserts.

Though all landforms of the karst surface, and most of the underground features, were formed by dissolution in the presence of carbonic acid derived from carbon dioxide, there is an important group of caves that were formed by sulphuric acid. The acid is produced by oxidation of hydrogen sulphide produced by microbial activity in hydrocarbon reservoirs; it then migrates from its basinal source into adjacent reef limestones. The large chambers of Carlsbad Caverns, in New Mexico, were excavated by reaction with sulphuric acid, under a limestone outcrop with minimal karst development (Hill, 2000). Acidic hydrothermal fluids have also created a number of caves scattered through eastern Europe; they are distinguished by their irregular chambers and fissure passages, the complex mineralogy of their decorations and their lack of correlation with surface landforms. Sulphuric acid probably plays a major role in the very early stages of cave development in many karst regions (before dissolution by carbonic acid dominates), but it also accounts for isolated and anomalous cave passages and chambers in relatively immature karst landscapes. Caves revealed during construction of roads and buildings in the deserts of Abu Dhabi were unexpected and isolated features in a very dry and immature karst, and were probably formed by ancient sulphuric acid dissolution.

1.4.1 Types of limestone karst landscape

Geomorphologists recognise some broad types of karst on limestones, and each is developed largely in a specific climatic regime. Sinkholes are a feature of all karst terrains.

Glaciokarst is distinguished by its limestone pavements, rock scars and deeply entrenched gorges. It occurs at high altitudes or latitudes, where it was scoured by the ice and meltwater of Pleistocene glaciers. Bare rock surfaces are formed by glacial stripping on any rock type, but they survive for longer on limestone where there is minimal development of post-glacial soils *in situ*. The open fissures that

Figure 1.7. Contrasting types of karst landscape (a) Well-developed alpine glaciokarst on the French side of the high Pyrénées, with dark grassy floors in solution sinkholes; the larger depression has some small subsidence sinkholes in its soil floor. (b) A dry valley in the fluviokarst of the Derbyshire Peak District, U.K.; it was last occupied by a perennial stream under periglacial conditions during the late Pleistocene. (c) Shallow sinkholes across the lowland polygonal doline karst of Kentucky's Sinkhole Plain. (d) The extensive series of conical hills and intervening dolines in the forest-covered fengcong karst of the Cockpit Country in northern Jamaica.
TW.

characterise limestone pavements were created almost entirely by post-glacial dissolution, and are noticeably wider where they are or have been under or close to an organic soil. Bedrock sinkholes are generally small and widely scattered, but areas of semi-permeable glacial drift are well known for their huge numbers and dense concentrations of small subsidence sinkholes. The high plateaux of the Austrian Calcareous Alps and the French Pyrénées (Figure 1.7(a)), some wilderness plateaux in Montana, and the limestone outcrops of the Yorkshire Dales, England, are classic examples.

Fluviokarst has extensive dendritic systems of dry valleys that were cut by rivers before they were captured by underground drainage into caves. It is therefore well developed on some of the less cavernous limestones, including the Chalk Downs and the Cotswold Hills in England. Most of it occurs in regions that experienced periglacial conditions during the cold stages of the Pleistocene, when meltwater stream courses were re-activated over ground temporarily sealed by permafrost. Sinkholes occur both in the valley floors and on interfluve plateaux. The limestone plateaux of the Causses in southern France, and the Derbyshire Peak District in England are fine examples (Figure 1.7(b)).

Doline karst has a polygonal network of interfluves separating the closed depressions that have replaced valleys as the dominant landform because all drainage is underground. It is a mature landscape, developed in temperate regions with Mediterranean climates, and sinkholes of all types occur within the larger solutional dolines. The classical karst of Slovenia and Croatia is an upland with considerable local relief and steep-sided dolines, while the Sinkhole Plain in Kentucky is a rolling lowland polygonal karst, with wide shallow dolines, some of which contain lakes or open shafts into the underlying caves (Figure 1.7(c)).

Cone karst (or fengcong karst) is dominated by repetitive conical hills between closed depressions that are either stellate dolines or larger alluviated poljes, with almost the only areas of flat ground on the polje floors. It is a very mature landscape, almost totally restricted to tropical regions. The fengcong areas of Guizhou, China, provide the classic examples, while the Cockpit Country of Jamaica consists entirely of conical hills with rather lower profiles (Figure 1.7(d)). An extreme variant in the wet tropics is the tower karst (or fenglin karst) with isolated, steep-sided towers rising above extensive karst plains that have thin spreads of alluvial sediments over rock floors that are planed across the rock structure but fretted into highly irregular rockhead surfaces. The Yangshuo region of Guangxi, China, is the prime example (Figure 1.8). In both these types of karst, sinkholes of all types are typically common and can be very large. Tall limestone towers are the extreme karst landforms, and the largest of them are mostly on very thick sequences of almost pure carbonates, where they could evolve through long periods of surface denudation before the entire limestone was removed. For the same reasons, many of the

Figure 1.8. The classical fenglin tower karst near Yangshuo, in Guangxi, China; the plain between the towers has a thin alluvial cover on cavernous linestone and is the location of numerous subsidence sinkhole events.
TW.

largest caves and sinkholes are also in the very thick limestone sequences, of which that in southern China is notable.

1.5 DOLOMITE AND ITS SINKHOLES

Strictly speaking, a limestone has calcite forming at least 90% of its carbonate minerals. With only 50–90% as calcite, and 10–40% as dolomite, the rock is correctly called a dolomitic limestone, though many are still described as limestones, either because the exact chemistry has not been determined, or because the difference is generally irrelevant to the construction engineer. A rock with dolomite forming more than 90% of its carbonate minerals is also known as a dolomite. One with only 50–90% as dolomite, and 10–40% as calcite, may be known as dolostone (Figure 1.9). Though this is the preferred term in some countries, the term dolomite is still used for a much wider range of mineral compositions in most parts of the world. Terminology in the pages of this book is simplified to either limestone or dolomite, and is qualified with reference to content of insoluble impurities and calcite/dolomite ratio only where necessary.

Primary dolomites (or dolostones) tend to be thinly laminated, are formed of extremely fine-grained crystalline dolomicrite (<20 μm) and are unfossiliferous (Folk, 1973). They originate in sabkha environments, so are commonly associated with evaporites, and may contain nodules or scattered crystals of gypsum or

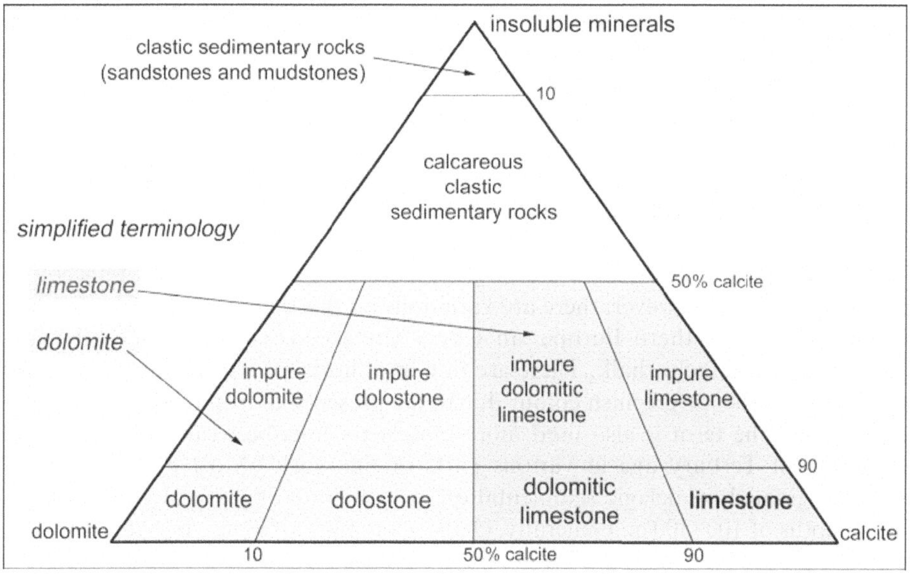

Figure 1.9. Relationships between limestones, dolomites and dolostones in a classification of pure and impure carbonate and clastic rocks.
After Burnett and Epps (1979).

anhydrite. Secondary dolomites are more coarse-grained or saccharoidal in texture with plentiful evidence of diagenetic replacement of calcite by dolomite, so that original textures and structures, as well as the fossil content, are obscured or lost (Zenger *et al.*, 1980). Selective dissolution of calcite in a dolomitic limestone can cause disaggregation to leave a weak dolomite sand. Partial dolomitization may create a patchy distribution of dolomite within the rock mass, whereas total recrystallization produces a medium to coarse, crystalline mosaic. The process of dolomitization leads to an increase in porosity of up to about 13%, which subsequently may be decreased by the processes of induration. Dedolomitization may follow, whereby dolomite is replaced by calcite, probably in response to the partial dissolution of gypsum (Williams and McNamara, 1992). Both dolomitization and dedolomitization can be responsible for brecciation of the rock mass because of changes in volume due to addition or removal of material. The mechanical properties of dolomites span about the same range as those of limestone, with unconfined compressive strengths of 30–100 MPa for the intact rock.

In many regions, including the classical karst of Slovenia, the main cliffs, crags and karren landforms are on the limestone, while the dolomite surfaces are mantled by broken rock debris and thicker soils. Caves and sinkholes lie in both rock types, and in Slovenia are especially concentrated along the dolomite/limestone boundaries where the chemical contrast has facilitated cave inception. In terms of their sinkhole hazards and construction difficulties, most limestones and dolomites are essentially the same. Though limestones are more widespread, some dolomites are famous for their karst landforms. Some of the largest and most catastrophic sinkholes in the world developed when dolomites in the Far West Rand, in South Africa, were dewatered to allow expansion of the gold mines beneath them (Box 8.1). Landscapes in the Dolomite Mountains of Italy are dominated by alpine glacial features, but they are formed of both limestone and dolomite, and areas on both rocks contain significant karst with many caves and sinkholes.

1.6 CHALK AND ITS SINKHOLES

Chalk is a soft, fine-textured, white or grey, marine limestone of late Cretaceous or early Tertiary age. However, there are variations on this definition. It describes most of the Chalk in northern Europe, an Upper Cretaceous stratigraphic unit with a lithology that is largely chalk. There are thin hard horizons within the Chalk, and it can have a reddish or greenish colour due to the presence of iron oxide or glauconite respectively. The term is also used more loosely to describe weak, soft and friable limestones of Tertiary age in various parts of the world. Most of the European chalks originated by pelagic sedimentation in temperate or tropical shelf seas with water depths of 100–300 m. Generally, chalk is a remarkably pure micritic limestone, containing over 95% $CaCO_3$, with >70% of grains composed almost entirely of coccolith skeletons that are $< 40\,\mu m$ in diameter.

Chalk varies in hardness. In the soft chalks of south-east England, the individual particles are bound together at their points of contact by thin films of calcite and aragonite, and there are only minute amounts of these cements. Early cementation prevented gravitational consolidation occurring in soft chalk, and helped retain its high porosity. In contrast, most of the chalk in northern England and in northern France has >50% of its voids occupied by cement due to overburden pressure bringing about pressure solution and reprecipitation of calcite (Bell *et al.*, 1999). In addition to effects of intrinsic diagenesis, chalk lithologies vary due to their history of tectonic compaction (Clayton, 1983). Alpine orogenic stresses induced mechanical disaggregation of the coccolith matrix, followed by reconsolidation, pressure solution and re-precipitation of the calcite under sustained pressures. In areas where the chalk dips at angles in excess of 30°, its density and strength properties are increased to those of the weaker limestones.

Although the porosity of chalk may be as high as 50%, the primary pores are very small. Most of the water that flows through chalk is therefore concentrated along systems of discontinuities. Dissolution along these creates open fissures, small cave passages, and the sub-vertical solution pipes that are a type of buried sinkhole (Chapter 5). The fissure systems account for high flow rates within chalk aquifers, 1,200 m per day in fissured zones in chalk in northern England (Pitman, 1983), and up to 3 km per day through well-developed conduit systems in chalk in southern England (Atkinson and Smith, 1974).

Many chalk terrains evolve into a subdued variety of fluviokarst, with dendritic systems of dry valleys largely inherited from periglacial erosion, but sinkholes are also a feature of chalk karst, as well as occurring in areas where chalk lies beneath a thin cover of clastic sediments or soils (Figure 1.10). Large, inactive sinkholes in the

Figure 1.10. Large buried sinkholes filled with clay soils (locally known as chalk pipes) exposed in a road cutting through the chalk in southern England.
Photo: Martin Culshaw.

sand-covered chalk heathlands of Norfolk and Dorset, U.K., evolved through a combination of rockhead dissolution and cover suffosion, but active subsidence sinkholes are small and of limited distribution. Buried sinkholes are common near the contact of the chalk and the Tertiary cover in England, and a few stream sinks are known in the same contact zones. Chalk is often described incorrectly as being non-cavernous, but small cave conduits are frequently encountered in deep excavations and in well drilling. Coastal cliff sections also expose caves, with one at Beachy Head, on England's south coast, having 350 m of mapped passages (Waltham *et al.*, 1997). The denser and stronger chalk of the Paris basin in France has at least four cave systems each with 1–2 km of mapped passages (Chabert, 1981). It is reasonable to assume that caves too small to be explored by man are probably much more common in chalk than previously has been appreciated (Lowe, 1992).

A special feature of chalk is that dissolution aids its disintegration by weakening its fabric and by emphasizing structural weaknesses, however slight, thereby reducing its mass strength. Frost shattered chalk, widespread within about 6 m of the ground surface as a feature inherited from cold stages of the Pleistocene, is prone to liquefaction when it is both saturated and disturbed. Concentrated water flows, from either soakaway drains or leaking pipes, can cause both saturation and enhanced dissolution, with catastrophic effect where the liquefied chalk has an outlet in an open space beneath. The classic subsidences in Bury St. Edmunds, England, were caused by soakaway drains inducing liquefaction failure of the chalk over old mine workings (Bell *et al.*, 1992).

Dissolutional weakening of chalk is analogous to the chemical effects in other soft limestones, where the pores are enlarged by groundwater flow, thereby enhancing permeability and water circulation, and encouraging further dissolution. In the unweathered, indurated oolitic limestone underlying much of Miami, Florida, intergranular porosity averages 15% above the water table, but is as high as 75% just below the water table, where groundwater circulation is greatest (Sowers, 1975, 1996). This dissolutional loss means that the rock has a compressibility similar to that of loose sand, and stress increase at grain contacts leads to increasing corrosion. Settlement of around 100 mm can take place rapidly, with most occurring during the construction period, as is characteristic of a sand.

1.7 EVAPORITE ROCKS AND THEIR SINKHOLES

Evaporite deposits are formed by precipitation from saline waters that are concentrated beyond the levels of mineral saturation by evaporation in lagoonal or lacustrine environments. The vast bulk of evaporites are the sulphates (gypsum, $CaSO_4.2H_2O$, and anhydrite, $CaSO_4$) and rock salt (halite, NaCl). Karstic conditions on both gypsum and salt create extensive areas of difficult ground for construction engineers (Johnson and Neal, 2003). Other quantitatively minor evaporites are valued as ore deposits, except for the small amounts of evaporite limestone. Gypsum, anhydrite and salt underlie more than 20% of the world's land surface (Kozary *et al.*, 1968), though more than 90% of this, including nearly all the salt, is buried beneath other rocks. Much of the world's gypsum karst is interstratal, whereby caprock sinkholes and breccia pipes are developed through surface outcrops of insoluble rocks (Section 3.5).

1.7.1 Dissolution and sinkholes on gypsum and anhydrite

Gypsum is rarely present in ancient evaporites at depths greater than about 1,000 m, where anhydrite is the stable calcium sulphate mineral. Gypsum forms as the original precipitate and survives the early stages of burial, before becoming unstable above a temperature of about 42°C, which occurs at a depth of about 1,200 m in areas with normal geothermal gradients. The influence of pressure appears to be slight, but the presence of salt lowers the transition temperature, so that transformation can occur at lesser depths in some environments. Conversely, gypsification of anhydrite occurs with exhumation, where evaporite beds near the surface as a result of uplift and erosion.

The extent of hydration of anhydrite may be directly related to depth, but is also dependent upon the ease with which water has access to anhydrite (Holliday, 1970). In a bed of anhydrite between two beds of impermeable mudstone, the likelihood of complete hydration is reduced appreciably. Hydration proceeds by dissolution of the anhydrite, followed immediately by re-precipitation of the calcium and sulphate ions as gypsum. The rate at which hydration proceeds, when water is available, depends on the extent to which anhydrite is out of equilibrium. The reaction is slow at depth, at a little below the transition temperature, but is more rapid nearer the surface where lower temperatures prevail. Gypsum with coarse porphyroblastic texture forms early in exhumation by hydration under near-equilibrium conditions, whereas finer alabastrine textures appear later, when the rocks are much nearer the surface (Mossop and Shearman, 1973). Although gypsification involves a large increase in volume in the solid phase, the volume of water necessary to hydrate anhydrite is greater than the additional volume of gypsum that is formed, and extra space for gypsum is created by parallel redistribution and dissolution (Holliday, 1970).

Total solubility and dissolution rate are both much greater for gypsum in water than they are for limestone. In massive gypsum (in water that contains no dissolved salts, is at 10°C and is under a hydraulic gradient of 0.2), a fissure 0.2 mm wide and 100 m long would have widened by dissolution within 100 years, so that a block 1 m^3 in size could be accommodated in the tapering entrance to the fissure (James and Lupton, 1978). A cave would have been formed. If the initial width of the fissure exceeded 0.6 mm, large caverns would form and a runaway situation would develop in a very short time. In long fissures, the hydraulic gradient is low and the rate of flow is reduced, so that solutions become saturated and little or no material is removed. A flow rate of 10^{-3} m/s is critical as turbulent flow then occurs; if it is exceeded, extensive solution of gypsum can take place (James and Lupton, 1978). Under the extremely high hydraulic gradients generated beneath dams, dissolution rates for gypsum may be as high as 10 mm/year (Dreybrodt *et al.*, 2002). Rates of gypsum dissolution in the Ripon area of England, imply that the walls of phreatic

Figure 1.11. One of the collapse sinkholes that house the "bottomless lakes" at Roswell, in New Mexico, with the bedded gypsum exposed in its walls.
TW.

caves could retreat by 0.5–1.0 m/year under favourable flowing water conditions, with significant impact on the scale of the sinkhole geohazard (Cooper, 1995), but dissolution rates measured for gypsum elsewhere are appreciably less than this (Klimchouk *et al.*, 1996).

Karst terrains on gypsum bear some comparisons with the types of karst developed on limestones, and all contain sinkholes (Figure 1.11). Gypsum is known to occur at outcrop in all climatic regimes except those with very high rainfall. Its most distinctive terrain type is polygonal doline karst, of which the finest examples lie in the interiors of Uzbekistan and Turkey where rainfalls are 200 to 500 mm/year. A different style of doline karst is created by the innumerable caprock sinkholes over the interstratal gypsum karst of the Ukraine (Klimchouk *et al.*, 1996). Glaciated gypsum terrains in northern Russia have caves, sinkholes and some bare crags (Waltham and Cooper, 1998), but the residue of insoluble interbeds creates a more extensive soil, which supports a forest cover in place of the pavements that typify glaciokarst on limestone. Cone and tower karsts do not exist on gypsum, as these landforms evolve during long periods of surface lowering, in which time gypsum beds are totally removed by dissolution.

Sinkholes and caves can develop in thick beds of gypsum more rapidly than limestone. In the U.S.A., such features have been known to form within a few years where beds of gypsum are located beneath dams, and extensive surface subsidence has occurred in Oklahoma and New Mexico due to the collapse of cavernous gypsum (Brune, 1965). The problem is accentuated by the fact that gypsum is weaker than limestone and therefore collapses more readily. The upward migration of voids through both gypsum and any overlying insoluble beds, creates breccia pipes filled with fallen debris, that may reach rockhead from depths of many hundreds of metres. Active pipes may express themselves as caprock sinkholes (Case study #1), while inactive pipes may act as foci for the development of subsidence sinkholes.

1.7.2 Dissolution and sinkholes on salt

The solubility of rock salt (halite, NaCl) in water is 35.5% by weight, and is therefore 7,500 times more soluble than limestone. Because it is so soluble it only survives at the surface in arid environments, and bare salt karst is restricted to Israel's Negev Desert, salt domes on some Iranian islands in the Persian Gulf and a few other small areas (Frumkin and Raz, 2001). Deep-seated dissolution can continue in salt buried beneath thick sedimentary sequences, and breccia pipes may rise through the cover rocks (Chapter 3), but it may be difficult or impossible to distinguish between active salt karst and its palaeokarst equivalent (Johnson, 1997).

In wetter climates, salt only survives beneath a cover of mudstone or drift, and its only karstic landforms may be shallow solution dolines in areas of permeable drift cover (Chapter 2). Beneath England's Cheshire Plain, the salt that occurs in the Triassic Mercia Mudstone terminates at a solution surface at depths of 70–150 m below ground. This rockhead is overlain by a solution breccia formed of collapsed, insoluble, interbedded mudstone, and that lies beneath permeable Pleistocene glaciofluvial sands (Bell, 1975). Natural solution at the rockhead has caused formation of the Cheshire Meres, lakes up to 1 km across that fill some of the larger, soil-mantled solution sinkholes. This type of subsidence, although it operates over large areas, takes place extremely slowly, as dense saturated brines sit on the rockhead and prevent aggressive freshwater reaching the salt until they slowly drain away to brine springs. Dissolution has been greatly accelerated where brine abstraction for the salt industry has effectively drawn in fresh supplies of aggressive groundwater, and subsidence has therefore been reactivated or induced anew (Bell, 1992).

1.8 MAN'S INFLUENCE ON DISSOLUTION AND SINKHOLES

Most of the world's sinkholes are inherited from natural erosion processes over geological time, and are therefore largely prehistoric features. However, the major geohazard is created by those sinkholes that are now active and therefore developing failure events. The vast majority of these are induced or accelerated by man's own activities (Newton, 1987). Nearly all of these are subsidence sinkholes that develop by soil failure within the lifetimes of engineered structures, or even during the period of construction. They are induced by increased drainage flows that washes any soil cover down into cavernous karst, and that is induced either by pumped declines in groundwater level, by redirected and concentrated flows of storm-water run-off, or by leaking pipelines (Figure 1.12). Sinkholes may cause problems in construction for engineers, but the geohazard is largely self-induced (Chapter 8).

Figure 1.12. A small suffosion sinkhole that has opened up beneath a leaking pipeline in the soil-covered dolomite karst south of Pretoria, South Africa.
Photo: Fred Bell.

There are thousands of sinkholes throughout the eastern U.S.A. Whereas it takes thousands of years to create natural sinkholes, those created by human activity have largely occurred since the early 1900s. More than 4,000 sinkholes have been catalogued in Alabama as being caused by human activities, with the great majority of these developing since 1950. In Shelby County, Alabama, more than 1,000 sinkholes developed between 1958 and 1973 in an area of $26\,km^2$ (LaMoreaux and Newton, 1986). These sinkholes have been caused largely by continuous dewatering in the carbonate rocks by wells, quarrying and mining operations and by surface drainage changes. Only China has a greater record of artificially induced sinkhole events, and those too are almost all related to mining or construction works.

A classic case of induced sinkholes was the result of dewatering the thick cavernous dolomites that overlie the gold-bearing reefs of the Rand in South Africa. Sinkholes formed concurrently with the water table decline in areas that formerly had been free from sinkholes, and these included some of the world's most catastrophic sinkhole events (Box 8.1). Some areas became unsafe for occupation, and entire villages and towns were moved, on a scale that has only been approached, with respect to sinkholes, at some locations in China's extensive karst. In both South Africa and China, the situation has been locally aggravated by the development of high-density, informal settlements that are poorly serviced so that waste water is disposed of directly into the ground. Any urbanisation can induce sinkholes and subsidence since it radically changes the way in which water enters the ground compared with an open, undeveloped site. Where housing, road surfaces and concrete structures render 50–80% of the surface area relatively impermeable, infiltration and percolation of precipitation occurs only in the remaining areas of open ground. Furthermore, much of the water draining from built-over areas is concentrated into soakaways, and backfilled service trenches also tend to act as preferred paths for water ingress (Chapter 8).

These few examples serve to demonstrate that, while sinkholes constitute a significant geohazard in the many parts of the world underlain by limestone and other cavernous rocks, much of the potential geohazard is manageable or controllable if the processes of sinkhole development are fully appreciated.

2

Sinkhole classification and nomenclature

2.1 SINKHOLES AND DOLINES

Karst is defined as a landscape that is distinguished by its underground drainage, so that its landforms evolve in response to rainfall and surface water flowing into the ground. Consequently, river valleys cannot develop in a mature karst. Instead, drainage and consequent fluvial erosion is centripetal towards points where the water sinks underground. Dendritic systems of linear valleys are replaced by internally drained closed depressions as the dominant surface landforms of karst. The erosion processes that create closed depressions are essentially downward, and therefore may involve subsidence that makes them a geohazard to engineers. The surface depressions are matched by the underground landforms, caves, which, by their very existence, create the second engineering geohazard in karst.

Closed depressions are often regarded as the diagnostic features of a karst terrain. Their closure refers to their rim of surrounding and higher land, so that no downhill surface drainage outlet exists. Closed topographic depressions may be created by tectonic sag or by glacial over-deepening, but then fill with water to become lakes, except in desert environments (or where incidental karstic drainage keeps them dry). Aeolian deflation may create closed depressions in sandy deserts. Volcano craters may also remain dry by underground drainage in very permeable, porous volcanic rocks that are non-karstic. All these are readily distinguished from karst features. Closed depressions with internal drainage, that exist in non-arid environments and do not contain lakes, are valid indicators of karst conditions. Karstic closed depressions exist in all sizes. Most are less than 1,000 m across, with a high proportion of these less than 100 m across. All of these are known as either sinkholes or dolines.

The term doline is derived from the Slovene word *dolina*. This translates as a valley, but really means just a negative landform; there are no true valleys in the limestone karst of Slovenia, where sinking drainage means that the *dolina* depressions are closed. The term entered scientific literature through the writings of Cvijic (1893), and is now the accepted term among geomorphologists worldwide.

The term sinkhole derives from the processes of its evolution. Water sinks into the ground of a closed depression (to prevent it becoming a lake), sediment sinks into the ground where it is washed down by water, the ground surface is lowered (and effectively sinks) by slow dissolution of its soluble rockhead, and the ground may subside (or sink) catastrophically in a collapse event. In different situations, sinkholes have been defined by any of these characteristics. In some other cases, a sinkhole has been defined as an open hole into which a stream sinks, not necessarily within a closed depression larger than the hole itself. Such discrete sink points are better known as sinks, stream sinks, swallow holes, swallets or ponors, and they may or may not be within the floor of a sinkhole depression.

There is therefore a history of minor confusion over what constitutes a sinkhole, which is why geomorphologists use the term doline. However, the North American engineering literature has seen wider use of the word sinkhole to describe karstic depressions caused by one or more of the above sinking processes. As that region has the lion's share of sinkhole subsidence and collapse events within the English-speaking world, the term sinkhole has become the accepted term in the engineering literature, where the term doline is often not understood.

Sinkhole is a very descriptive term for a site where ground or water is sinking. The term has therefore been over-used by application to ground failures that are clearly not karstic, especially in emotive circumstances where property is lost to "sinking ground". In South Africa and elsewhere, ground failures by migrating roof collapse over mine workings have been described as sinkholes, or sink holes, though many of these are correctly and more widely known as crown holes. Substantial flows of water can also sink into open rock fractures or very porous rocks, neither of which is soluble in water or karstic; these features are better described as elements of pseudokarst.

For all practical purposes, sinkholes and dolines are the same, and the former is used in these pages.

2.2 CLASSIFICATION OF SINKHOLES

Sinkholes develop by a cluster of inter-related processes, including bedrock dissolution, rock collapse, soil down-washing and soil collapse. Any one or more of these processes can create a sinkhole. The basic classification of sinkholes has six main types that relate to the dominant process behind the development of each, the main characteristics of which are shown in Table 2.1 and are further considered below. There are also long spectra of sinkholes that classify between the main types because they form by combinations of processes. These include two notable groups. First, thousands of sinkholes are formed in soils by successions of slumps and collapses that make them intermediate between suffosion and dropout sinkholes; these are collectively and non-specifically known as subsidence sinkholes. Second, many sinkholes form by the combination of dissolution and small-scale collapse that typifies most karstic erosion.

The classification and terminology of sinkholes in Table 2.1 has evolved from

Table 2.1. The six types of sinkholes, with typical cross sections and major parameters for each type; the dropout and suffosion sinkholes may be described as forms of subsidence sinkholes.

Solution sinkhole 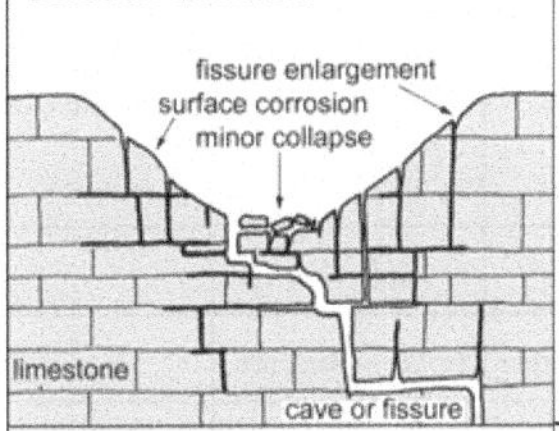	*Formation process*	Dissolutional lowering of surface
	Host rock types	Limestone, dolomite, gypsum, salt
	Formation speed	Stable landforms evolving over >20,000 years
	Typical max size	Up to 1,000 m across and 100 m deep
	Engineering hazard	Fissure and cave drains must exist beneath floor
	Other names in use	Dissolution s/h, cockpit, doline
Collapse sinkhole 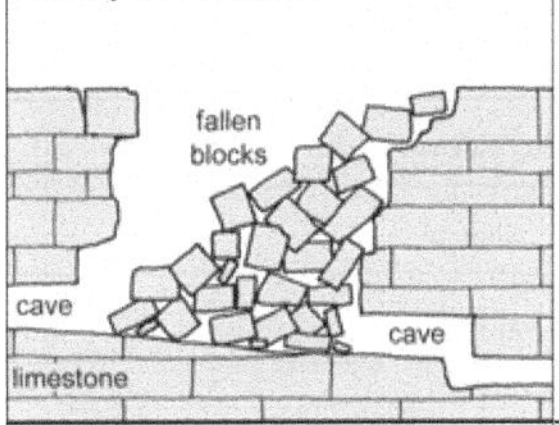	*Formation process*	Rock roof failure into underlying cave
	Host rock types	Limestone, dolomite, gypsum, basalt
	Formation speed	Extremely rare, rapid failure events, into old cave
	Typical max size	Up to 300 m across and 100 m deep
	Engineering hazard	Unstable breakdown floor; failure of loaded cave roof
	Other names in use	Cave collapse s/h, cenote
Caprock sinkhole 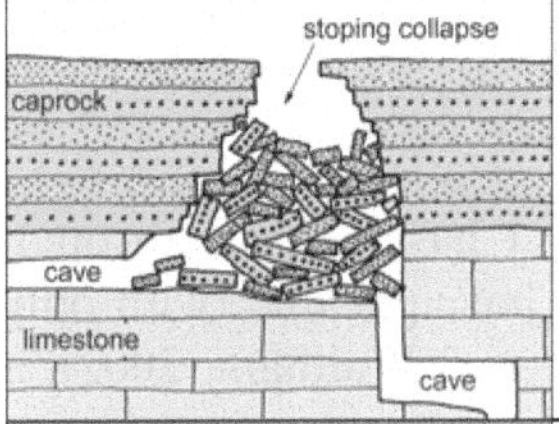	*Formation process*	Failure of insoluble rock into cave in soluble rock below
	Host rock types	Any rock overlying limestone, dolomite, gypsum
	Formation speed	Rare failure events, evolve over >10,000 years
	Typical max size	Up to 300 m across and 100 m deep
	Engineering hazard	Unastable breakdown floor
	Other names in use	Subjacent collapse s/h, interstratal karst
Dropout sinkhole 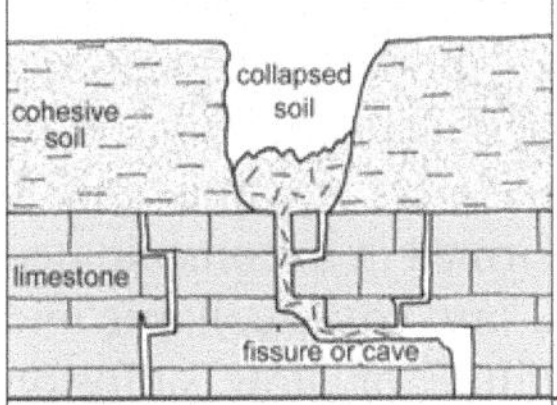	*Formation process*	Soil collapse into soil void formed over bedrock fissure
	Host rock types	Cohesive soil overlying limestone, dolomite, gypsum
	Formation speed	In minutes, into soil void evolved over months or years
	Typical max size	Up to 50 m across and 10 m deep
	Engineering hazard	The main threat of instant failure in soil-covered karst
	Other names in use	Subsidence s/h, cover collapse s/h, alluvial s/h
Suffosion sinkhole 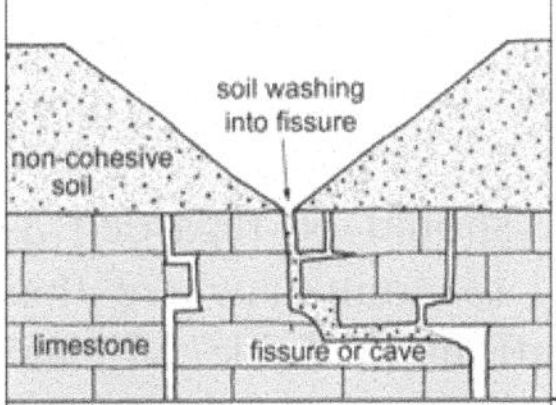	*Formation process*	Down-washing of soil into fissures in bedrock
	Host rock types	Non-cohesive soil over limestone, dolomite, gypsum
	Formation speed	Subsiding over months or years
	Typical max size	Up to 50 m across and 10 m deep
	Engineering hazard	Slow destructive subsidence over years
	Other names in use	Subsidence s/h, cover subsidence s/h, alluvial s/h
Buried sinkhole 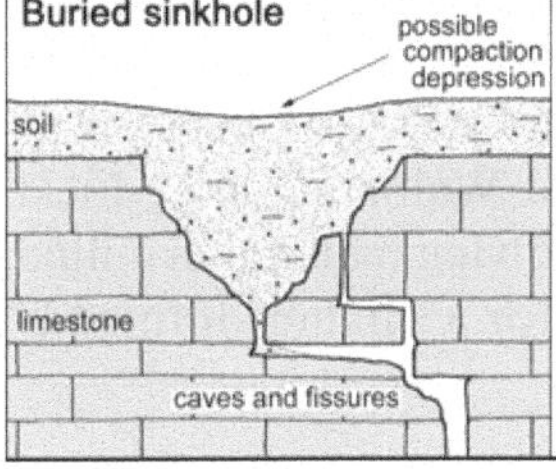	*Formation process*	Sinkhole in rock, soil-filled after environmental change
	Host rock types	Rockhead depression in limestone, dolomite, gypsum
	Formation speed	Stable features of geology, evolved over >10,000 years
	Typical max size	Up to 300 m across and 100 m deep
	Engineering hazard	Local subsidence on soft fill surrounded by stable rock
	Other names in use	Filled s/h, compaction s/h, paleosinkhole

Table 2.2. Sinkhole terminology as applied by this and earlier classifications.

Waltham *et al.* (2005)	Solution	Collapse	Caprock	Dropout	Suffosion	Buried
				Subsidence		
Williams (2004)	Solution	Collapse	Caprock collapse	Dropout	Suffosion	Buried
Lowe and Waltham (2002)	Dissolution	Collapse	Caprock	Dropout	Suffosion	Buried
Ford and Williams (1989)	Solution	Collapse			Suffosion	–
White (1988)	Solution	Collapse	–	Cover collapse	Cover subsidence	–
Culshaw and Waltham (1987)	Solution	Collapse	–	Subsidence		Buried
Beck and Sinclair (1986)	Solution	Collapse		Cover collapse	Cover subsidence	–
Jennings (1985)	Solution	Collapse	Subjacent collapse	Subsidence		–
Bögli (1980)	Solution	Collapse (*fast*)/Subsidence (*slow*)		Alluvial		–
Sweeting (1972)	Solution	Collapse	Solution subsidence	Alluvial		–
Other terms in use	Cockpit	Cave collapse Tiankeng Tumour Cenote	Interstratal collapse Breccia pipe		Shakehole Ravelling	Filled Compaction Pipe

various predecessors (Table 2.2). Over the last few years it has matured through discussions among international karst specialists, so that it has appeared (with only minor variations) in the current dictionary of karst (Lowe and Waltham, 2002), a new encyclopedia of karst (Williams, 2004) and the engineering classification of karst (Waltham and Fookes, 2003). The terms for the sinkhole types in Tables 2.1 and 2.2 have borrowed the best from earlier classifications, and are both descriptive and distinctive. Caprock collapse is a usefully descriptive term (Williams, 2004), but is abandoned in favour of the one-word term. The concept of cover sinkholes (Beck and Sinclair, 1986) is also useful with reference to all those formed in unconsolidated soil cover over karstic bedrock; however, these were divided into cover collapse and cover subsidence sinkholes, forming by rapid dropout failure or slow suffosion subsidence respectively. The single term retained outside the six main types is the subsidence sinkhole; it is widely used, and it also encompasses very successfully the many sinkholes in soil cover that are intermediate between the dropout and suffosion types and may become indistinguishable once weathered and degraded.

The seven terms for sinkholes, as used in Table 2.2, should be regarded as standard within the English language; they apply equally to dolines, where that term is preferred. Abandoned terms in the older classifications in Table 2.2, and many other more obscure and sometimes confusing individual terms, should therefore be avoided. A selection of other terms that appear in past and present literature related to sinkholes are defined briefly in the glossary, and many more terms are defined in available karst dictionaries (Lowe and Waltham, 2002; Field, 2002). Dissolution is the correct term for the process of dissolving material in a fluid, while that fluid is a solution. The technically correct name for a sinkhole formed by dissolving the rock away is therefore a dissolution sinkhole. This became fashionable for a short time, but has proved cumbersome and unpopular, being dropped in favour of the solution sinkhole.

2.3 SOLUTION SINKHOLES AND SOLUTION DOLINES

On outcrops of soluble rocks without significant soil cover, solution sinkholes in various shapes and sizes are the dominant karst landform of intermediate size. They are the surface depressions that range between 1 m and 1,000 m across. Together with the smaller karren features and the larger landforms of cones, towers and pavements (Chapter 1), solution sinkholes characterise the landscape of karst. As they are slowly evolving features of a karst terrain, they are long-term geomorphological landforms; they are perhaps better known as solution dolines, and that term is used in these pages. An American may describe a closed basin some kilometres across within a karst as a sinkhole; but this term would jar with any foreigner, who would describe the large feature as a doline or basin, and would refer to individual small depressions and sink features within it as sinkholes. Many very large karst depressions are often referred to just as dolines, with their solutional origins implied. Among all the types of dolines/sinkholes, the solution features dominate in most karst terrains, but they are the least important with regard to engineering geohazards.

Solution dolines are formed where surface water and/or soil water dissolves bedrock at the surface or rockhead, as it flows towards points where it can sink into the fissured and/or cavernous ground. The doline is then self-exaggerating, as increased centripetal drainage inside the enlarging depression increases local dissolution and consequent surface lowering within it. Solutional dolines enlarge so that their diameters are orders of magnitude greater than the mean fracture spacing within the bedrock. Rainfall input then sinks at many points into the numerous fissures that are either exposed or at rockhead beneath a soil cover. A solution doline has numerous points where water sinks into bedrock fissures (Box 2.1). There may be no sign of centripetal surface drainage, and this diffuse infiltration deepens the doline across its entire floor so that it opens out to a saucer profile (Figure 2.1). Where a larger proportion of rainfall input drains over the soil or rock surfaces towards a single central sink, the doline develops a steeper, conical profile.

Dissolution of soluble rocks takes place on all exposed surfaces, at rockhead beneath any soil cover and also on the walls of discontinuities that are widened to fissures below rockhead. Most dissolution by aggressive soil water takes place within a zone of bedrock that lies just below the surface. This is known as the epikarst (Klimchouk, 2000) or the subcutaneous zone (Williams, 1983). Over time, the bedrock within the epikarst is reduced to a 3-D fissured mass and then to a layer of separated blocks. On block dimensions of about a metre, this instigates repeated small rock collapses into enlarging fissures and caves. In this respect, all solution dolines involve some elements of collapse in their evolution. Dissolutional removal

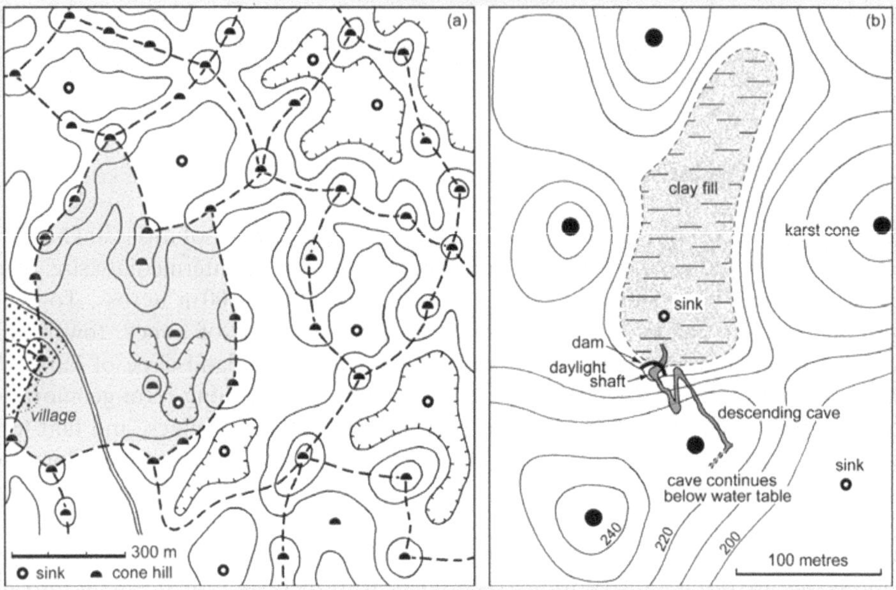

Figure 2.1.1. (a) Part of the cone karst of Gunung Sewu, with broken lines marking the doline boundaries, and shading in the Puleirang doline; (b) the lowest part of the doline, with its alluviated floor and the cave system that is on its southern side.

Figure 2.1.2. The Puleirang doline as modified in 1982, with its mortar-sealed slopes, graded clay floor and the concrete dam around the open sink.
TW.

BOX 2.1. SOLUTION DOLINE

An example – Puleirang, Java, Indonesia

Along the south coast of the island of Java, Gunung Sewu (meaning Thousand Hills) is formed of massive Miocene reef limestones, and is eroded into one of the world's finest examples of cone karst. Conical or hemispherical hills, each 30–100 m high, are separated by dolines (or sinkholes), where all drainage sinks underground (Waltham *et al.*, 1983). On the eastern side of Puleirang village, a doline of about 35 ha forms a polygonal basin with eleven hills at its corners (Figure 2.1.1(a)). The slopes of the doline have about 60% rock outcrop as small irregular pinnacles and scars, and about 40% cover of patchy clay soils some of which is artificially retained by stone walling to create small sites for growing vegetables. The floor of the doline has a thicker soil of alluvial clay over an area of about 2.5 ha (Figure 2.1.1(b)). Much of the wet-season rainfall sinks straight into the ground, but there is considerable surface run-off during intense rainstorms. Originally, most of the run-off gathered into streams that crossed the doline floor to sink into an open shaft on the south side (Figure 2.1.2). The shaft system is open down to the water table at 190 m below the level of the doline floor.

Drought-season water storage is needed for the villagers. In 1982 a concrete dam was built round the open shaft, the clay floor of the doline was reworked to seal the short stream channels, and mortar was hand-placed to seal the limestone in the doline walls above the clay floor (Figure 2.1.2). In the first wet season after construction, water ponded on the doline floor, but almost immediately drained out through a new suffosion sinkhole (Figure 2.1.3). The water flowed into bedrock fissures, most of which linked across to the main shaft about 20 m below ground level. The new sinkhole through the soil was blocked with clay and mortar in the next dry season, but all bedrock fissures could not be sealed, and it failed again in the following wet season. The reservoir is now abandoned, the doline floor is used for agriculture, and all drainage sinks through the floor before reaching the lip of the shaft. Multiple fissure flows ensure that the lower parts of the shaft system are well watered in the wet season, but the dam remains high and dry. As a reservoir, Puleirang was not a success, but the site has revealed a fine example of ground conditions beneath a solution doline (or a solution sinkhole).

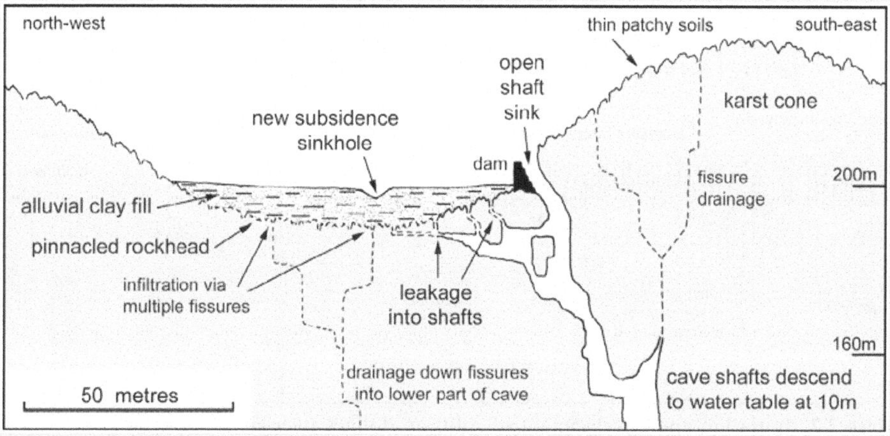

Figure 2.1.3. Profile showing known details of the Puleirang doline floor and its open sink.

Figure 2.1. Large solution dolines in limestone above Malham in the Yorkshire Dales karst, U.K. The person is standing near the lowest point of the nearest doline, but the far doline is deeper. Neither doline has surface stream courses nor discrete sinks, and all rainfall infiltrates through the soil of the doline slopes and their wide floors.
TW.

of the limestone is accelerated as fissure widening permits increased infiltration, and is concentrated in the centres of dolines by the net centripetal flows (Figure 2.2) (Williams, 1983, 1985).

A thickness of the epikarst cannot be precisely defined, as its base grades into relatively intact rock, though a figure of 15–30 m is related to stress relief opening of rock fractures (Klimchouk, 2000). The upper part of the epikarst in massive

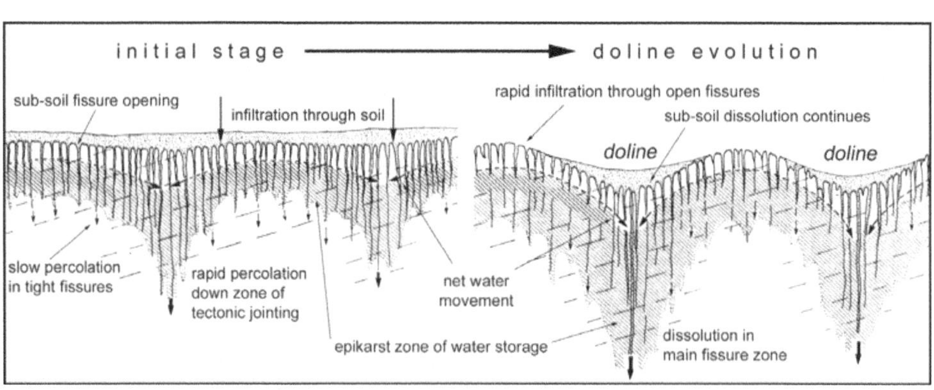

Figure 2.2. Initiation and development of solution dolines over sites of increased infiltration via zones of greater tectonic jointing.
After Williams (1983).

Figure 2.3. A steep-sided solution doline in steeply dipping limestone beneath an almost flat plateau surface in the classical karst of Slovenia.
TW.

limestone has widely spaced fissures, which have been greatly deepened by high rainfall inputs, to produce a pinnacled rockhead. Observations of quarry faces and borehole records in the Korean karst indicate an epikarst that reaches depths about four times the rockhead relief, below which the limestone is relatively tight and dry. This picture matches the concept (to which karst is no exception) of dendritic drainage convergence at depth, so that below the epikarst there are just a few isolated large conduits. In general, the epikarst reaches deeper, with greater local relief to the rockhead, and with taller pinnacles, beneath the floors of solution dolines.

Sizes and shapes of solution dolines vary enormously. A typical size is 20–200 m across. If they are less than a few metres across they may not be recognisable landforms, but some can be well over 1,000 m across. Their profiles may be anything between those of gentle saucers or steep cones, so depths vary from 1 m to well over 100 m (Figure 2.3). Many solution dolines have a rounded plan shape, though they may be elongated by valleys carrying streams into internal sinks. In mature karst regions, valleys pre-dated the development of efficient karstic drainage, and were subsequently broken into chains of solutional dolines where water sank underground at various points along the valley floors. In the evolving karst, the dolines expanded and the valley systems were destroyed, but linear patterns in the dolines may reveal the ancestral valley patterns. Uvalas are distinguished by their lobate shapes where dolines have coalesced, but this term is now rarely used.

Shafts or potholes with steep or vertical sides are surface depressions formed by bedrock dissolution in karst, normally where discrete streams drain into cave systems. They are also known as sinks, stream sinks, swallow holes or swallets, but are not solution sinkholes or dolines in the strict sense. Open shafts or swallow holes may swallow all or some of the drainage in the floors of solution

Figure 2.4. Closely packed solution dolines with polygonal inter-basin ridges in the Sinkhole Plain of Kentucky.

dolines (Box 2.1), but they may equally lie in valleys or on slopes away from dolines or in the sides of large dolines. Shaft diameters increase by dissolutional wall retreat, but rarely exceed 10 m across at a single stream sink. Larger potholes have generally evolved by progressive wall collapse, perhaps with the merging of adjacent shafts, and these are effectively collapse sinkholes.

Where solution dolines expand across a karst surface so that they all meet their neighbours, they create polygonal karst. This is entirely occupied by polygonal dolines, each typically some hundreds of metres across, between a net of interfluve drainage divides (Figure 2.4). As this matures by deepening of the dolines, the interfluve nodes are left as remnant hills that take on roughly conical shapes. Within such a cone karst, many of the intervening dolines then have stellate floor shapes with blunt arms extending between the conical hills that stand along a polygonal topographic divide. Dolines in this style are also known as cockpits, after the Cockpit Country of Jamaica that consists entirely of conical hills and cockpit dolines (Figure 1.7(d)). Dayas are formed where dissolutional removal of a thin limestone caprock creates a very wide shallow solution doline (up to 1 km across but only a few metres deep) in the arid karst of the African Sahara Desert (Conrad *et al.*, 1967). Dayas have internal drainage into porous underlying rocks; some of these are insoluble sandstones, while others are calcareous sandstones where ground has subsided due to partial removal in solution. These very large dolines, cockpits and dayas, and the even larger poljes and karst basins, are major landforms where the term sinkhole is inappropriate; they commonly have sinkholes within them, in either their soils or their rock floors. In the Chinese karst, the larger features are known as karst depressions, while smaller negative features of solutional origin are known simply as dolines.

Large solution dolines are long-term landforms. Though they are continually evolving, they have stabilities and lifespans approaching those of mountains and valleys. In some high altitude karsts, solution dolines a few metres across have developed within the 10,000 years since Pleistocene glaciation, but most solution dolines evolve over ten times as long, and many can be dated to over a million years old. Most solution dolines occur in strong, cavernous limestone. They can form in the softer limestones, including chalk, but these typically behave as more diffuse aquifers and the scarcity of open conduits limits the scope for doline evolution around single efficient drainage points. Solution dolines in chalk are generally the most gently graded and shallow of closed bowls, created by diffuse infiltration that is only slightly concentrated by centripetal overland drainage. They are common in gypsum and salt, where the much greater rates of rock dissolution do make their evolution significant within engineering timescales (Sections 2.3.2 and 2.3.3). They cannot form in the pseudokarsts of cavernous basalt and loess.

2.3.1 Solution dolines and engineering

The very slow dissolution of carbonate rocks in the natural environment precludes any geohazard from the formation of solution sinkholes or dolines during the lifetime of an engineered structure. Failure events are not a feature of these landforms, though the same does not apply to sinkholes in gypsum and salt (Sections 2.3.2 and 2.3.3).

The geohazard that solution dolines do present to foundation engineering is the higher degree of fissuring and cavitation that is likely to occur beneath a doline floor. The lowest point of a doline floor is the natural drainage collector, and is therefore a point of enhanced dissolutional erosion of the immediately underlying bedrock. The increased cavitation and water movement also increase the potential for rock collapse and for soil down-washing to form subsidence sinkholes (within the floor of the larger solution doline). A construction project should regard a doline or sinkhole apex as a site to be avoided wherever possible, besides taking into account any effects of blocking natural drains and thereby increasing flooding. Beneath the floors of many dolines, dissolutional development has widened fissures so that they are open and accessible cave passages and shafts, which create a potential geohazard. The main open caves may however descend obliquely, so that they are not directly below the doline floor (Box 2.1). Excavations in the floors of many other dolines have revealed nothing more than networks of narrow fissures, which create a lesser geohazard to foundation engineering (Figure 2.5). All solution dolines have floors so permeable that they present major difficulties for reservoir construction and landfill retention (Chapter 12).

Absolute values for the increased ground fissuring and cavitation, and therefore the potential hazard, are impossible to define in a general statement. At many sites,

Figure 2.5. A fresh road cut in limestone karst in eastern Korea that sections a shallow filled solution doline, with a zone of widened and soil-filled fissures diminishing with depth beneath its floor; the face in the image is about 15 m wide and 8 m high.
TW.

local data, ground mapping and/or drainage observation should indicate the potentially difficult zones that are best avoided. Most very large, saucer-profiled solution dolines and karst basins have one or a few point outlets that create small hazard zones between large areas of relatively stable ground. A polygonal karst such as Kentucky's Sinkhole Plain (Figure 2.4) has 100% of its area occupied by sinkholes (or dolines), though less than 10% of this (as small areas at the lowest point of each sinkhole) is highly fissured ground, in contrast with strips of ground along the inter-basin ridges that are commonly underlain by rock that is almost non-cavernous. Perhaps the most unpredictable sites are the large flat dayas in the deserts of North Africa; their flat floors, and the scarcity of modern rainfall, give little or no indication of the extent or locations of any underlying dissolution cavities that were formed in past times with wetter climates.

There are positive aspects to solution dolines. In mature karst, their alluviated floors offer the only ground suitable for agriculture, and thousands of families around the world survive purely by farming on doline floors (Figure 2.6). However, this land may be prone to flooding where the floor is at a level within the zone of annual fluctuation of the local water table. Seasonal flooding of deep dolines in China has been controlled by cutting drainage tunnels into neighbouring dolines at lower levels. Doline flooding is also caused by farming practices that increase soil erosion and subsequent blocking of the cave drains below the dolines, or by inappropriate urbanisation. One example of an unusual benefit of a doline is

Figure 2.6. A solution doline in the Xingwen karst of Sichuan, China, where the alluviated doline floor and a narrow zone on one side have been graded by a local farmer as they offer the only soils for farming in the rocky karst terrain.
TW.

the Arecibo doline in Puerto Rico which matches the shape of the huge dish reflector required for the ionosphere observatory (Monroe, 1976).

On a broader scale, the combination of conical hills and deep solution dolines in a mature cone karst creates an unusually complex topographic relief. The substitution of valleys by polygonal dolines leaves a conspicuous shortage of linear corridors through the karst, making road and rail construction expensive. New railways through the karst of China have incorporated numerous tunnels between viaducts and embankments that have to avoid flood-prone doline floors.

2.3.2 Solution dolines on salt

Rock salt, or halite, is so rapidly and highly soluble in natural waters that its dissolution can cause ground subsidence at rates significant to engineered structures. A conduit 1 m in diameter can be dissolved out of salt within a few years, and surface lowering by rockhead dissolution can achieve rates of >10 mm/year.

Salt does not survive at outcrop except in deserts with negligible rainfall and negligible dissolution. Where salt (or a saliferous bed of interbedded salt and mudstone) is exposed at the surface or at rockhead, its rapid dissolution leaves an insoluble residue as a layer of collapse breccia. This is sometimes known as "wet

Figure 2.7. Subsidence across the active solutional depression that houses Elton Flash in the salt karst of the Cheshire Plain, U.K.; the centre of subsidence is left of the railway, causing both lateral ground strain and downward movement towards it.
TW.

rockhead", and overlies an epikarst of fissured salt where dissolution continues unless it is protected by a layer of saturated brine. Where dense brine almost stagnates on the rockhead, the impermeable salt cannot be reached or dissolved by fresh and chemically aggressive groundwater. Large caves are rare in salt as the ductile material commonly permits void closure at rates matching dissolutional enlargement.

Where groundwater flow patterns towards natural brine springs allow continuing salt dissolution over a limited area of rockhead, the resultant areal subsidence may be some metres deep and in the order of a kilometre across. In the English saltfield that underlies the Cheshire Plain, the Elton Flashes are active areal subsidences of this size (Waltham, 1989). These are effectively very broad solution dolines, though their dimensions are more comparable to karst basins and poljes in limestone karst. Under natural conditions of groundwater circulation, these areal subsidences on salt typically subside at rates of up to 15 mm/year. Short-term subsidence rates of up to 100 mm/year, recorded over small areas, may relate to small deep-seated collapse events within the rockhead collapse breccia. Solution mining involves various methods of pumping brine from wellfields for purposes of salt production (Waltham, 1989). One method is uncontrolled "wild brining", where brine is abstracted from shallow wells sunk into zones of wet rockhead. This greatly increases water circulation, drawing renewed flows of aggressive fresh water onto the salt, and thereby increasing dissolution rates. While brining was taking place nearby, the Elton Flashes subsided at rates of up to 900 mm/year, necessitating ongoing engineering works on the railway that crosses them (Figure 2.7). Since the brining operations have ceased, the subsidence has reduced to the minimal rates that can be ascribed to natural dissolution in the undisturbed drainage regime.

Without disturbance by brining, solution dolines on salt may be largely self-stabilising. The numerous "meres" of the Cheshire Plain are stable lakes within old

Figure 2.8. Houses in Sandbach subsided and tilted due to dissolution and removal of salt from the rockhead zone beneath 20 m of drift cover, in the Cheshire Plain, U.K.
TW.

solution dolines. They are freshwater, with their surfaces at the regional water table within the glaciofluvial sediments that veneer the Plain. Salt removal beneath the surface dolines has created rockhead depressions that now hold ponded brine porewater and are lined with insoluble clay residues, so that further dissolution is minimal or absent.

Though natural dissolution and subsidence rates on salt are high compared with those on limestone, the geohazards and engineering difficulties on salt are almost entirely related to the vastly enhanced dissolution rates induced by brine pumping (Bell, 1975; Waltham, 1989). Subsidence patterns can be recognised from historical records and sometimes from detailed geomorphological mapping, notably in the linear subsidences (or solution troughs) that form over lines of concentrated groundwater flow (known as brine streams) at rockhead. Distribution of isolated non-linear subsidence zones (that could be described as shallow solution dolines or sinkholes) are not so easily predicted, and all areas over salt rockhead (or outcrop) must be treated as zones of potential subsidence hazard (Figure 2.8).

Beds of salt at depth may be totally dry where they are interbedded with impermeable mudstones. This is the situation in the Cheshire Plain, and the dry salt beds overlain by mudstone (in areas known colloquially but incorrectly as "dry rockhead") cannot cause solutional subsidence. These areas are exploited by controlled brining operations, which create stable dissolution caverns with no subsidence risk. Isolated surface collapses over these solution mines have the appearance of sinkholes (Walters, 1977) but are, quite simply, mine failures, outside the scope of these pages. However, salt beds within sequences of permeable rocks may be prone to dissolution in circulating groundwater at depths of many hundreds of metres, creating a type of interstratal karst beneath insoluble caprock. Wide subsidence basins may therefore be created, but may become inactive when the remaining salt is too deeply buried (Johnson, 1989a), and more localised ground subsidence may develop over deep-seated breccia pipes (Section 3.4.3).

2.3.3 Solution dolines on gypsum

The solubility of gypsum and its dissolution rate in natural water lie between those of limestone and salt. Total removal of gypsum can take place within engineering timescales where flows of aggressive surface or ground water are localised and continuous. In a worst-case scenario, where major leakage under a dam is driven by a steep hydraulic gradient, fissures may widen in gypsum by more than 10 mm/year, which is over ten times faster than in limestone (Dreybrodt *et al.*, 2002). This dissolution rate could not be matched across an entire solution doline, but it does indicate the potential for very localised gypsum removal at rates that could make surface subsidence significant.

Worldwide, the most common solution dolines on gypsum are over 100 m across and only a few metres deep (Sauro, 1996), so that they are often described as subsidence basins or areal subsidences. The low profiles are a function of diffuse infiltration and gypsum dissolution across the entire doline floor, with less central concentration of surface lowering as in many limestone dolines. They have evolved through gypsum dissolution at the rockhead and within the epikarst, but there is debate over the details of the mechanisms of ground lowering. Dissolution may create tight networks and mazes of small cave passages within the top few metres of the bedrock, and these then collapse across large zones due to the low mechanical strength of gypsum. This process of localised cavitation and collapse is included within the concept of the subsidence doline or sinkhole (see above), but may be notably more extensive in gypsum than in limestone.

This mechanism has been invoked in the case of many shallow depressions in the gypsum karst around Ripon in northern England (Cooper, 1986). Though many of these features are actively subsiding, little has been seen of their subsurface structure. Others of the broad and shallow depressions around Ripon are interpreted as overlying active breccia pipes, and these would therefore be classified as collapse sinkholes. There are also smaller features that are simple collapse or caprock sinkholes (Chapter 3). The town of Ripon is built on the gypsum outcrop, and many houses are damaged by ongoing, slow, active subsidence (Figure 2.9). This may be due to rapid dissolution of the gypsum along localised drainage zones at, and close to, the rockhead, which lies beneath some metres of alluvial cover. However, the localised surface lowering may be all or partially due to compaction of partly drained sediments, including peat, that fill old and relatively stable solution

Figure 2.9. A terrace of houses in Ripon, U.K., subsided due to a combination of rockhead dissolution of gypsum and compaction of soft sediments filling an old buried solution doline. TW.

depressions in the rockhead (Cooper, 1998; Waltham and Cooper, 1998); in this case, the features would be classified as buried sinkholes.

Solution dolines, each 100–500 m across, can be so closely packed that they constitute very spectacular polygonal karst on some gypsum outcrops (Figure 2.10). Very fine examples are known in the dry uplands of eastern Turkey (Waltham, 2002) and Uzbekistan, where low rainfalls ensure low rates of dissolution. In wetter climates, gypsum does not survive at outcrop (and only to a limited extent at rockhead), where it is replaced by breccias of insoluble residue. In these cases, doline karst cannot evolve, but there remains a significant engineering geohazard from dissolution of the last traces of gypsum and collapses of small multiple ground cavities (Cooper and Saunders, 2002). The most extensive infrastructure damage by solution doline growth on gypsum takes place in areas of drier climates, such as Sicily, Spain (Gutierrez and Cooper, 2002) and interior U.S.A. (Rahn and Davis, 1996).

2.4 COLLAPSE AND CAPROCK SINKHOLES

Both these types of sinkhole are defined by fracturing, breakdown and collapse of unsupported bedrock slabs, beams and arches that are left around dissolutional

Figure 2.10. Polygonal karst with a network of solution dolines, each floored by soil that is dark where ploughed, on the gypsum outcrop near Sivas, Turkey.
TW.

cavities in karst. They vary in size, but few are more than 100 m across, and are generally distinguished by steep rocky profiles, unless degraded by subsequent erosion.

The processes behind the formation of collapse sinkholes are inextricably linked with the dissolution of bedrock, which enlarges fractures into open fissures and then into shafts and potholes. The same processes of fissure opening also undermine small blocks of limestone, that are therefore broken and subsequently eroded, and thereby contribute to the surface lowering that also characterises solution sinkholes. Collapse and solution sinkholes are linked by a spectrum of processes and morphologies between end-members that are extensively collapsed or are almost wholly dissolutional. Even the clear cases of massive rock collapse, where roof rock fails into cave chambers, are preceded by the dissolution that created those chambers. Once a cave chamber is formed, roof stoping may cause upward void migration – ultimately causing ground surface failure in a collapse sinkhole, without any further dissolution. Caprock sinkholes are formed where similar roof stoping migrates up through overlying insoluble rocks. Breccia pipes are formed by similar roof failure, and now appear as columns of failed debris with or without modern surface expression; they are more common over gypsum and salt than over deeply buried limestone.

Collapse sinkholes are not common, and though the final rock collapse may be almost instantaneous, natural collapse events are extremely rare. The main geohazard posed by rock collapse is roof failure where engineering loads are imposed over unseen, shallow caves, thereby artificially inducing a collapse sinkhole. Natural sinkholes, both collapse and caprock, are considered further in Chapter 3, and the geohazards of induced rock collapse are assessed in Chapter 7. It is also significant that collapse sinkholes can develop in a variety of insoluble rocks, notably basalt (Chapter 6).

2.5 SUBSIDENCE, SUFFOSION AND DROPOUT SINKHOLES

The most widespread geohazard in karst of limestone or any other rock is not due to the collapse or failure of the soluble rock. Instead it is due to the failure of ground as soil is rapidly washed down into bedrock cavities that have formed over geological timescales; the rock cavities remain stable while the soil fails into them. The effect on the ground surface is rapid and cause severe localised subsidence – in the form of subsidence sinkholes. In non-cohesive soils the subsidence process tends to take place over months or years, and the results are known as suffosion sinkholes. More cohesive soils can develop larger voids that subsequently collapse, and actual failure of the ground surface can be instantaneous and catastrophic, as a dropout sinkhole event. These events are referred to in China as karst collapses (which do not distinguish between dropout sinkholes in the soil cover and collapse sinkholes that originate in the rock beneath). The infinite variations of soil properties accounts for a spectrum of failure processes and rates. Both suffosion and dropout sinkholes, along with all those sinkholes with intermediate properties, are conveniently grouped under the name of subsidence sinkholes.

Suffosion and dropout sinkholes provide many of the major problems for engineering works on cavernous karst, and also account for the great majority of sinkhole damage to roads and buildings (Figure 2.11). They may be 1 m across or 100 m

Figure 2.11. A house in Centurion, South Africa, destroyed by development of a large subsidence sinkhole when the soil beneath its foundations was washed into the dewatered underlying limestone.
Photo: Fred Bell.

across, but even the smallest can deprive a column base of its integrity. Both types are considered further in Chapter 4. Their key formation process is soil removal by downward drainage, and the great proportion of subsidence sinkhole events are induced by the disturbance of natural drainage patterns during engineering activity, as further assessed in Chapter 8.

2.6 BURIED SINKHOLES

Karst depressions of any type, size, shape and morphology that are features of the rockhead, wholly or partly filled, buried or obscured by the soil cover, are best known as buried sinkholes. Most originate as solution sinkholes, with or without modification by collapse, and are subsequently buried or filled by sedimentation induced by climatic or environmental change. They may be 1–1,000 m across, with gentle or steep profiles. They include filled caves, often known as pipes or pocket deposits, which are disproportionately abundant in chalk karst. Deep-seated breccia pipes may be viewed as filled sinkholes, but they develop upwards by roof failure, and are therefore more closely related to collapse sinkholes (Chapter 3).

Buried sinkholes constitute a geohazard for engineers where their soft fills create areas of unstable ground over very uneven rockhead. They can also lose their fill material by suffosion into underlying voids, so that they become the sites of new subsidence dolines. Compaction sinkholes are shallow depressions formed by compaction of the soil within buried sinkholes. All types are considered further in Chapter 5.

2.7 KARST TYPES AND SINKHOLE DISTRIBUTION

Sinkholes occur in all types and varieties of karst. They tend to be larger and/or more numerous in the more mature karst terrains, which are generally those in the hot wet tropical environments. This reflects the higher rainfall and greater availability of biogenic carbon dioxide that together maximise the processes and rates of dissolution of carbonate rocks, including limestone (see Section 1.4). The largest collapse sinkholes are features of the largest cave passages, which are mainly found in the tropical regions of cone and tower karst.

There is however infinite variety in the distribution of caves and sinkholes within karst terrains, and isolated features that are anomalously large can occur scattered across terrains of otherwise smaller landforms. Much depends on local details of geological structure. The only predictable aspect is that caves and sinkholes are likely to be larger and more numerous in the karst adjacent to outcrops of insoluble rocks – which can supply large inputs of dissolutionally aggressive water from allogenic surface catchment basins. Drainage from shale caps provides long lines of stream sinks, caves and sinkholes along the outcrop boundaries of the limestone in Britain's Yorkshire Dales karst. Both the geomorphic history and the paleoenvironmental climatic record are also significant, in that they can account for isolated remnants of past drainage regimes that now appear as scattered caves or sinkholes out of scale with more widespread modern features. Desert regions on carbonate rocks are notable for their minimal modern karst, but also for isolated features that formed in pluvial stages within the climatic oscillations of the Pleistocene.

Scale increases of carbonate dissolutional features in hot and wet climatic regimes are not matched by the landforms on evaporite rocks. Conversely, salt karst mainly becomes a geohazard in desert regions. The world's only extensive open caves and modern collapse features in salt lie in outcrops in Central Asia and the Middle East, notably in the Israeli Negev Desert and on some of Iran's islands in the Persian Gulf. Salt has been completely removed by dissolution at outcrops and shallow depths in all regions with significant rainfall, so that it survives to create dissolutional subsidence features only beneath thick drift and residual insoluble breccias, as in England's Cheshire Plain, or as deep-seated interstratal karst, as under the Canadian prairies.

The correlation between rainfall and sinkholes on gypsum is different again. The largest and most numerous caves and sinkholes on gypsum occur in continental interiors with low rainfalls, mainly of 200–500 mm/year. In drier regions, dissolution is limited by the scarcity of water. In wetter regions, gypsum has been largely or entirely removed at and close to outcrop, though it does survive at depth to create extensive geohazards over interstratal karst. In all but desert terrains, the sinkhole geohazard on gypsum is enhanced by its high rates of natural dissolution.

2.7.1 Sinkholes in the engineering classification of karst

Karst ground conditions have been divided for engineering purposes into a progressive series of five classes (Waltham and Fookes, 2003). These are represented in Figure 2.12 by their typical morphological assemblages. The five classes provide the basis of an engineering classification that characterises karst in terms of the complexities and difficulties to be encountered by the foundation engineer. Description of karst ground conditions by a single class label creates concepts of the scale of anticipated foundation difficulties, but the variations that are typical of karst demand a more specific and more detailed definition. A full description of karst ground conditions embraces four descriptors; "Karst class + sinkhole density + cave size + rockhead relief" (Waltham and Fookes, 2003). The classification is not quantitative, as a points-scoring system comparable to those used for rock mass rating has not yet been developed. A description such as "a mature karst of class kIII, with about 3 solution sinkholes and over 40 subsidence sinkholes per km^2, numerous cave passages mostly <3 m wide and narrow fissures creating observed rockhead relief of 3 m", focuses attention on the potential geohazards. Every karst site must be assessed individually with respect to engineering development, though Waltham and Fookes (2003) give outline concepts of appropriate ground investigation approaches and likely foundation concepts.

Sinkhole numbers constitute one of the key parameters in the engineering classification. The sinkhole density may be a simple number per unit area, but is

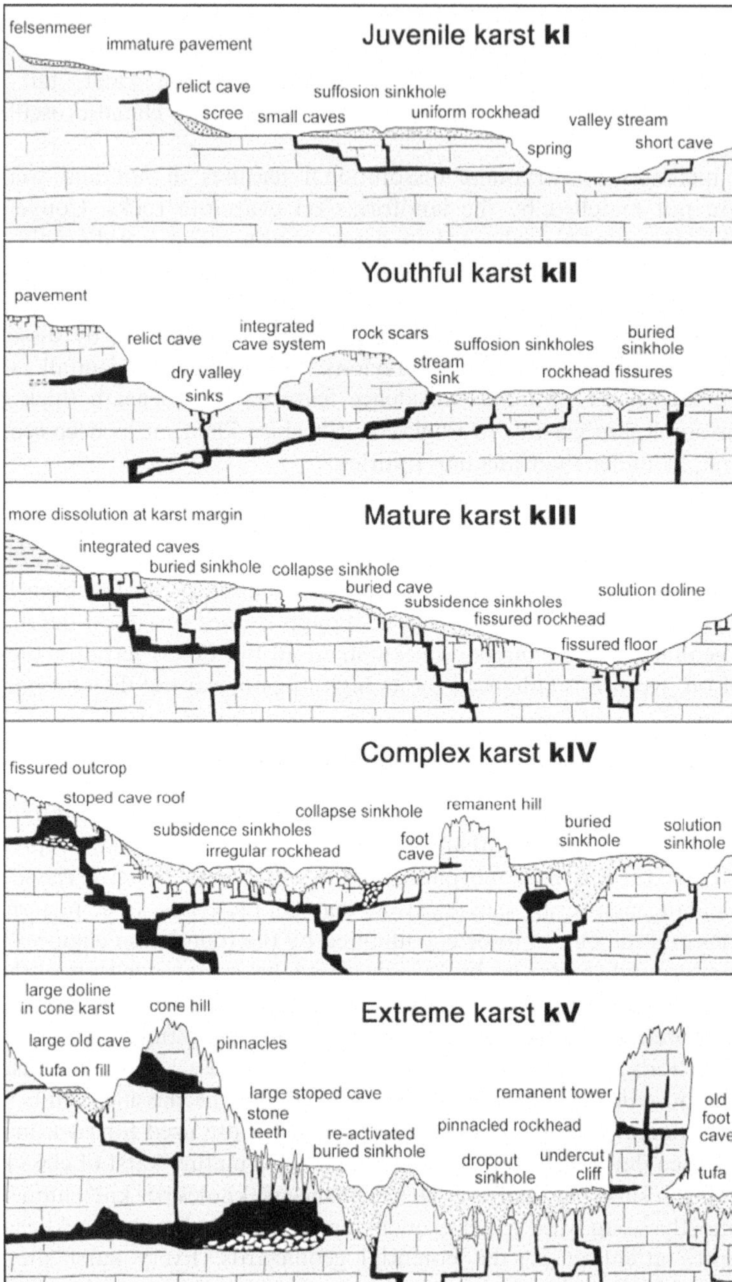

Figure 2.12. Diagrammatic representations of the five classes in the engineering classification of karst, with various sinkholes and solution dolines forming one of the components of all karst terrains.

From Waltham and Fookes (2003).

qualified if densities are low because the sinkholes are large, and is improved if the types of sinkholes are recorded. In areas with subsidence sinkholes in drift cover, it is useful to define the rate at which new sinkholes failures (NSH) occur, expressed in events per km^2 per year, but records are rarely adequate for anything better than a broad generalisation. This is also qualified if the NSH rate is temporarily enhanced by engineering activities including any water abstraction or drainage modifications.

The abundance and scale of sinkholes may be broadly characterised within each of the five classes of karst. The following notes are expanded from the tabulated definitions by Waltham and Fookes (2003).

Juvenile karst (class kI) occurs mainly in desert and periglacial zones or on impure carbonates. It has few sinkholes, with NSH <0.001. Dissolution is so restricted that solution sinkholes have barely developed beyond the opening of fissures where small streams sink. Few caves are more than a metre wide, so collapse processes have not been initiated. Scattered suffosion sinkholes may have formed in any soil cover, but they are unlikely to be more than a few metres across. Buried sinkholes may exist as relics from wetter or warmer climates in the past. Limestones in some modern deserts contain truncated fragments of large old cave passages and also hide isolated large buried dolines, all relics from wetter Pleistocene climates. Unless these features dominate the landforms of a given area, a desert limestone outcrop with such isolated old features is most likely to be classified sensibly as juvenile karst.

Youthful karst (kII) has the minimum scale of dissolutional features normally encountered in temperate regions. Suffosion and dropout sinkholes are common, but typically, <10 m across, and NSH is 0.001–0.05. Streams flow into open sinks in solution sinkholes that are also generally <10 m across. Collapse sinkholes are untypical and then only a few metres across, but scattered larger buried sinkholes can occur as in the juvenile karst.

Mature karst (kIII) is the normal on carbonate rocks in temperate and Mediterranean climates. In many regions, solution sinkholes are up to 100 m across, and may be so frequent that they create a polygonal system within a doline karst terrain. Suffosion and dropout sinkholes may number >100 per km^2, with NSH of 0.05–1.0, though their abundance and size also depend on the extent and thickness of any soil cover (Chapter 4). Collapse and buried sinkholes may both be significant, and the presence of large caves may allow caprock sinkholes to develop in suitable geological situations.

Complex karst (kIV) is more typical of the tropical regions, but can be developed in some more temperate climates. Large solution dolines may be the dominant landform, with individual features up to 1 km across. Solution and collapse sinkholes can occur in their floors. Subsidence sinkholes are numerous in any soil cover, with NSH of 0.5–2.0, though individual sizes are still restricted where the soil cover is thin. Buried sinkholes may be numerous and large as past climates are also likely to have been warm and wet.

Extreme karst (kV) has the largest of all dissolutional features, including ground cavities, and is only found in the wet tropics. The landscape has cone and tower hills interspersed with solution dolines and karst basins kilometres across, with all types of large active sinkholes in their floors; NSH is normally much greater than 1, including dropout sinkholes that can occur almost anywhere in organic soils overlying heavily fissured bedrock. Buried sinkholes are so large and deep that soil compaction within them, either naturally, enhanced by dewatering or under imposed load, may cause differential surface subsidence that is significant to engineering.

As a guide to sinkhole distribution and geohazards, the engineering classes of karst can only offer broad indications, but they do emphasise the enormous variations of the sinkhole geohazard in different karst regions.

3

Rock failure in collapse and caprock sinkholes

3.1 KARSTIC COLLAPSE

All ground voids constitute elements of weakness within a rock mass, and karst is distinguished by having the largest natural voids, where roof failure can create a significant geohazard. The natural consequence of progressive roof failure is upward void migration, which may reach the surface where it causes instantaneous major subsidence in the form of a collapse sinkhole. Where the roof failure migrates up through non-karstic rocks, the surface failure in an outcrop of insoluble rock creates a caprock sinkhole. Both these forms of sinkhole involve failure and collapse of bedrock, and are therefore distinct from the subsidence sinkholes where soil cover is flushed into stable rock fissures (Chapter 4).

Collapse and caprock sinkholes are initiated where cave passages or cave chambers are enlarged beyond the limits of their own roof rock stability. Roof collapse is a natural and automatic process in all karstic caves, though its development to a scale that influences surface stability is dependent on the necessary geological structures and the long periods of geological time for processes to mature. Failure of a cave roof can be initiated or accelerated by imposed loads from construction works, and is therefore a major geohazard where large caves exist at shallow depths (Chapter 7).

Processes of cave roof collapse within gypsum and salt are comparable to those in limestone, but are distinguished by the much lower mechanical rock strengths. Collapse sinkholes are formed in both rocks, and caprock sinkholes are proportionately more important than they are in limestone. Breccia pipes are related structures that originate with deep-seated dissolution in these rocks, but are rare over limestone. Collapse sinkholes also develop over lava caves, but the origins, structures and formative processes of both the caves and the sinkholes are very different from those of karst landforms, as they are unrelated to post-genetic dissolution (Chapter 6).

3.2 COLLAPSE OF CAVE CHAMBERS

The stability of a limestone cave is a function of its unsupported span and the structural integrity of its roof rock. Most caves are structurally sound; relationships between cave widths and the rock mass quality of cavernous limestones suggests that the majority of caves would require little or no support if they were regarded as engineered structures (Waltham and Fookes, 2003). A tubular tunnel dissolved out of massive limestone far beneath the surface is extremely stable. In contrast, even the smallest caves can collapse in zones of heavily fractured rock or beneath very thin roof slabs (Figure 3.1).

Cave passages are typically less than 10 m wide in most temperate regions, though chambers formed at optimum sites or by passage coalescence are common to about 50 m across, and isolated larger caverns do exist. In tropical regions, caves 30 m across are not unusual, and numerous chambers (often known as caverns) are over 100 m wide. Surface collapse is related to cave size, and there is therefore a climatic influence on the size and frequency of collapse sinkholes in limestone karst. (Parameters for caves in gypsum and salt are different, and are discussed below.)

The positions, shapes and sizes of both cave passages and cave chambers are guided by the structural and stratigraphic features of the host rock. The geological influences can be recognised in nearly all accessible caves, but the multiple choices offered by complex rock structures mean that the positions of neither caves nor enlarged chambers can be predicted ahead of exploration. Large caverns tend to

Figure 3.1. Collapse of the thin limestone roof over a small cave in the side of a wadi in Jordan; the two beds that roofed the cave, and partially survive in the entrance arch, are probably an indurated duricrust over a more easily eroded blocky limestone.
TW.

Figure 3.2. The world's largest known cavern – Sarawak Chamber, in the Mulu karst on Borneo. Eight cavers with very large flashguns just light the multiple arches that form the roof, with the far wall 300 m from the camera.
Photo: Jerry Wooldridge.

form where a weak, thinly bedded or densely fractured limestone is underlain by a stronger or more massive unit, so that rapid erosion can progress beneath a stable roof (Gilli, 1986). Favourable sites for cave chambers occur where larger flows of water enter the karst, from rivers off insoluble rocks or where flows are concentrated through breaches in shale beds within the limestone sequence. Many chambers have formed by coalescence of adjacent passages and shafts with intervening wall collapse, by lateral wall undercutting, or by enhanced mixing-corrosion (Bögli, 1964) at the junctions of major passages. The world's largest known cavern, 700 m long and 300 m wide in the Mulu karst of Sarawak (Figure 3.2), is formed in massive limestone with bedding planes 15–20 m apart, where a major cave river cut laterally on a favourable geological structure (Waltham, 1997). In contrast, the famous Carlsbad Cavern, in New Mexico, was cut in almost structureless reef limestone by slow-moving water enriched in sulphuric acid that was derived from hydrocarbons in adjacent basins.

3.2.1 Cave roof breakdown by bed failure

Roof collapse by progressive roof failure is widespread in limestone caves, and can be very conspicuous in any zones of more thinly bedded limestones (Figure 3.3). These contrast with roof profiles that retain their original dissolutional features in massive limestones or within single very thick beds, and can span large voids with no sign of breakdown (Figure 3.4). The simplest analysis of a cave roof in bedded limestone is to treat the beds as beams failing under their own weight between the cave walls.

For an unsupported beam of span length L, of unit width, of depth d, of unit weight γ and of weight W ($= Ld\gamma$), the bending moment $M = WL/12 = L^2d\gamma/12$, and the section modulus $Z = d^2/6$.

Assuming elastic behaviour of the rock, failure occurs when the tensile strength T is reached at M/Z in the outer surface of the deforming beam.

Consequently the stable beam depth (or bed thickness) $d = L^2\gamma/2T$.

This assumes that the beam is constrained at its ends, which is the normal situation in a rock mass. If the beam is unconstrained, due to the presence of open fissures, the bending moment is $WL/8$ and the stable beam depth $d = 6L^2\gamma/8T$.

For limestone, the unit weight γ may be taken as $26\,\mathrm{kN/m^3}$.

The critical factor is therefore the tensile strength T of the limestone, and, for a typical cavernous limestone of unconfined compressive strength around 100 MPa, this may be taken as about 5 MPa. The stability envelope for constrained limestone roof beams of this tensile strength may therefore be determined (Figure 3.5).

This representation of cave roof stability by beam integrity is only a simplification of reality, though it has been used to design safe working under bedded rocks in

Figure 3.3. Cave passages with flat limestone roofs developed by progressive failures along bedding planes; on the left in Mammoth Cave, Kentucky, and on the right in Ogof Agen Allwedd, U.K.; in both caves the roof is some metres above the original dissolutional passage. TW.

Figure 3.4. Massive limestone forming the roof of the Niah Great Cave, Sarawak, with no signs of breakdown to destroy its dissolutional sculpture.
TW.

British iron ore mines (Whittaker and Reddish, 1989). Factors that reduce stability include rock jointing that weakens or destroys the beam, any overburden load that bears on the beam, dipping geological structures that complicate the roof profile and time that allows strength decrease from the short-term levels. Factors that increase stability include cohesion within the shale that occupies most bedding planes in unweathered limestone, irregular fractures that allow blocks to lock against each other and any development of compression arches through multiple bed sequences.

Data on approximate bed thicknesses and roof spans, from numerous caves around the world that have either failed or remain stable are included in Figure 3.5. The observed data show a reasonable correlation with the calculated beam envelope, although a rather better limit is given by the simple relationship of $L = 17d$. Earlier analyses of roof beam failure (Davies, 1951; White and White, 1969; White, 1988) used shear strength in place of tensile strength for failure and therefore achieved an unrealistically large stability envelope (Figure 3.5). Rock strength decreases with time due to fracture propagation from stress points at grain boundaries on critical surfaces, and it is realistic to cite a long-term decrease of tensile strength to about 30% of the short-term value (Tharp, 1995). However, the stability envelope based on this reduced tensile strength is small in relation to observed data (Figure 3.5). Unconstrained displacement of the rock beam ends also reduces the stability envelope, and appears to be unrealistic as it has even less correlation with observed data. Analysis of mine roofs in strong bedded limestone was correlated with monitoring data from an experimental mine room (Merrill, 1957) to define a stability envelope very similar to that proposed above for cave roofs (Figure 3.5).

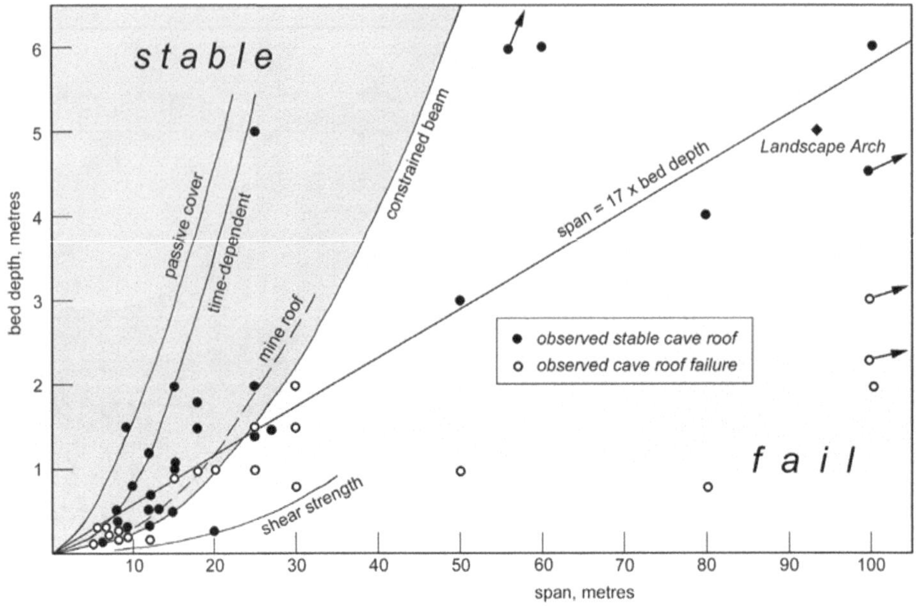

Figure 3.5. Correlation between bed thickness and chamber span width with respect to the failure of intact rock beams over wide caves. The shaded stability envelope is calculated for constrained beams of rock with a tensile strength of 5,000 kPa. The other envelopes are (from the left) calculated for a passive cover as thick as the cave width, based on long-term tensile failure of limestone at 30% of its short-term strength (Tharp, 1995), determined for roof failure in limestone mines (Merrill, 1957), the simplistic relationship of $L = 17d$ (span width $= 17 \times$ bed depth) and calculated for beam failure by employing its shear strength (White, 1988). Point data refer to failed and/or stable observed roofs in 43 caves and also the sandstone Landscape Arch, Utah.

Where a cave roof is reliant on support by a single thick bed of rock, any cover of more thinly-bedded rock and/or unconsolidated soil adds a distributed load to the roof beam, and thereby reduces the stability envelope (Figure 3.5). The lack of correlation between this envelope and the observed data suggests that this is not the general case. Most roof rock masses appear to be internally supportive, with each bed or element providing its own support.

Cave roof failures that lie inside the stability envelope (Figure 3.5) may be ascribed to rock fractures that destroy beam integrity. Cave chambers only survive in rock with relatively intact beds. Heavily fractured rock does not appear in Figure 3.5 because chamber roofs collapse before they can become accessible and observable, and erosion is directed to create tall and narrow fissure caves instead of chambers. Stable caves that lie outside the stability envelope appear to gain their support from compression arches, and not purely from rock beams.

3.2.2 Stable arch development in cave roofs

Distortion of gravitational stress around a ground cavity creates an arched compression zone over the roof and into the walls, with a tension zone in the roof immediately beneath the arch. Most natural cave passages and chambers evolve towards an arched profile as rock falls away from the tensile zone where it is immaterial to the total roof stability. The large entrance chamber of Tham En (Figure 3.6), in the karst of Laos, has evolved to a stable arched roof profile in massive rock after stoping up through thinly bedded limestone (Waltham and Middleton, 2000). A cantilever is weaker than a beam, but there are zones within the compression arch where beds remain in cantilever and contribute to support of a wider span. Within almost any arch there are zones where bedding planes and fractures are oblique to the compressive stress and the underside is unconfined. Shear failure in these zones leads to development of higher arches to achieve stability. Uncontrolled breakdown of mine roofs generally leads to the development of a stable arch profile when the arch rise equals about half the span width (Franklin, 1989), a profile that is close to that of the theoretical tension zone above a cavity. Even in well-bedded limestones, most cave chambers are observed to have arched roof profiles, and most of these have a span much greater than the rise of the arch; the Tham En roof arch rises by less than a third of its span.

Natural cavern roofs that have evolved slowly and only carry their self-load are stable in very low arch profiles in the strong limestones that can contain large caves. For a given roof thickness and span, there is a range of possible arches that balance arch height against arch thickness. A higher parabolic profile develops in weak rocks, notably mine roofs in thinly bedded coal measures, but strong limestones are stable in much lower arches, especially those that are laterally constrained by high horizontal stresses in the wall rocks; furthermore, the thick roof over a deeply buried cave chamber has space to develop an arch of almost any profile. In practice,

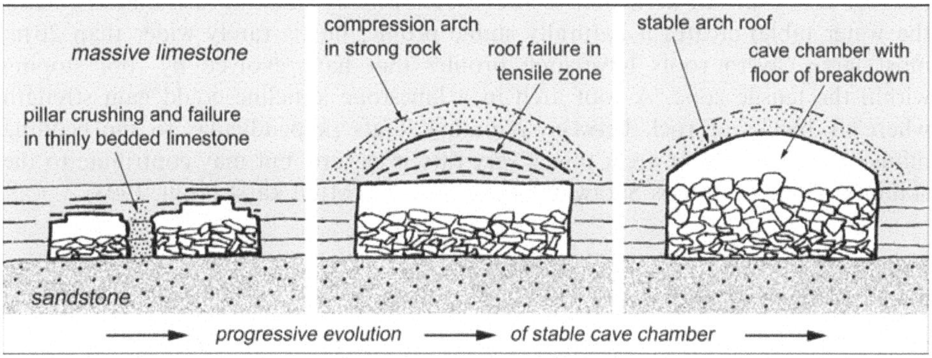

Figure 3.6. Development of a compression arch in massive limestone over the southern entrance chamber of Tham En, Laos; roof failure and passage enlargement in thinly bedded limestones was followed by evolution of an arched cave roof by stoping of rock from the tension zone.

Figure 3.7. The stable, low-profile, compression arch that forms Landscape Arch in the sandstone of Arches National Park, Utah, spanning 93 m and only 5 m thick.
TW.

the arch profile is determined by the details of its geological structure, and these remain unseen in an intact arch. Analyses of arched cave roofs are therefore very approximate, as they have to rely on estimates of the reduction of rock mass strength due to the unseen fractures. These become critical to ground stability where the rock may be stressed by imposed construction loads (see Chapter 7).

A compression arch is much stronger than a beam in the same material, and natural arches can be very thin. The spectacularly thin Landscape Arch (Figure 3.7), in Utah, demonstrates how strong rock is under compression. It is weathered out of sandstone that is weaker than most cavernous limestones, and is not devoid of fractures. Regarded as one bed, it plots on Figure 3.5 very close to the stability limit of $L = 17d$. An arched cave roof developed by phreatic dissolution (below the water table) creates a naturally stable profile, but is rarely wider than 20 m; most large cavern roofs have arch profiles that have evolved by roof stoping within the tensile zone. A roof arch in a limestone anticline could gain strength where its blocks of rock between radiating joints perpendicular to the bedding mimic a stable voussoir arch; this is very rare in nature, but may contribute to the stability of the enormous Sarawak Chamber in the Mulu karst (Gilli, 1993).

3.2.3 Breakdown processes

Cave roof breakdown is generally initiated by lateral stream undercutting that is excessive for the roof span strength. This becomes self-propagating where a stream is then diverted into an undercut around a pile of fallen breakdown. Roof collapse may also be initiated where a phreatic chamber (formed below the water table) is first drained by rejuvenation. Buoyant support in water takes nearly 40% of the rock load, but this is imposed immediately on drainage, when it probably creates the largest short-term stress increase in the lifetime of many chambers. Hydrostatic pressure due to the aquifer head is not relevant as it is applied equally as joint water pressure within roof fissures. Mineral wedging (notably sulphate growth) can contribute to roof breakdown in some environments (White and White, 2003), while frost action becomes significant in caverns at shallow depths, including entrance chambers, outside the tropical regions.

While the final collapse of a cavern roof is an instantaneous event, progressive stoping and cavity migration may extend over geological timescales, and there is almost no available data that records roof evolution. Roof stoping in abandoned mines has been known to migrate through tens of metres of cover to cause surface crown holes (also known as chimney subsidences) within only months of the loss of support in the underlying mine. Surface failures have occurred within 3–10 years of the initial roof collapse in ironstone mines at depths of 100 m in Britain (Whittaker and Reddish, 1989). This rapid stoping only occurred where water entered from overlying aquifers within sequences of weak rocks, so the analogy to cavernous limestone is limited. In contrast, some crown holes have developed hundreds of years after mine abandonment, but these data apply mainly to weak sedimentary rocks of the Coal Measures. In Quebec, Canada, roof failure of a mined cavity at a depth of 70 m reached the surface by uncontrolled progressive stoping within less than two days (Franklin, 1989). Major surface collapses over large brining cavities in salt have followed stoping failure through hundreds of metres of rock within only months or years (Section 3.4.3), but these are also failures in weak cover rocks.

In the Pokhara basin of Nepal, natural cave excavation and subsequent total collapse of a roof about 20 m thick appears to have occurred all within about 500 years (which is the likely age of the host limestone), but this is an exceptional site of powerful erosion in weak rock (Waltham, 1996). Most cave chambers can only be dated as many thousands of years old, and early roof failures are likely to be followed by more stable conditions as arched profiles evolve. Cave roof stoping rates remain unknown, but are likely to be much lower in strong limestones than in the situations cited above. Roof reduction by progressive stoping failure presents a karst geohazard that is negligible when compared to rock failure accelerated by imposed loads from construction works over an unknown cavern.

There is limited available data on cavern roof collapses in gypsum and salt, both of which are mechanically weaker than most limestones. Large chambers in salt are created by brining operations, and, unless carefully controlled, many do eventually collapse to form breccia pipes and/or sinkholes (see below). Cave chambers wider than about 25 m are almost unknown in gypsum. Roof collapse is complicated by the material's plastic deformation and its transitions to and from anhydrite, so that beds can bend and curve away from cave roofs without breaking. Dissolutional undercutting and fissure opening is so rapid in gypsum that cave streams can promote collapse of the weak roof rock within only years or decades of flow past an originally stable site. This is especially important where streams are diverted by piles of collapse debris so that they undercut chamber walls in gypsum and cause a renewal or widening of the collapse within only tens of years.

3.3 COLLAPSE SINKHOLES

Where cave roof failures propagate through to an exposed karst surface, they become collapse sinkholes. These appear as surface depressions from 1 m to 300 m across, typically with some rock outcrops, scars or walls in their perimeter remaining from the collapse processes that modified the initial dissolutional cavity or cavities within the bedrock. Collapse sinkholes have very variable depth/width ratios, commonly >1, thereby distinguishing them from most solution sinkholes. They may or may not have a soil cover, but their key factor is the collapse of bedrock and not only of the soil, as in some subsidence sinkholes.

3.3.1 Collapse sinkholes in limestones

The simplest type of collapse sinkhole is created by the total failure of the roof of a shallow cave. A young collapse sinkhole has almost vertical rock walls and a debris floor sloping down into an open cave passage (Figure 3.8). Over time this degrades, so that an old collapse sinkhole has flared and broken walls, is likely to have accumulated a wind-blown or in-washed soil on its floor and probably has no open entrance into the cave below (Jennings, 1975). Further degradation leaves it inactive (Figure 3.8), with an alluviated floor, a rounded plan-shape and only traces of rock wall that distinguish it from a solution sinkhole. In contrast, an active collapse sinkhole is likely to have a jagged outline that reflects the original cave shape, and it may have almost no debris on its floor, where the fallen roof breakdown has been removed by the stream or river that flows across it. The Slovenian karst at Skocjan provides very fine but very large examples of collapse sinkholes (Box 3.1), and their morphological features could be reduced by scale factors of 5 or 50 to match numerous other collapsed sinkholes in karst regions around the world. Cenotes and many blue holes are cliff-ringed lakes that are collapsed sinkholes into flooded caves (Figure 3.9); some lakes are floored by debris piles which block access to the outlets, but many cenotes in Mexico's Yucatan karst are windows into networks of flooded caves that extend for many kilometres (Beddows, 2004). As open windows into highly productive aquifers, these cenotes can constitute valuable water resources; the Otjikoto cenote lake

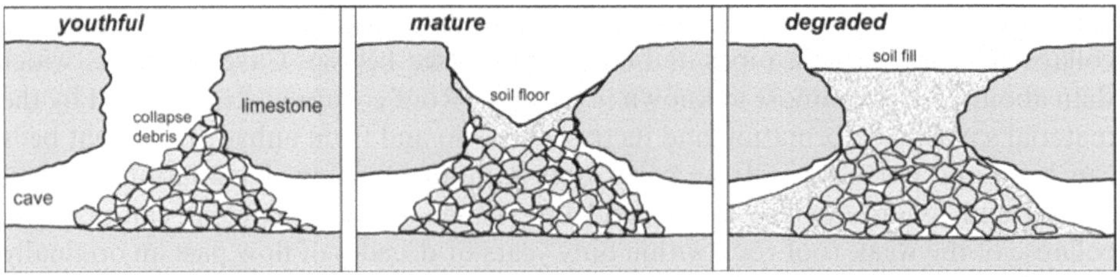

Figure 3.8. Three profiles that demonstrate the evolution of collapse sinkholes from youthful to mature to degraded.
Partly after Jennings (1975).

Figure 3.9. The splendid Otjikoto cenote in the Otavo karst of Namibia; the water is 120 m deep, but there is no known access to the flooded caves that extend either way from the cenote floor.
TW.

(Figure 3.9), in Namibia, has been pumped for many years without any decline of its level.

While collapse sinkholes may form by the failure of individual cave chambers, many are formed by multiple and progressive collapses over a zone of shafts, caverns and passages. Cave passage and shaft widening can cause adjacent features to coalesce as intervening walls of rock are thinned until they collapse and take with them the parts of the roof that they supported. In this manner large collapse sinkholes can be created in areas of heavily fractured and fissured rock where the roof of a single large chamber could not have survived even temporarily. Major collapse features are more common than large chambers inside cave systems, and this also accounts for the numbers of significant collapse sinkholes along the outcrops of faults. In the Slovenian karst, Rakovska kukava is a collapse sinkhole 240 m across and 70 m deep with steep rock sides. It is more than 15 times larger than the largest known cave chamber in the area, in limestones so fractured that larger underground voids are inconceivable, and the sinkhole is considered to have formed gradually and progressively by serial collapses into an underlying cave. As it is believed to have developed from a small hole that has grown steadily larger, it has been called a tumour doline (or sinkhole) that is a variety of collapse sinkhole where no large cave chamber ever existed (Šušteršič, 1998). Tumour sinkholes fall within the spectrum of features transitional between collapse and solution sinkholes, where dissolutional enlargement of numerous fissures and small caves is followed by repeated small-scale collapses.

BOX 3.1. COLLAPSE SINKHOLE

An example – Skocjanske Jama, Slovenia

The river cave of Skocjanske Jama is cut into Cretaceous limestones on the south-east side of the Kras plateau in southern Slovenia. The River Reka sinks into the cave's massive entrance, and beyond 200 m of passage it flows across the floors of two very large collapse sinkholes (Figure 3.1.1). Mala (*Small*) Sinkhole is 130 m in diameter, while Velika (*Great*) Sinkhole is more than 150 m by 250 m in plan (Habič *et al.*, 1989). These both have walls that are a combination of bare rock cliffs, undercut scars and very steep rocky slopes that now support a dense cover of trees (Figure 3.1.2). They both formed by collapse into complexes of large passages and chambers along the underground river course. High-level galleries were abandoned and undercut by newer river passages, and the Velika sinkhole is also at the junction of a major old branch passage towards the north. The dimensions of the sinkholes are the same order of magnitude as those of the largest chamber, 145 m long and 125 m wide, that survives intact downstream in the Skocjan cave (off the map, Figure 3.1.1) beneath a roof that is 100 m thick. It is not known how large were any individual chambers that suffered massive roof collapse, but the two sinkholes most probably developed by progressive roof and wall collapses into multiple passages. Lisicna is a third large

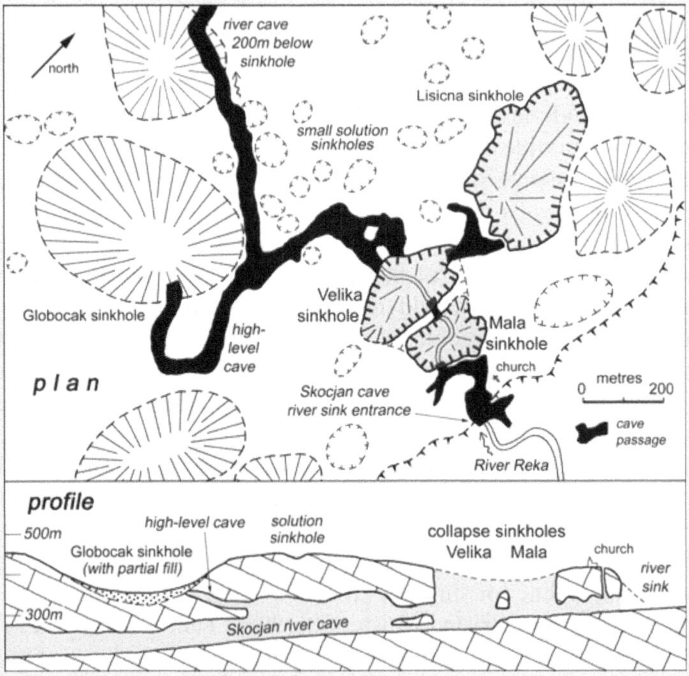

Figure 3.1.1. Map and long profile of the Velika, Mala and Lisicna collapse sinkholes at the entrance to the Skocjan caves in Slovenia.

collapse sinkhole, that appears to be failed into an old outlet cave passage whose northern continuation is hidden by collapse debris. Now lacking its cave river, it is a little more degraded, with a floor that is an inverted cone in a debris fill, but it still has cliffs around most of its perimeter.

The karst plateau above the Skocjan cave is pitted with numerous large and small solution sinkholes (Mihevc, 1998). One of the largest is the Globocak sinkhole, 500 m across and 90 m deep (Figure 3.1.1). A high-level cave passage in Skocjan is blocked by breakdown debris (but breached by a small mined tunnel) under its slopes, and appears to be an old trunk passage that once continued further west. The Globocak sinkhole is almost conical except for its aggraded debris floor, with no cliff margins that compare it to Velika and Mala. It appears to be a solution sinkhole, where collapse of the cave has made only a modest contribution to its development. The topography immediately around Velika and Mala suggests that a shallow solution sinkhole was developing above the caves, and its deepening contributed to the thinning of the limestone, and therefore to the ultimate collapses. The timing of this event, or of these events, remains unknown, but is not within the age of historical records.

Figure 3.1.2. The view eastwards across the Velika and Mala collapse sinkholes. The village houses and church stand above the cliff that drops into the Mala sinkhole breached by the upstream segment of the Skocjan cave, and the wooded ridge in the foreground separates the two sinkholes.
TW.

Figure 3.10. A small collapsed cave in Penyghent Gill, in the English Pennines, where a roof about 1 m thick has collapsed along the length of a low cave that was 5–10 m wide.
TW.

Total roof collapse over a length of shallow cave passage can create a very elongate collapse sinkhole lined by large slabs of undermined rock (Figure 3.10); failure of these wide caves in Penyghent Gill has created the largest area of undisturbed cave collapse in Britain (Waltham *et al.*, 1997). On a much larger scale this process can create a gorge or ravine (Figure 3.11); the Patale Chhango gorge in

Figure 3.11. The chaos of large fallen blocks that defines the collapsed cavern at the mouth of the Patale Chhango cave in lowland Nepal; the short gorge is 55 m deep, and most of the collapse blocks are of the conglomerate that roofs the visible cave openings.
TW.

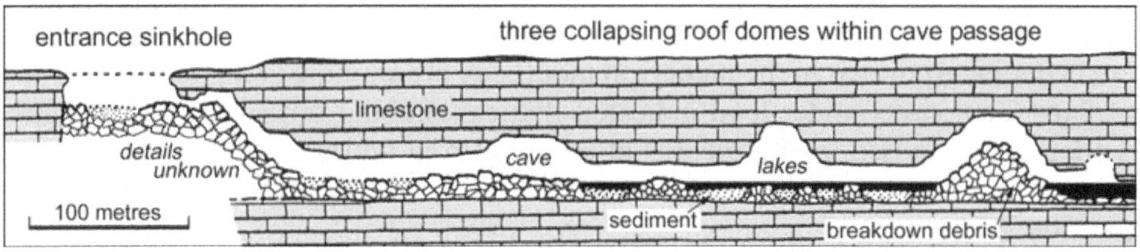

Figure 3.12. Long profile of Koonalda Cave, in Australia's Nullarbor Desert, with roof collapse forming the three high chambers and the entrance sinkhole; the rock floor of the main passage is unseen and its position is therefore conjectured, and the breakdown pile obscures any rock details below and left of the entrance sinkhole.
After survey by J. Hinwood *et al.*

Nepal is one of the very few in the world that have been created by massive cave collapse (Waltham, 1998). Nearly all karst gorges that are popularly described as "collapsed caverns" are actually subaerial fluvial features.

A collapsing chamber within a cave evolves into a collapse sinkhole when the original cave roof is thinned to the point of failure. This may develop entirely by roof stoping within the cave, or may be a feature of gross surface lowering. Both processes are natural and automatic in active karst terrains, and most collapse sinkholes owe their evolution to a combination of both processes. Roof stoping has largely or entirely formed the wide sinkholes into isolated large caves in the almost static surface environment of the Nullarbor Desert in Australia (Lowry and Jennings, 1974). Koonalda Cave has a number of breakdown domes along its large trunk passage, whose original dissolutional roof is about 60 m below ground (Figure 3.12). One dome is now within 15 m of the ground surface, and another has broken through to create the large collapse sinkhole at the entrance. In contrast, more than 20 large collapse sinkholes in the karst of Belize have been formed by surface denudation breaching an old high-level series of large cave passages, while large stable chambers still lie in the modern river caves at greater depths (Miller, 1987). It is also significant that the roof of a collapsing chamber may be thinned to failure by deepening of a solution sinkhole directly above it, guided by the same fracture zone. This is recognised as significant over Slovenia's Postojna Jama (Šebela, 1996), where deep solution sinkholes overly some of the cave's collapse chambers while large collapse sinkholes appear to overlie passage continuations now blocked by breakdown debris. The double process may also account for the formation of tiankengs (see Section 3.3.2).

The critical roof thickness for collapse was investigated by numerical modelling of the chamber that is 130 m wide in the cave of Brezno pri Medvedovi Konti in Slovenia (Kortnik, 2002). Local failures started to appear when the roof was thinned to about 20 m. The roof developed major failures when it was down to 6 m thick, and it had fully collapsed before the thickness reached 4 m. The collapse was modelled twice, with values of 1.85 and 4.0 MPa for the tensile strength of the limestone. Though pre-failure displacements were greater for the weaker material, the increase in minor failures and the ultimate collapse occurred at the same roof thickness for each material. The results broadly conform to field observations of cave chambers in the region, but are only regarded as preliminary until joint patterns are better incorporated in the numerical models. A programme of numerical analyses of flat-roofed caves under load indicated that caves about 50 m wide may collapse without any imposed load at depths of about 10 m in typical karst limestone, while caves 30 m wide collapse naturally when their roof is reduced to about 2.5 m thick (Section 7.2.2).

The Bahamian island of San Salvador has numerous small collapse sinkholes that are known as banana holes, after the crop that is grown so easily on the soils capping the collapsed rock debris on their floors. They are mostly <12 m across and <4 m deep with vertical or overhanging walls, but lack accessible openings into continuing caves. The small chambers originally formed in the zone of preferential dissolution at the top of a freshwater lens graded to an old higher sea level (Wilson *et al.*, 1995). Where the roof rock was only about a metre thick it failed with the help of subaerial fissure development and dissolution from above, but it is thought that many other similar caves at slightly greater depth present a significant collapse hazard.

3.3.2 Tiankengs

The very largest collapse sinkholes are also known by the Chinese term, tiankengs (meaning *sky holes*). They are typically more than 250 m deep and wide, with vertical walls forming a large part of their perimeters, and the finest examples lie in the very mature karsts of China and New Britain. China's Xiaozhai Tiankeng is 660 m deep and 600 m across, with vertical walls above and below a sloping terrace at mid-depth (Figure 3.13); below the terrace a debris cone on one side of the hole descends to the river that crosses the floor of the tiankeng between cave passages only about 20 m wide (Senior, 1995; Zhu and Zhang, 1995; Zhu *et al.*, 2003).

The most likely origin of a tiankeng appears to be the deepening of a solution sinkhole directly above the stoping of a cave roof. Both features are likely to develop on a zone of locally increased rock fracturing. Deepening of the sinkhole concentrates infiltration, and thereby accelerates roof stoping below, where the cave river efficiently removes the breakdown debris. This model (Figure 3.14) was first based on the giant sinkholes of the Nakanai karst in New Britain (Maire, 1981), and may be applied on a smaller scale to many collapse sinkholes. Once its vertical walls are exposed, a tiankeng would evolve towards a rounded plan-form by face retreat due to weathering and small-scale breakdown. Distinct from this collapse origin, a second type of erosional tiankeng has been proposed (Zhu, 2001) where a large sinking river initiates the shaft development. However, waterfall retreat in a river sink tends to create an elongate slot that lacks the rounded form of the larger tiankengs, and waterfall erosion would appear to be only contributory to the collapse process. Tiankeng evolution is demonstrated by three sites within China's Xingwen karst (Waltham *et al.*, 1993). Xiaoyanwan is a massive collapse sinkhole ringed by vertical walls that truncate giant cave passages, while Dayanwan is an older degraded tiankeng, and the large chambers of Zhucaojing will eventually coalesce and collapse into a third tiankeng (Figure 3.15).

Figure 3.13. The giant collapse sinkhole of Xiaozhai Tiankeng, near Chongqing, China; the footpath winds across the wooded ledge before descending 300 m down a talus slope in the inner shaft.
Photo: Zhu Xuewen.

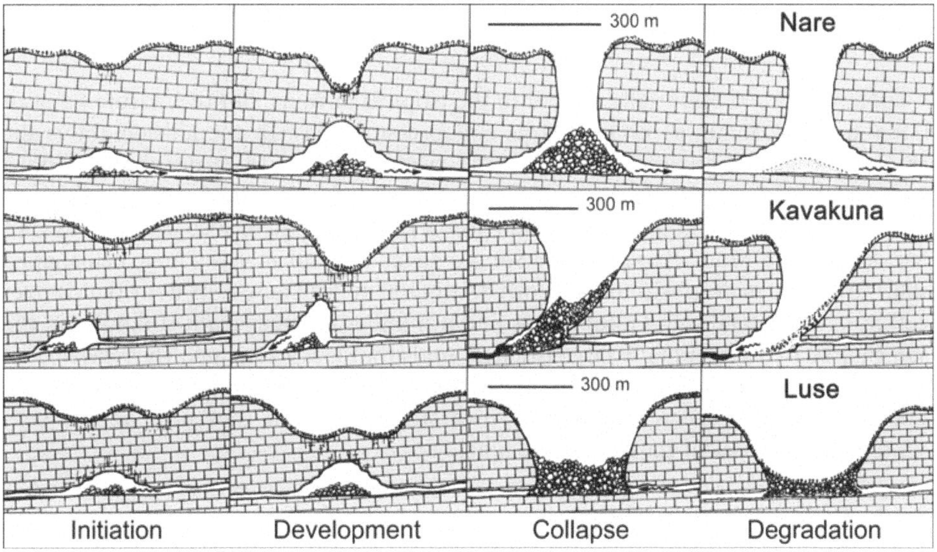

Figure 3.14. Four stages in the conceptual evolution of three of the tiankengs in the Nakanai karst of New Britain.
After Maire (1981).

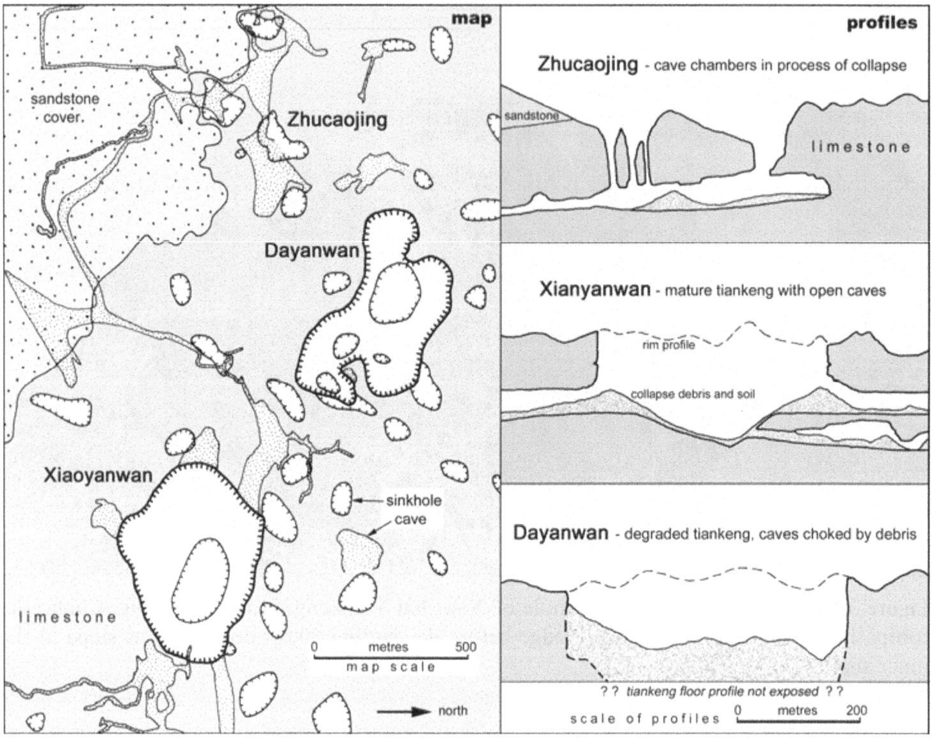

Figure 3.15. Map and profiles of the two tiankengs and the Zhucaojing cave that will eventually collapse into a third tiankeng in the Xingwen karst of China.
After surveys by the China Caves Project.

Tiankengs are essentially limited to the tropical terrains, but the mature karst of Croatia does contain some very large collapse sinkholes. In the hillside above the Imotski polje, the adjacent sinkholes of Crveno Jezero, 518 m deep, and Modro Jezero, 288 m deep, are each about 300 m across and half full of water; both these and the Skocjan sinkholes in Slovenia (Box 3.1) could be described as tiankengs. In Mexico's Sierra Madre Oriental, the Sotano de las Golondrinas is a huge bell-shaped shaft, over 500 m deep with breakdown debris 100 m deep across its floor that is over 200 m in diameter. Both Golondrinas and a number of other giant sinkholes in the same karst are exceptionally large isolated voids that lack associated river caves, and they may owe their initiation to hydrothermal sulphuric acid corrosion. Subsequently, they have been modified by roof stoping so that they now appear very similar to the classical tiankengs in China.

3.3.3 Collapse sinkholes in chalk

Collapse sinkholes are rare in chalk, as the weak rock contains few large cave passages. Though small-scale collapse does occur, it is generally subordinate to dissolution during landform evolution in chalk karst. An even more rarely observed collapse event created the Bîme des Enfants in France's Aube department in January 1948. It left a sinkhole 16 m deep and 10 m across, with walls in typically well-fractured chalk, in a gently sloping hillside (Figure 3.16). There was no cave accessible around the edge of the mound of fallen debris, but the immediate area is distinguished by having some long stream caves with passages generally 2 m high and 1 m wide, which are large by the standards of the European chalk. The limestones of Australia's Nullarbor karst are described as chalky, and have porosities normally >25%, but they do have large cave passages slowly developed in the desert climate, and consequently have large collapse sinkholes totally unlike anything in the European chalk.

A special feature of chalk is its susceptibility to liquefaction when it is both disturbed and saturated (Burland *et al.*, 1983). Disturbance may be due to engineering handling, when wet chalk is well known to liquefy into a slurry, but may also be due to stress within a cavity roof, or due to frost shattering. The latter is ubiquitous across the outcrops of southern England and reached depths of up to 10 m during the Pleistocene cold stages. Saturation is easily achieved at point inputs of drainage off a soil cover or by concentrated engineered drainage, and increased dissolution is then an inevitable side effect. A site with drainage from soakaway drains directed into chalk that has inherited frost shattering and stands over ground cavities is optimal for liquefaction and consequent ground failure. The well-known collapses (Figure 3.17) in a new housing estate at Bury St. Edmunds, U.K., remain one of the most spectacular consequences of chalk liquefaction (Waltham, 1989; Bell *et al.*, 1992). However, these developed around drains over old flint mines close to the depth limit of Pleistocene frost action, and strictly they are therefore crown holes. Comparable liquefaction over natural cavities is limited by the scarcity of adequately large caves, though sinkholes of this type have been known to develop over gull cavities along cambered escarpments. It is debatable as to how much liquefaction contributed to the roof stoping that formed the Bîme des Enfants (Figure 3.16). There were no indications of point drainage input that could have locally saturated the chalk, but the failure was during a cold and wet winter, and its exposed walls are in putty chalk and rubble chalk very similar to that in the collapses at Bury St. Edmunds.

3.3.4 Collapse sinkholes in gypsum

Due to the relative lack of gypsum at outcrop, collapse sinkholes in gypsum are not very widespread, as opposed to the abundance of caprock sinkholes over interstratal gypsum karst (Section 3.4.2). However their role as a geohazard is increased by their potentially rapid development under gypsum's high rate of dissolution. New collapse sinkholes occur almost annually in the forested gypsum karst at Pinega in northern Russia (Waltham and Cooper, 1998). Houses have been damaged in Rapid City, South Dakota, by the sudden development of collapse sinkholes up to 10 m across that have formed in the brecciated outcrop zones of gypsum beds up to 9 m thick (Rahn and Davis, 1996). Just over the state line into Wyoming, the Vore Buffalo Jump is a much larger collapse sinkhole in the same gypsum. Nearly 60 m across and

Figure 3.16. The collapse sinkhole of the Bîme des Enfants in the chalk karst of northern France, 26 years after its sudden appearance.
TW.

Figure 3.17. One of the collapse sinkholes formed by liquefaction failure of the chalk over old mines in Bury St. Edmunds, U.K., seven years after their failure destroyed the road.
TW.

20 m deep, it was used by native Americans around 400 years ago to stampede buffaloes to their deaths over its precipitous walls, but it is now degraded to a rocky bowl (and partly filled by thousands of buffalo bones) adjacent to the interstate highway.

Small collapse sinkholes are lost within the rapid denudation of gypsum outcrops, but larger features survive, especially in dry climates where dissolution is reduced. The nine "Bottomless Lakes" at Roswell, New Mexico (Martinez *et al.*, 1998), lie in sinkholes 50–100 m wide ringed by walls of broken gypsum up to 40 m high above water that is actually only 5–25 m deep (Figure 1.11). The sinkholes are cut into a dissolutionally cambered escarpment, and were largely formed by undermining and collapse around sites of rising artesian water (a process that is limited in limestone karst by the need for meteoric carbon dioxide to provide dissolutional aggressivity).

Isolated collapse sinkholes are scattered across the exposed gypsum karst near Sivas in eastern Turkey (Waltham, 2002a). Some of these are up to 400 m across and 50 m deep; they are orders of magnitude larger than gypsum caves observed locally or anywhere else. One sinkhole is still very active, as it is partially flooded with water

Figure 3.18. An evolutionary sequence of collapse sinkholes in the gypsum karst near Sivas, Turkey. (a) Active dissolutional undermining and rock collapse in a small corner of the riverside Bielekbasi sinkhole. (b) An almost static lake in the mature Kizilcam sinkhole. (c) A large degraded sinkhole on the hills above Mahmutaga.
TW.

that circulates rapidly to and from the adjacent Kizilirmak River. Its margin is an almost complete line of dissolutional undermining with small collapses producing talus slopes of gypsum blocks into the lake on its floor. In one corner, the collapse is into larger blocks (Figure 3.18(a)), and indicates a maximum failed span that could have reached about 25 m – compatible with the maximum sizes of gypsum caves recorded in many karsts. It appears that numerous small collapses accumulated to form the large Sivas sinkholes. An evolutionary sequence can be recognised (Figure 3.18) from the active riverside collapses, to the mature collapse sinkhole of Kizilcam far from the river's circulation and therefore less active, to the unnamed degraded sinkholes high on the plateau and fossilised by fluvial rejuvenation and water table decline. The same sequence may also apply to collapse sinkholes in limestone, but over much longer timescales.

3.4 CAPROCK SINKHOLES

Where roof stoping and cavern collapse migrate up through overlying non-karstic rocks, any ultimate failure of the surface creates a caprock sinkhole. These sinkholes are totally dependent on the initial formation of significant voids within the interstratal karst that is developed in the underlying soluble rock. Like collapse sinkholes, many caprock sinkholes have steep rocky walls as their surface features typically develop by rapid or instantaneous failure events. They may be only a few metres across, as is the example at Dankivsky (Box 3.2). However, they can be much larger, and the two examples of Dankivsky and Skocjan (Box 3.1) lie at opposite ends of the spectrum of sizes that can be found in both collapse and caprock sinkholes. The steep walls of fresh caprock sinkholes degrade to lower profiles over time, especially in the weaker caprocks, and the morphology of caprock sinkholes in poorly consolidated clay rocks is very similar to that of dropout sinkholes formed in weak but cohesive clay soils (Chapter 4). In plan shape, most caprock sinkholes tend to be more circular than most collapse sinkholes in exposed karst. This is because the stress-controlled stoping processes evolve as the void migrates up through the caprock, and smooth out any fracture pattern irregularities that guided dissolution in the original cave beneath.

Though caprock sinkholes may have diameters up to many hundreds of metres, their depths are generally limited by the lack of dissolutional removal of the breakdown debris. Their debris piles may reach considerable depths, and their morphology grades into that of deep-seated breccia pipes, but few caprock sinkholes are open to more than 20 m deep. Collapse features of larger areal extent in caprock are recognised as grabens and areas of foundered strata where the interstratal karst beds have been totally removed by dissolution, but these cannot be described as sinkholes.

3.4.1 Caprock sinkholes over limestone

The dependence of efficient carbonate dissolution on soil-derived carbon dioxide restricts extensive development of interstratal karst in limestones beneath cover rocks of low permeability. The optimum situation is a gently dipping limestone bed with drainage from an outcrop supplied by allogenic streams that pass beneath a tilted plateau of cover rock and out to a deep valley. This creates potentially large caves where collapse can be initiated. If the caprock is a permeable

Figure 3.19. A large caprock sinkhole over the interstratal karst of Mynydd Llangattwg in Wales, U.K.; all the exposed and collapsed rock is strong gritstone, and the top of the limestone is about 10 m below the floor of the sinkhole.
TW.

sandstone, any localised concentrations of downward infiltration to the buried limestone provide sites for collapse initiation, that ultimately stope upwards to create the caprock sinkholes.

Dip slopes of gritstone in southern Wales provide Britain's finest interstratal karst, expressed by hundreds of caprock sinkholes (Thomas, 1963, 1974). Mapping of the grit outcrops revealed 437 sinkholes, with average dimensions of 29 m across and 7.5 m deep (Figure 3.19). Small cavities are bridged by the strong grit, which therefore lacks the profusion of smaller sinkholes that characterise limestone outcrops. In contrast, there are numerous subsidence depressions and outcrops of foundered strata, each 100–300 m across, caused by dissolutional removal of the buried limestone in areas much larger than those causing the sinkholes. The maturity of the caprock sinkholes is reflected in more than a quarter of them having almost perfectly circular outlines, and also by aprons of talus and soil that drape most of the gritstone walls forming their perimeters.

Only in some of the Welsh interstratal karst are there known cave systems that can be correlated with the caprock sinkholes. Most of the sinkholes appear to have been caused by dissolution of the limestone from just beneath the permeable but insoluble grit cap, while some appear to relate to collapse zones and boulder chokes in large old caves far below. Sinkholes on the Llangattwg moorland appear to form an evolutionary sequence, and clastic cave sediments derived from the caprock grit are evidence of connections between the sinkholes and the caves (Bull, 1980). A caprock sinkhole, first developed by dissolution at the top of the limestone, may direct drainage towards a deeper cave, and thereby promote roof collapse that may in turn stope upwards to deepen the sinkhole by deep-seated undermining. This

BOX 3.2. CAPROCK SINKHOLE

An example – Dankivsky, Ukraine

The world's most cavernous interstratal gypsum karst underlies the Dniester Valley of the western Ukraine and northern Moldova. On 11 January, 1998, a caprock sinkhole developed in a clay hillside at Dankivsky when a small instantaneous surface collapse broke through from an underlying gypsum cave. The new sinkhole was 5 m across, with vertical sides that descended 22 m to a sloping floor of fallen debris. It was in the previously unbroken grass cover of a field (Figure 3.2.1), and the collapse was heard at a farm one kilometre away. At the foot of the newly opened shaft (Figure 3.2.2), there was an opening over the debris cone and into the domed roof of a cave 9 m wide (Fig. 3.2.3). This dome had formed by wider collapse of the gypsum roof beds of a cave passage that continued about 5 m high but totally underwater where its roof was intact. The cave passage is very similar to those, just 12 km to the south, that extend for 92 km in the maze cave of Zolushka, which has been drained by pumping activity in an adjacent quarry (Klimchouk and Andrejchuk, 2003). Both the gypsum and the caprock clays are of Miocene age.

Figure 3.2.1. The freshly collapsed clay walls of the Dankivsky caprock sinkhole, Ukraine, with the way into the underlying cave lost in the darkness.
Photo: Alexander Klimchouk.

Figure 3.2.2. The grass-covered clay slope at Dankivsky broken by the new caprock sinkhole.
Photo: Alexander Klimchouk.

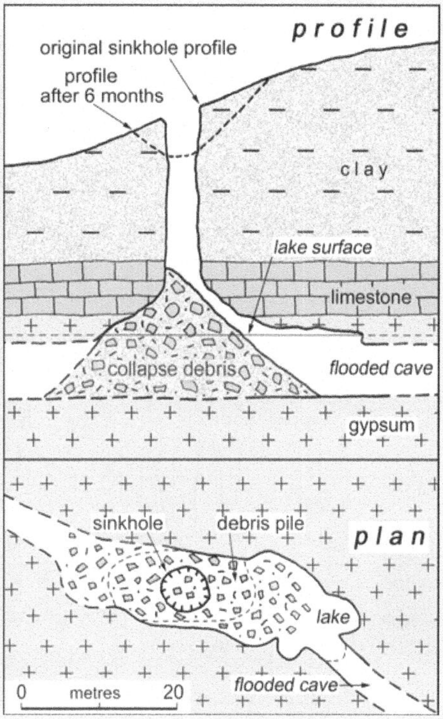

Figure 3.2.3. Plan and profile of the Dankivsky caprock sinkhole; the extensions of the cave underwater and behind the debris cone are unknown.
After Klimchouk and Andrejchuk (2003).

The roof dome at Dankivsky had developed on a slickensided fault zone that had provided a focus for upward dissolution of the gypsum, and probably also a site of downward percolation of water through the cover rocks. A nearby surface stream lies 19 m above the standing water level in the cave, indicating the presence of groundwater perched above the gypsum. By a combination of dissolution and collapse, the cave roof dome enlarged upwards through 3 m of gypsum. It then stoped upwards through 6 m of limestone and 17 m of clay before breaching the surface. Fallen debris from the shaft walls blocked the way into the cave four months after the initial collapse, and the shaft had degraded into a bowl-shaped sinkhole just 4 m deep within nine months of the event. Though the Dankivsky sinkhole is quite small, it demonstrated the morphology of both a fresh caprock sinkhole with vertical sides, and also a degraded caprock sinkhole that is barely distinguishable from other types except for its occurrence in the outcrop of an insoluble rock. Because the upper caprock is a clay, it also resembles a dropout type of subsidence sinkhole.

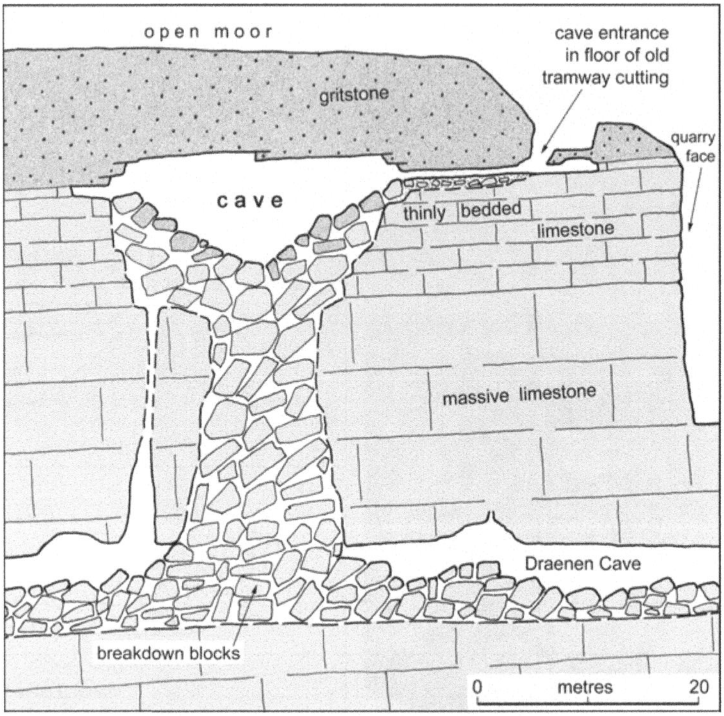

Figure 3.20. Profile through Siambre Ddu, a cave that has already migrated from the limestone up into the overlying gritstone and will ultimately form a new caprock sinkhole in the interstratal karst of South Wales, U.K.
Draenen passage after survey by J. Stevens *et al.*

process accounts for the Waen Rûdd sinkhole, 100 m above a collapse in the cave passage, but caprock sinkholes are unrelated to collapse zones in the caves at greater depths.

At the eastern end of Wales' interstratal karst, the cave of Siambre Ddu appears to offer a model for formation of the caprock sinkholes. It has a single rounded chamber, 25 m across and 9 m high, with walls and roof entirely in the gritstone (Figure 3.20), entered through a low collapse-modified passage that formed in the top bed of the limestone. The roof is about 6 m thick, and there is as yet no surface sinkhole. Breakdown blocks of grit line a subsided cone in the floor of the chamber, which descends well below the original limestone/gritstone contact. About 20 m below, a passage 10 m wide in the cave of Ogof Draenen is nearly blocked by a cone of gritstone blocks that descend from totally choked shafts (Figure 3.20); an adjacent clean shaft carries drip-water from Siambre Ddu. Any collapse zone that may have stoped upwards from the lower cave is obscured by debris, but the massive lower limestone is more likely to contain just the dissolution shafts that are now tapping the pile of gritstone breakdown blocks from below. It appears that Siambre

Ddu originated by dissolution forming a wide cavern near the top of the thinly bedded upper limestone, and it will ultimately develop into a caprock sinkhole.

Britain also has numerous sinkholes in the weak sedimentary rocks that provide a caprock on parts of the chalk dip slopes (Sperling *et al.*, 1977). These are better described as subsidence sinkholes (Chapter 4), as there is no evidence of collapse into large cave chambers in the chalk, and the poorly consolidated sands and clays of the cover appear to have behaved as soils that have been washed down into networks of fissures within the bedrock.

3.4.2 Caprock sinkholes over gypsum

Interstratal karst is extensive in gypsum that is highly soluble in groundwater not dependent on subaerial processes for its aggressivity. Dissolution of gypsum buried at shallow depths ultimately causes vast numbers of caprock sinkholes, and dissolution at greater depths creates many deep-seated breccia pipes (see Section 3.5).

Sinkhole densities on interstratal gypsum karst can rise to 200 per km^2, with new sinkholes appearing at rates of 0.01 to 3.0 per km^2 per year. These are mainly caprock sinkholes but include significant proportions of subsidence sinkholes where the gypsum is capped by thin and poorly consolidated sedimentary materials that behave as soils. The best-documented cases are in the Ukrainian karst, where the sinkholes can be correlated with breakdown features in extensive networks of accessible caves (Klimchouk and Andrejchuk, 1996, 2003). Roof failures in the caves develop mainly at drainage points (mostly inlets, but also where outlets existed before the cave was drained by local rejuvenation), which are mainly on joints or faults. Most passages are about 3 m wide, and wider chambers are not major collapse zones as they are mostly formed in stable zones of less fractured rock. Roof stoping then proceeds through the upper gypsum and a thin limestone, and then through the capping clays and sands until a collapse sinkhole develops (Figure 3.21). The sinkhole may initially be steep-sided (Box 3.2), and later degrades to a gentler bowl, or it may start with a lower profile where liquefiable sands in the cover sequence fail by suffosion instead of collapse (Chapter 4).

Caprock sinkholes decrease in numbers where the caprocks are thicker. The critical cover thickness relates to the material properties of the cover and the sizes of the caves, and is specific for each region. In Ukraine, critical thicknesses beyond which sinkhole densities decrease noticeably range from 20 to 40 m. The passages of Mlynki cave contain 144 roof breakdown features, but there are only two sinkholes above, as the cover is a clay 25–30 m thick. There is a close correlation between sinkhole distribution and the underlying Kungur Caves, in the Russian Urals (Figure 3.22), where the caprock is down to about 25 m thick, but sinkholes also exist in smaller numbers where the caprock is over 60 m (Klimchouk and Andrejchuk, 1996).

Roof collapse in interstratal gypsum caves creates columns of largely insoluble breakdown debris (Figure 3.21) that are small versions of the breccia pipes formed over sites of deep-seated dissolution of either gypsum or salt (Section 3.5). These features constitute a significant geohazard due to their potential failure at the surface

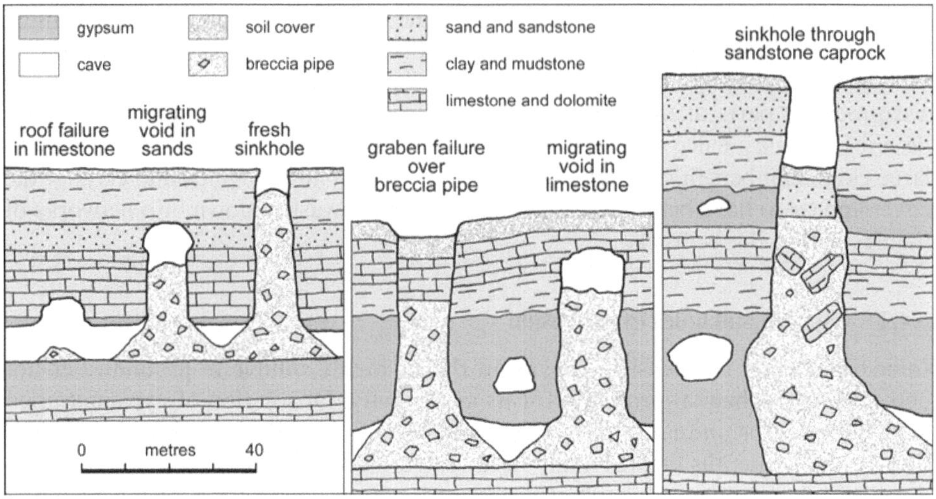

Figure 3.21. Evolving and maturing caprock sinkholes over the gypsum karst of Ukraine and England.

After Klimchouk and Andrejchuk (1996) and Cooper (1998).

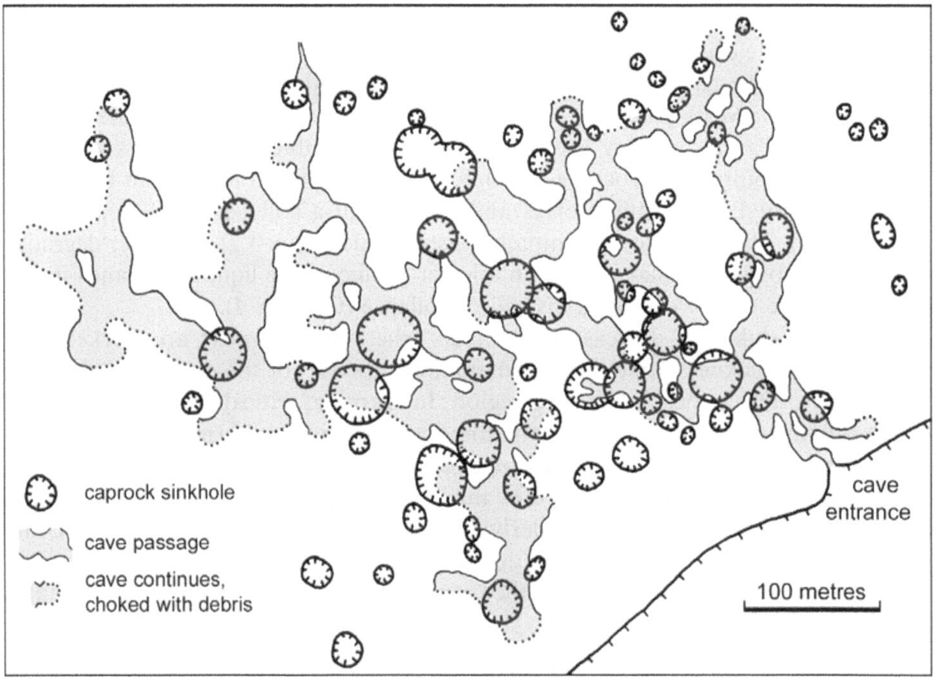

Figure 3.22. Correlation between caprock sinkholes and cave passages in the buried gypsum at the Kungur Caves in Russia.

After survey by K. Gorbunova *et al.*

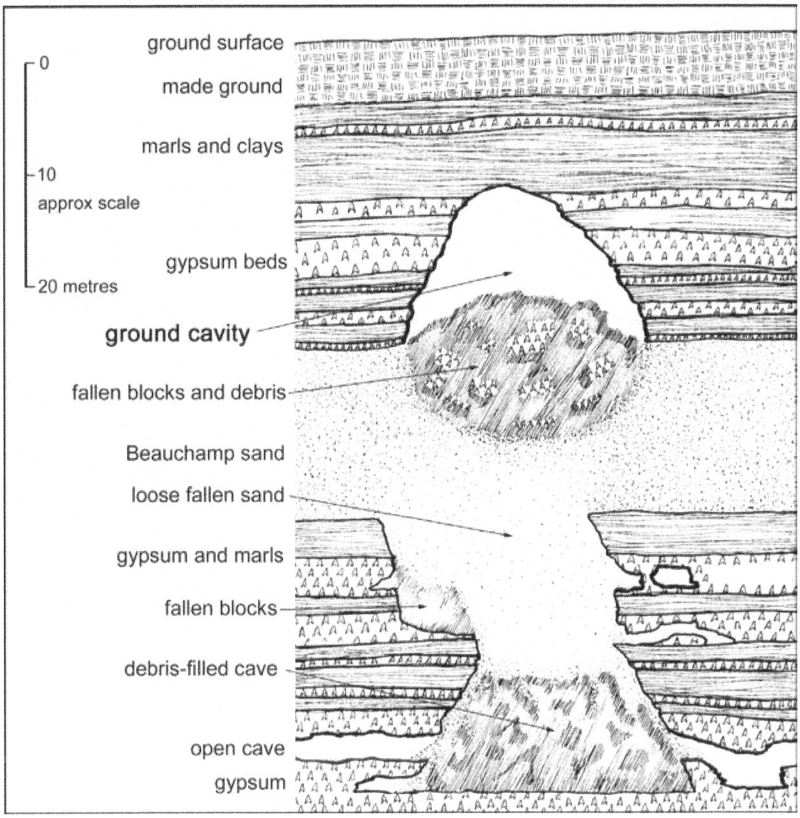

Figure 3.23. Schematic section through a concealed void of 2,500 m^3 found entirely within cover rocks over gypsum and beneath a Paris railway station.
After Toulemont (1987).

as caprock sinkholes, unless the bulking of the breakdown material fills the void and provides roof support. Many surface failures occur over the gypsum beneath Paris, France (Toulemont, 1987), and investigation of a cavity found in 1975 beneath railway engineering works revealed a failure migrating up through the cover rocks from dissolution cavities in a number of gypsum horizons (Figure 3.23). Numerous caprock sinkholes at Ripon, U.K., lie on top of breccia pipes that are reactivated by ongoing dissolution of the gypsum lying beneath 40–60 m of mudstone, limestone and sandstone cover (Cooper, 1998; Cooper and Waltham, 1999). Surface collapses are generally 10–30 m across, and cause extensive and repeated damage to the town's roads and buildings (Case study #1). Over time, most caprock sinkholes in weak cover rocks degrade to gentle depressions that may be confused with solution sinkholes unless the subsurface structure is investigated or exposed. Those in stronger rocks survive as deep circular pits, as seen at a number of sites in Canada where dolomite overlies gypsum within the Prairie Evaporite Formation under northern Alberta (Figure 3.24).

Figure 3.24. Caprock sinkholes rimmed by dolomite over the Wood Buffalo interstratal gypsum karst of northern Alberta, Canada.
Photo: Derek Ford.

3.4.3 Caprock sinkholes over salt

Once salt is reached by circulating groundwater, its dissolution can be both rapid and extensive. In some situations, caprock sinkholes can be formed, but total removal of the salt and consequent regional subsidence are more common. The McCauley Sinks, in eastern Arizona, include twenty caprock sinkholes that are each more than 100 m across and 30 m deep (Neal, 1995). They are formed in sandstone over thick salt beds at a depth of about 300 m, due to discrete cavity migration and breccia pipe development through a remarkable thickness of cover; there are no local indications of diapiric rises of soluble salt that could pre-date pipe formation. These sinkholes lie just ahead of a dissolution front that is migrating down dip and causing more widespread removal of the salt. Ultimately, the breccia pipes and their caprock sinkholes will be incorporated into the disturbed rock structure of a bowl of regional subsidence.

In March 1879, a sinkhole developed catastrophically near Meade, in western Kansas (Figure 3.25). It was 52 m across, and initially 27 m deep though filled with water to within 4 m of its rim. It overlies salt at a depth of about 150 m and is clearly a caprock sinkhole, though the surface feature may have been widened by slumping of the cover soils in the manner of a subsidence sinkhole. Significantly, the salt below

Figure 3.25. The caprock sinkhole that developed in 1879 over deeply buried salt at Meade, in Kansas.
Photo: US Geological Survey.

Meade is broken by faults that allow groundwater circulation to and from permeable sandstones and thereby encourage localised dissolution (Frye and Schoff, 1942). A century later, the Wink Sink formed in northern Texas (Case study #11). It was a caprock sinkhole 34 m deep and 110 m across, and the timing of its development is uniquely well known (Johnson, 1989b). The roof of a new cavern collapsed, and formed the sinkhole after void migration through 400 m of cover within less than 50 years – at a mean rate of at least one metre every six weeks.

Even more remarkable is the caprock sinkhole that developed in 1986 over a new breccia pipe above the Berezniki potash mine in the Russian Urals (Andrejchuk, 2002). Brine leakage into the mine, at a depth of 400 m, warned of massive dissolution of the 90 m of salt overlying the mined potash, though the extent of older cavities remains unknown. Only seven months later, cavity migration reached the surface through 300 m of limestones, mudstones and sandstones with the instantaneous appearance of a new caprock sinkhole 150 m deep and 40 × 80 m across at the top of a breccia pipe of the same plan dimensions. Roof stoping over the pipe took only about 12 days through the last 100 m of mudstone, but this very high rate was aided by structural weaknesses in a fracture zone along a local fold axis. The dissolution cavities in the salt were clearly very large, but the rate of cavity migration through the caprock has implications for ground stability in many terrains of interstratal karst.

While these caprock sinkholes have developed over breakdown sites that created breccia pipes, Crater Lake in south-east Sakatchewan, Canada, appears to have formed by block subsidence within ring faults (Christiansen, 1971). The sinkhole is 300 m across, largely filled by a lake with thick sediment above a rockhead that lies 30 m below the surrounding ground level. Salt in the Prairie Evaporite Formation lies at a depth of 900 m, and it is unknown to what extent a classic breccia pipe is developed at depth beneath the ring faults that breach the upper cover. Though the structure beneath Crater Lake may be more akin to that of a volcanic caldera, it is essentially a variety of caprock sinkhole, albeit an unusual variety.

3.5 BRECCIA PIPES

Collapse and stoping of the roof of a cavern created by dissolution can migrate upwards through considerable thicknesses of cover rock to create columns of fallen breakdown. These are generally known as breccia pipes. They may be only a few metres across and a few tens of metres deep, as are the debris piles that underlie all caprock sinkholes (and most collapse sinkholes); these features are effectively small-scale breccia pipes (Figure 3.21). But they may be much larger, extending up through many hundreds of metres of cover rocks, and proportionately many tens or hundreds of metres across. These deep-seated breccia pipes may or may not reach the surface. Where they do crop out, the initial collapse or caprock sinkhole may subsequently be degraded to a very modest surface depression. Alternatively, they may just appear as small circular breccia outcrops with no specific topographic expression. Settlement of the breccia within a pipe does constitute a potential subsidence hazard, but surface collapse events are extremely rare.

Breccia pipes are not a major feature of limestone karst. Strong limestone tends to develop stable arched roofs over caverns of the size normally created by cave river erosion. Typically, cavity migration is terminated with an arched roof over a modest pile of breakdown within a chamber only slightly modified by the collapse process. Exceptions do occur in thinly bedded or heavily fractured limestone. A breccia pipe within thinly bedded limestone, fortuitously exposed in an island cliff in Vietnam's Halong Bay, is at least 20 m high (its floor is below water level) and about 5 m across (Figure 3.26). It still has an open cavity on top of the column of limestone blocks, with about 5 m of overlying limestone ready to fail in order to create a collapse sinkhole. However the breccia pipe now appears to be inactive, with stalagmite growth above it, some cementation of the breccia and no signs of recent settlement within the pipe. Deep inside Switzerland's Hölloch cave, the Schwarzer Dom is the top, open part of a breakdown-filled shaft, 225 m high, with its roof still nearly 400 m below the ground surface (Figure 3.27). This style of rather short breccia pipe is only recognised in limestone caves where there happens to be access to both the top and foot of the debris pile, but it does demonstrate the process of progressive roof stoping in limestone.

Larger breccia pipes in limestone occur as features of paleokarst that have developed over very long timescales. These are generally associated with mineralisation, with the implication that dissolution by hydrothermal fluids may be essential to their development on this scale. A number of breccia pipes are exposed in the walls of the Grand Canyon, Arizona, with bases in dissolution zones in the Redwall Limestone, and rising through about 600 m of shales and sandstones, so that some reach the Kaibab Limestone at canyon rim level. Some younger caves have exposed the breccia pipes underground (Wenrich and Sutphin, 1994), and renewed settlement within a pipe 80 m in diameter has left an open cavern with walls of insoluble clastic

Figure 3.26. Limestone breccia pipe with an open cave at its top, exposed in the cliff of an island in Halong Bay, Vietnam.
TW.

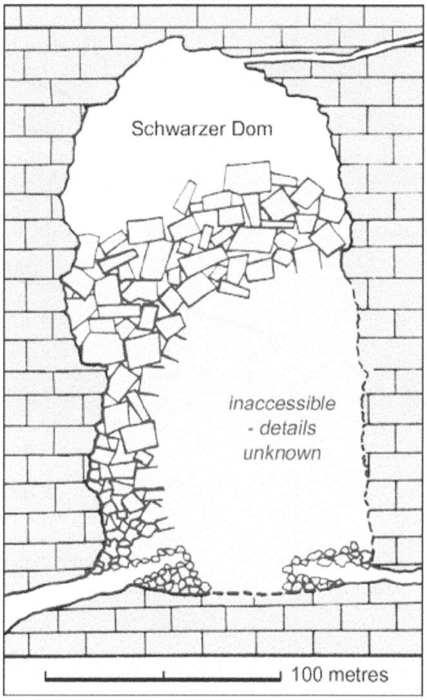

Figure 3.27. Cross section through the Schwarzer Dom in Hölloch, Switzerland, where progressive collapse of a cave roof has produced a short breccia pipe.
After Bögli (1980).

rocks in Paiute Cave (Hose and Strong, 1981). The roof of this has then stoped obliquely into the wallrock, to create an entrance through a collapse sinkhole in the Kaibab Limestone. The nearby Ah Hol Sah is a collapse sinkhole, 150 m across and 50 m deep, in the Kaibab Limestone outcrop. It appears to overlie another deep-seated pipe, but its breakdown has destroyed or obscured any exposures of breccia.

Nearly all very deep breccia pipes originate from dissolution of gypsum or salt. These materials can host more extensive cavities, that are formed more rapidly by deeply circulating groundwater at greater depths than comparable features in limestone, and they also lack limestone's strength to span the larger voids. It is estimated that there are more than 5,000 breccia pipes over gypsum and salt in North America (Quinlan *et al.*, 1986). Diameters range up to 1,000 m, and they propagate from depths as great as 1,200 m, but the total number includes features down to just 1 m in diameter that are more comparable with the cave roof collapses and caprock sinkholes described in Section 3.4.2.

Probably the finest exposures of deep-seated breccia pipes have been achieved in the coal mines and boreholes that intersect them in China. Some 2,875 breccia pipes

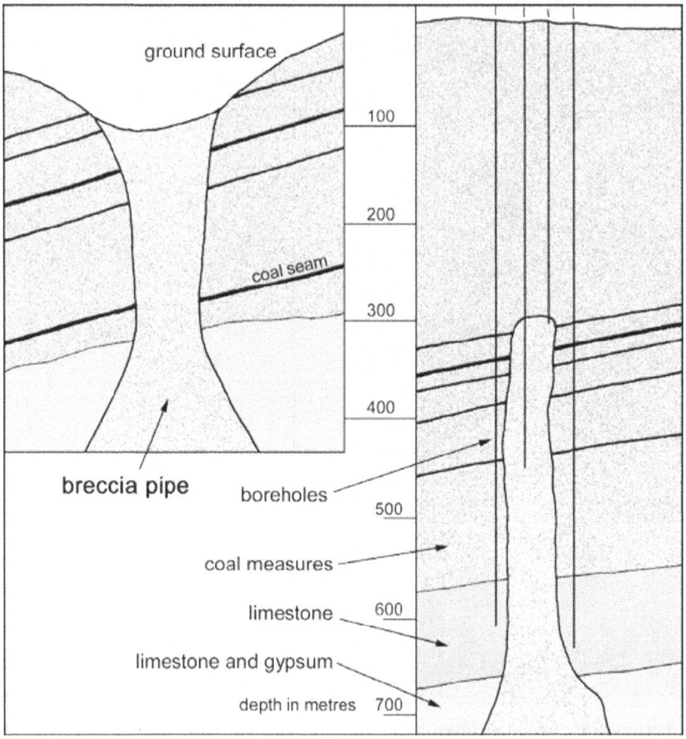

Figure 3.28. Two large breccia pipes exposed in coal mines over interstratal gypsum karst in northern China.
After Lu and Cooper (1997).

have been recorded, mainly in the mines of Shanxi and Hebei (Lu and Cooper, 1997). They originate on dissolution cavities within Ordovician gypsum that is interbedded with limestone, and they pass up through the Carbo-Permian coal measures (Figure 3.28). The vertical transmissivity of the pipes creates a major mining hazard from large flows of water that pass either up or down them. The pipes vary in diameter from tens of metres to a few hundred metres, and the larger ones pass up through 500 m of cover. Some reach the surface and contain a loose upper fill, but these are paleokarstic pipes that have been breached by surface lowering, and they never were active caprock sinkholes. Others are topped by stable bedrock arches. Near the northern margin of the Shanxi coal basin, 1,300 pipes are known in the Xishan mine, with a pipe density reaching 70 per km^2. It is significant that the breccia pipes are located mainly around the margins of the coalfield, and are not known in the deeper centre of the basin.

Breccia pipes comparable to those in China have been encountered in mines in Germany, Belgium and Canada, and have been mapped at outcrop in many other countries. Large breccia pipes over salt have also been induced artificially by uncontrolled brining operations that have caused ground failure in large sinkholes. Though

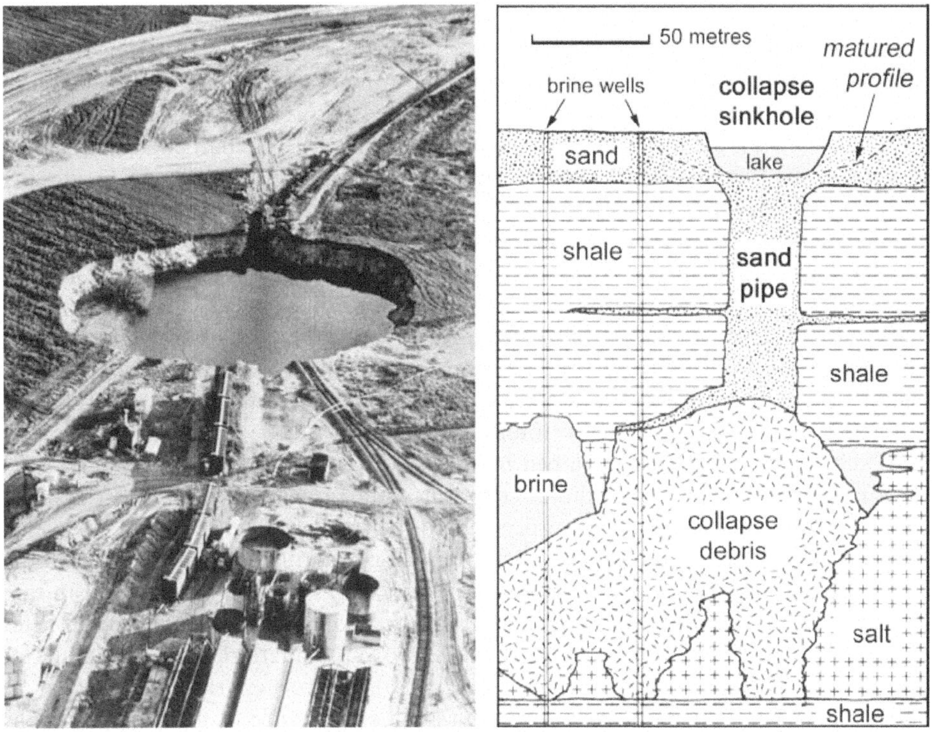

Figure 3.29. Oblique air photograph and conjectured profile of the caprock sinkhole that developed over a breccia pipe rooted in a large void created by brine pumping from a salt bed 130 m below ground level, near Hutchinson, Kansas.
Profile after Walters (1997). Photo: Hutchinson News.

these are effectively mine roof collapses, the ground structures revealed by post-failure investigations (Walters, 1977; Wassmann, 1979) provide useful models for the development of breccia pipes and caprock sinkholes (Figure 3.29). They are comparable with the Wink Sink in Texas (Case study #11), whose failure through 400 m of cover in about 50 years provided a unique indication of long-term stoping rates from depth. However this stoping was through very weak rocks, and stoping rates in cavernous limestones are orders of magnitude greater.

3.6 THE COLLAPSE GEOHAZARD

Though collapse landforms are a characteristic of most karst landscapes and collapse processes are integral to the evolution of most caves, natural events that involve surface collapse of bedrock are extremely rare. Collapse and caprock sinkholes rarely occur at densities higher than one per square kilometre in limestone landscapes that have matured through perhaps a million years. In such situations, the chance of a new sinkhole developing within a single construction site of one hectare during a lifetime of 100 years is therefore one in a million. Small caves and collapses can achieve similar densities in young coastal limestones within timescales that are an order of magnitude less, implying an equally increased probability of new sinkholes (Wilson *et al.*, 1995). Either of these figures presents a very small engineering risk, which may be further greatly reduced by prior ground investigations to locate any hazardous voids close to the surface by geophysics and/or shallow boreholes. In most karst terrains, collapse and caprock sinkholes are much more widely spaced, and the hazard is reduced. Higher densities of these types of sinkholes are largely restricted to some very mature karsts where landforms have evolved over much longer timescales, and the implied geohazard is therefore comparable to that in less mature karsts. In all cases, the risk of ground failure induced by imposed load due to engineering works is slightly greater (Chapter 7). The extremely small geohazard from rock collapse is totally independent of the much greater hazard from subsidence sinkholes developed in any soil cover (Chapter 4).

Collapse and caprock sinkholes constitute a far greater geohazard on gypsum karst due to its faster dissolution processes. In the English town of Ripon, 43 events of subsidence or collapse in the caprock over the gypsum have been recorded in the last 160 years, within an area of about 7 km^2 (Cooper, 1998). This gives a mean rate of one new sinkhole every 26 years in each square kilometre. The highest event rates are found in the thin and weak clay caprocks above the interstratal gypsum karst of the Ukraine (Klimchouk and Andrejchuk, 1996), where new sinkholes appear at rates of 0.01 to 3.0 per year per km^2. This implies a chance of greater than evens that any misguided one-hectare construction site in the worst areas will develop a new caprock sinkhole during its lifetime of 100 years. Many of these events are induced by engineering activity, so the natural event frequencies are lower, but these figures are pertinent to sites where development is taking place. However, the implication remains that appropriate engineering can reduce the hazard (Chapter 11). Both the soft ground and the high hazard level in the Ukrainian karst approach the conditions of subsidence sinkhole development in soils (Chapter 4).

Collapse sinkholes, caprock sinkholes and breccia pipes that reach outcrop all offer ground conditions that are essentially columns of broken rock. Many of these offer a subsidence potential by settling and compaction, either under load, or by down-washing and suffosion of their fines, and/or by continuing dissolution of their breakdown blocks. These hazardous sites are normally small and easily identified, and are best avoided by construction where they occupy only a small proportion of the karst area.

4

Soil failure in subsidence sinkholes

The vast majority of ground failures within karst terrains are due to the erosion, transport and failure of the soils that overlie cavernous bedrock. Dissolution and removal of enough limestone to form a cave of significant size requires geological timescales that cover tens of thousands of years. In contrast, a comparable volume of soil can be removed during a single rainstorm, where there is a stable, old karstic cave somewhere beneath that can accept the displaced material. The rock left over a limestone cave is generally strong enough to stand for very long periods of time. In contrast, a soil arch over any void is inherently unstable, and may fail immediately or during a subsequent rainstorm. For these two reasons, soil failures are much more common than rock failures in karst. The chances of an engineered structure being damaged or destroyed by sinkhole development due to soil failure, during its design lifetime, are orders of magnitude greater that the chances of rock collapse.

These karstic subsidences and collapses caused by soil failure are collectively known as subsidence sinkholes. They can form in any unconsolidated soil that overlies a karstic limestone – the only requirements are caves or networks of fissures that can accept the inwashed soil, a fissure to rockhead as a corridor for the soil transport and percolation water (naturally from rainfall) that becomes the transporting agent. In soil-mantled karst terrains, they occur in their thousands, most of them just a few metres across, but some reaching 100 m in diameter. Most subsidence sinkholes are permanent or maturing features of the landscape, but new sinkhole events are a significant geohazard in many karst regions.

4.1 SUBSIDENCE SINKHOLE MORPHOLOGY

The classic profile of a subsidence sinkhole is an inverted cone, but this may vary to a rounded bowl or to a vertical cylinder. The slope angle is a function of the soil cohesion and the sinkhole's maturity. Those in sandy soils tend to have slopes

Figure 4.1. A typical small, fresh subsidence sinkhole formed in till overlying cavernous limestone in the Yorkshire Dales, U.K.; the new sinkhole is only about 3 m wide and 1 m deep, and it does not expose the limestone as rockhead is at a depth of about 3 m.
TW.

close to their angles of rest around 35°. A fresh subsidence sinkhole in a clay soil can have vertical or even overhanging sides, but degrades over time into an ever-wider bowl. In 1994 a tropical storm over Georgia, U.S.A., triggered a population of new subsidence sinkholes in alluvium and residuum overlying the karstic Ocala limestone. Morphometric analysis of more than 300 new sinkholes and also over 300 old features (Hyatt *et al.*, 1999) showed that depths of both groups were comparable, but the older sinkholes were much wider due to slow degradation of their side slopes.

The depth of subsidence sinkholes is limited to that of the soil thickness. In a soil 2 m thick, a sinkhole cannot be more than about 8 m in diameter over a single input point to the bedrock, and many will be less in both depth and diameter. In the Yorkshire Dales, U.K., and in many other karst regions, depths of 1–5 m and diameters of 3–15 m are typical for the many hundreds of small subsidence sinkholes (Figure 4.1), commonly known as shakeholes, that pockmark benches of glaciokarst where the limestone lies beneath a veneer of till. In contrast, the Winter Park sinkhole in Orlando, Florida, was 30 m deep and just over 100 m across, because it developed in soils that are 45 m thick (Jammal, 1984). The Winter Park sinkhole is a fine example of a large single-event subsidence sinkhole (Figure 4.2), though its failure was induced by water table decline (Chapter 8). Most sinkholes

Figure 4.2. The large subsidence sinkhole in Winter Park, Florida, just after its initial dropout in May 1981; water is ponded within the steep-sided throat down through the clay soils, while the upper slopes were at the time still actively slumping back to form the wide bowl in the sand cover.
Photo: Orlando Sentinel.

that are much larger than these dimensions are either solution features eroded into bedrock or collapse features also into bedrock, though a soil mantle within the former or a soil cover slumped into the latter may create the false impression of a subsidence sinkhole.

Clearly, the volume of a subsidence sinkhole must be matched by the volume of pre-existing solutional voids within the bedrock, into which the soil can be washed, but there is no implication that bedrock caves should match the sinkholes in size. Most soil washed from sinkholes is lost into fissure networks. Even when a cave is associated with a subsidence sinkhole, its passages are normally much smaller than the surface feature (Box 4.1). Collapse of the Winter Park sinkhole in Florida involved nearly 150,000 m^3 of cover sediment disappearing into the bedrock voids, with a large proportion of this going in the few days of the main collapse event. No caves associated with this sinkhole are known, but passage sizes in the Floridan karst are commonly 5–10 m in diameter. The giant sinkhole that formed in the tailings

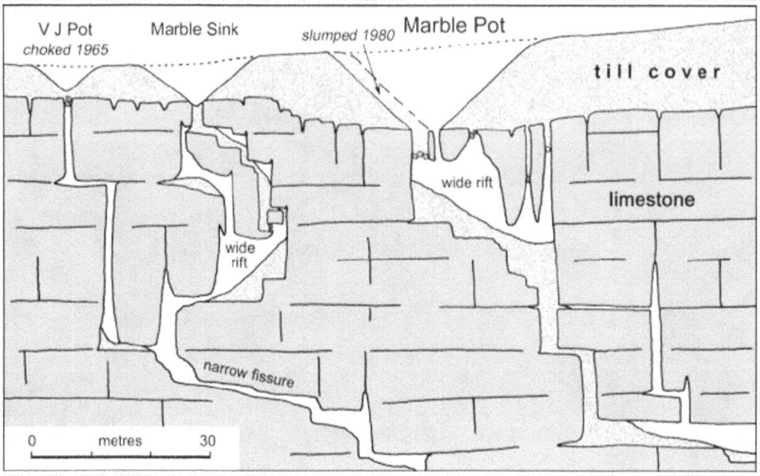

Figure 4.1.1. Profile through the three subsidence sinkholes at Marble Pot (simplified, with some passages omitted for clarity). Details of the rockhead and some of the totally choked cave passages are conjectured, but the sinkholes and main caves have been mapped by direct exploration.
After surveys by Dave Brook *et al.*

Figure 4.1.2. The subsidence sinkhole of Marble Pot, soon after its far side slumped into a newly open fissure in 1980.
TW.

BOX 4.1. SUBSIDENCE SINKHOLE

An example – Marble Pot, Yorkshire Dales, U.K.

Within the Yorkshire Dales glaciokarst, many of the broad limestone benches are mantled by till sediments that are pocked by thousands of subsidence sinkholes. Marble Pot is one of the larger sinkholes in the area, 40 m across and 15 m deep, and circular except for a stream trench entering on one side. A small stream drains into it, so that the underlying fissure system has evolved to be large enough to be accessible, at least in parts. This has allowed the underground features of Marble Pot and two more adjacent sinkholes to be mapped by cavers (Figure 4.1.1). If this profile had its scale adjusted so that most of the fissures were only 100–200 mm wide, it would suffice as a model for the karst morphology beneath subsidence sinkholes worldwide, where significant surface depressions appear unrelated to any visible or accessible voids in the bedrock.

Marble Pot has been developing ever since the mantle of till was left over the limestone outcrop on the retreat of the late Pleistocene glaciers about 12,000 years ago. The fissures and cave passages that constitute the constricted cave system beneath it have some earlier origins, as some of the lower passages contain stalagmite deposits that appear to be interglacial. Once the till had been deposited, seepage water (derived both from direct rainfall and small flows of allogenic surface drainage off the adjacent shale slopes) fed through to the rockhead fissures, and suffosion by this water carried sediment down into the limestone voids. The cohesive nature of the Yorkshire Dales till meant that the sinkhole then developed in a series of dropout events, each one followed by slumping of the sinkhole sides so that they flared out into the characteristic, steep, inverted cone. The most recent event was in 1980, when the whole of one side of the sinkhole failed (Figure 4.1.2). The soil all slumped into a newly opened fissure, which had been previously seen from below when cavers had looked up at a hanging choke of boulders with water dripping from between them. This temporary plug had failed during a winter storm event, and many tonnes of debris had fallen into the underlying rift. This totally choked the route down, except for the stream that could filter through the rubble and sediment.

Two other adjacent sinkholes lie over the same set of roughly parallel fissures that follow joints in the limestone (Figure 4.1.1). Marble Sink has its main route choked by suffosed sediment, but an open loop passage (not shown on the profile) provides access to the passages below. V J Pot is a smaller sinkhole that has had its outlet blocked since its sides degraded in 1965. All three sinkhole drains consist of systems of fissure passages mostly less than 1 m wide, linked by vertical shafts 1–3 m wide, and including some rift chambers up to 5 m wide. Their volumes are smaller than those of their respective sinkholes, but large amounts of sediment have been washed through them and into deeper caves during their long periods of evolution.

lagoon over Zambia's Mufulira mine in 1970 had a volume of 700,000 m^3 (Spooner, 1971). The failure was partly induced by the mining, and 300,000 m^3 of sediment filled the mine below the inrush point 530 m below the surface. But 400,000 m^3 of sediment were lost into the unseen network of fissures and caves between there and the surface. The sinkhole that formed in this event was less than 20 m deep, but it grew to 300 m across in the saturated and very unstable mine tailings.

Subsidence sinkholes are typically almost perfectly circular in plan outline. This is because each one is formed by slumping of the soil cover into a single opening in the bedrock. This opening may be a local widening of a single fissure at rockhead, but is more commonly a solutional enlargement at the intersection of two joints (or faults or dipping bedding planes); larger openings may be discrete but totally choked cave passages or shafts. It is very unusual for soil to be washed down any length of a fissure long enough to impose an elongate shape to the surface depression. Ashtree Hole, in the English Pennines, is 25 m across and 10 m deep, and is almost perfectly round, with a rounded soil floor, even though it loses its sediment into a major fault fissure that can be seen in a cave passage 40 m below (see Section 4.4). Those sinkholes that do have more complex or elongate shapes are almost invariably formed by the coalescence of two or more circular features, each over their own bedrock entry point. Alternatively, a subsidence sinkhole may become elongated by surface erosion where a stream happens to drain over one margin into it.

Though subsidence sinkholes are formed by loss of the soil cover into bedrock, it is unusual to have the bedrock exposed in their floor. Slumping of the sides, after the initial more localised collapse, normally leaves a sediment floor in the sinkhole. This commonly blocks the bedrock conduit, to the extent that ponds are common in sinkholes, even where they are perched far above the regional water table (Figure 4.3). Where subsidence sinkholes capture surface streams, they develop active stream sinks or swallow holes in their floor, but these may be either open or choked where the stream drains through boulders. Only a few subsidence sinkholes have open caves at their throats. A cave of accessible size may be in proportion to a very large sinkhole (Box 4.1), but may also lie at the apex of a sinkhole only a few metres across where a thin soil has slumped through an opening in the roof of a cave passage that drains from elsewhere.

Subsidence sinkholes can form in any type of soil. They are perhaps most abundant in till, notably where Pleistocene glaciers left a veneer of sediment over a limestone surface with fissures already enlarged by pre-glacial or sub-glacial dissolution. Impermeable till with a clay matrix can prevent water reaching the underlying rock, but most till has enough permeability to allow some infiltration and consequent initiation of subsidence sinkholes. The disorganised topography of a sheet moraine tends to inhibit efficient rainfall run-off, so that areas of till on almost level ground, on either valley floors or plateau benches, are commonly pocked by hundreds of usually small subsidence sinkholes, forming one of the distinctive aspects of glaciokarst terrains (Figure 4.4). In these cases, all the sinkholes have formed in post-glacial time, about 12,000 years in the case of Devensian till.

Comparable sinkholes also form in alluvium, in colluvium and in the red soils

Figure 4.3. Variations in morphology of subsidence sinkholes; (a) – a large and old sinkhole in till, that holds a lake in wet weather but drains out in dry conditions, in the Yorkshire Dales karst, U.K.; (b) – a small new sinkhole in clays that floor a valley in Pennsylvania that loses all its water underground; (c) – an old, degraded sinkhole with its fissure drain sealed by inwashed clay so that it has reed grass growing in a small pond, in the Yorkshire Dales, U.K.; (d) – a newly reactivated sinkhole with fresh scars in its soil slopes, on the limestone of County Cavan, Eire.
TW.

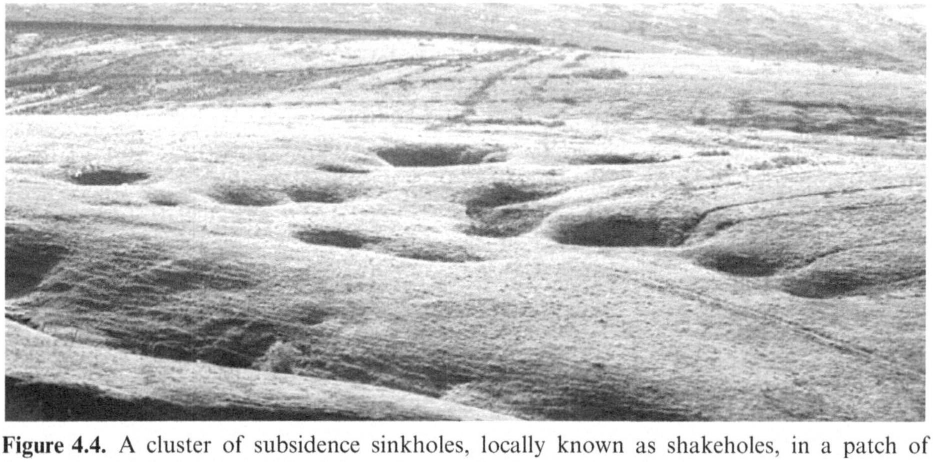

Figure 4.4. A cluster of subsidence sinkholes, locally known as shakeholes, in a patch of moraine till in the English Pennines.
TW.

that are common in karst as a combination of insoluble residues and loessic silts blown in on the wind from beyond the limestone outcrops. Total numbers of sinkholes are often smaller than those on till, but can be large where rejuvenation has followed tectonic uplift. Beside the west coast of Florida, Citrus County has new sinkholes forming in thin cover sands at a rate approaching one per week (Beck, 1986); each sinkhole is typically less than 3 m wide and deep, and their frequency represents a new sinkhole rate (NSH) of 0.05 per km^2 per year, typical of a mature covered karst of class kIII.

4.2 TYPES OF SUBSIDENCE SINKHOLES

Ranges of morphologies and of processes are encompassed by the concept of the subsidence sinkhole. The key process that defines all subsidence sinkholes is suffosion – the transport of disaggregated soil or sediment into fissures in the underlying bedrock. In a karst, these fissures have been enlarged by dissolution at some time in their history. The transport is almost invariably by water, though it can theoretically be solely by gravity in some sandy desert soils. The resultant sinkhole is therefore matched by sediment that is displaced underground. This may appear as a debris cone within an open cave (Figure 4.5), it may be dispersed through an inaccessible network of narrow fissures or it may ultimately be transported out of the aquifer by a resurging cave stream.

Time determines the subdivision of subsidence sinkholes into the dropout sinkholes where the ground failures are rapid collapses, or the suffosion sinkholes where the ground subsides slowly (Figure 4.6). The overall process may be described as suffosion in both cases, though the dropout term is introduced as a graphic description of the surface impact where a large soil cavity fails in the one type. The contrasting types relate largely to soil lithology. Only a cohesive soil, that is clay-bearing or indurated, can bridge over a void for long enough for the cavity to grow to a size (by wall ravelling and then debris removal from its floor) large enough for its subsequent and inevitable roof failure to have significant impact as a surface collapse. In contrast, the theoretically perfect suffosion sinkhole develops by particulate tapping of a non-cohesive, purely sand soil into a narrow underlying fissure, matched by slumping and subsidence of the soil profile (Figure 4.7). Though these can develop to very large conical profiles, most of this type that occur in the non-cohesive soils of Florida are only a few metres deep and wide even in soils that are 15–30 m thick (Sinclair and Stewart, 1985).

Nearly all soils have enough cohesion provided by clay within their matrix (or by clay-rich horizons within a predominantly sandy profile) to make dropout sinkhole events far more common than the slow subsidences. Instantaneous or rapid surface failure of the dropout sinkhole is the main geohazard in terrains of soil-covered karst. The slow subsidence of suffosion sinkholes is sometimes recognised, but these sites commonly develop into dropout sinkholes by intermittent stages of rapid soil collapse. More confusion between the two types is created by the fact that steep-sided dropout sinkholes generally degrade by wall failure until they

Figure 4.5. A small subsidence sinkhole in the karst of Indiana, and the cone of debris emerging from a fissure directly beneath it in Blue Spring Cave.
Photos: Art Palmer.

develop the larger diameters and the shallow conical profiles of suffosion sinkholes. This would appear to account for the questionable deduction that a part of Georgia, U.S.A., is dominated by new dropout sinkholes while the same area has older sinkholes of both suffosion and dropout types (Hyatt *et al.*, 1999).

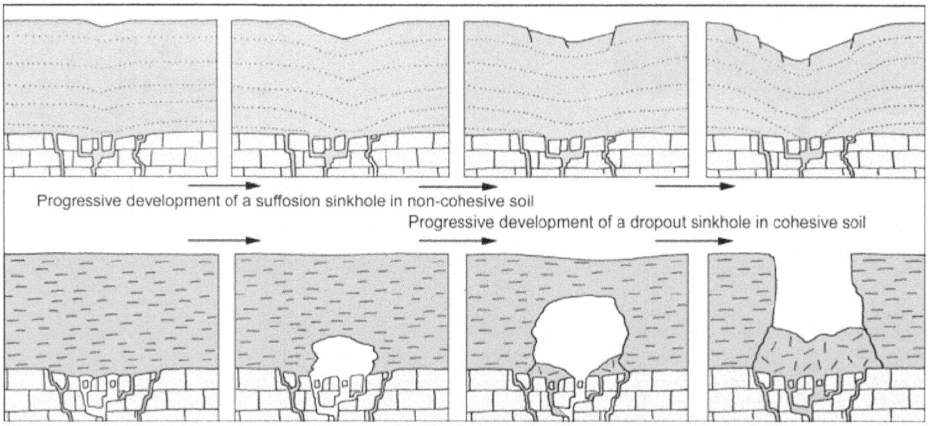

Figure 4.6. Sequences of progressive development of subsidence sinkholes, represented by stages in the two end members of morphologies, forming by perfect particulate suffosion in a non-cohesive sandy soil (above) and by dropout over an expanding soil cavity in a cohesive clay (below).

Figure 4.7. A small suffosion sinkhole newly developed in sandy soils near Kerman, Iran. Photo: Habibeh Atapour.

The morphological variety and wide distribution of subsidence sinkholes have led to an extensive and sometimes conflicting collection of terms for them (Table 2.2). Subsidence sinkholes (Jennings, 1985) usefully describe all those formed within the soil profile. Suffosion sinkholes (Ford and Williams, 1989) accurately describes those formed by the slow downward removal of the soil. Dropout sinkholes grew in common parlance from the visual impact of the rapid surface failure. These three terms are now widely accepted (Williams, 2004). Cover collapse and cover subsidence sinkholes (Beck and Sinclair, 1986) are also useful and descriptive terms (popular in the U.S.A.), but are less favoured because they each require three words. Also the word collapse is undesirable in this context, as its use in karst is normally retained to describe rock collapse, which significantly is not involved in subsidence sinkholes. Ravelling sinkholes, soil piping sinkholes and shakeholes are all less satisfactory terms for subsidence sinkholes. Description as alluvial sinkholes is unacceptable, as large numbers of them occur in till soils that are not alluvial.

Poorly consolidated clays of Tertiary age (old enough to be called rocks by geologists, but treated as soils by engineers) can develop ground collapses that are described as either subsidence or caprock sinkholes. The distinction should be that subsidence sinkholes form by particulate suffosion of the soil down narrow fissures in the bedrock, while caprock sinkholes involve intact plugs of the clay dropping into voids created by rock collapse of a cave roof. Surface morphologies may be indistinguishable. Sinkholes in the clay-mantled gypsum karst of the Ukraine (Klimchouk and Andrejchuk, 2003) are clearly of both types, but are only distinguished where the nature of the failure can be seen in the caves beneath. Transitional morphologies are also displayed by some of the larger sinkholes in the chalk karst of southern England. Culpepper's Dish and others have long been known (Sperling *et al.*, 1977) but more have been recognised by subsequent geological mapping. Many of these are formed in Paleogene sediments, but are almost certainly subsidence sinkholes, because large dissolutional voids that could collapse to create caprock sinkholes generally do not exist within the chalk. However, others are known to have formed by settlement into sediment pipes and filled pre-Paleogene sinkholes (Chapter 5).

4.3 DROPOUT SINKHOLES

Unheralded ground collapses, seemingly random in time and space, form one of the most widespread geohazards in karst, and dropout sinkholes are distinguished by their rapid development. A sinkhole just a few metres across can form in a virtually instantaneous event. In Pennsylvania, a young dogwood tree, growing about 5 m tall in a house front lawn, dropped almost out of sight as a man walked past (Figure 4.8). The hole was only 3 m across, and was soon filled with soil to prevent its sides flaring out, but the residents of the house promptly decided to move out. Rarely can a dropout sinkhole much wider than this be described as instantaneous. Though the hazard to property is very real, the threat to life is greatly reduced where the timescale is more than a few minutes. The most dramatic failure occurs where a thin, cohesive surface layer fails over a relatively large soil cavity that has taken many months or years to enlarge by slow ravelling and suffosion. An asphalt road surface provides the worst case, and many roads in China and the eastern U.S.A. have developed dropout sinkholes without any warning. At least one in China has

Figure 4.8. A dropout sinkhole in the front lawn of a house in Pennsylvania; the hole was originally 5 m deep but was filled with soil to prevent the sides flaring out, and the soil fill had compacted and settled by 0.7 m by the time this photo was taken.
TW.

opened rapidly enough to swallow and kill a person, and at least one in America has caused the death of a car driver who could not avoid the new hole.

A small dropout sinkhole developed in a highway across gypsum karst in eastern Michigan (Benson and Kaufmann, 2001). Though the hole in the road surface was only 1 m across, it opened below to about 1.5 m across, and appeared to be floored by a plug of soil that had dropped its full depth of about 1.5 m. Though the sinkhole was in sand, the dropout style was due to a clay soil at a depth of 1.2–4.8 m. Beneath the clay, sand extended to rockhead at about 12 m deep, and no large voids were known in the bedrock gypsum. The lower sand appeared to have been suffosed into fissures in the gypsum over an unknown period of time; this created a cavity bridged by an arch in the clay, which flaked away to the point of failure, when the remaining clay and upper sand collapsed into the lower void, leaving the road asphalt unsupported over the upper void until it broke through.

In contrast to the narrow fissures at the Michigan site, the Boxhead sinkhole in the Yorkshire Dales karst, U.K., was unusual in that its opening into the bedrock limestone was a metre wide, and over twice as long. This was revealed, overnight, when the floor of an older, shallow sinkhole dropped out (Figure 4.9), at the top of a vertical shaft 95 m deep, that flared out to a diameter of 8 m at its base. This open hole had been a significant pre-glacial stream sink taking drainage off the adjacent shale outcrop. Pleistocene glaciers then overrode the shaft, retreating to leave

Figure 4.9. The Boxhead sinkhole in the Yorkshire Dales karst, U.K. On the left, very soon after the first dropout revealed the deep open shaft in the bedrock limestone, rockhead is visible beneath the till on the left of the shaft. On the right, after more till had slumped into the shaft, the metre-diameter pipe was installed to retain access, and the sides had flared to a stable profile.
TW.

boulders jammed across the top and supporting a cover of till over 5 m thick. Subsequently, postglacial percolation drainage washed the fines out of the bridge of soil and boulders until it became unstable.

Truly instantaneous failures of large dropout sinkholes are rare. The most infamous example is the 1962 sinkhole in South Africa that swallowed a mine building, killing the 29 people inside it (Box 8.1). The Golly Hole developed on a forested hillside in Alabama in 1972. A local resident was disturbed by a house-shaking rumble and the sound of breaking trees, so it appears that there was a major instantaneous collapse. Two days later, a search of the forest found a subsidence sinkhole 35 m deep and 100 m in diameter, with steep soil slopes that must have degraded and widened to some extent in the two days since the main event.

Development of the Winter Park sinkhole in Florida was observed and documented (Jammal, 1984), because it was in a suburban district of Orlando. The ground was first broken when a large tree disappeared with a swishing noise into a new hole, and soil was then lost into it progressively over the next two days, when the sinkhole grew to a depth of 30 m and then had its sides flare out to a diameter of 106 m (Figure 4.10). Subsequently it achieved some sort of stability, before partially filling with water, and was later artificially modified to remain as a permanent lake. It appears that a large cavity had migrated up through the cohesive sandy clays of the Hawthorn Formation over an unknown length of time; its clay roof finally failed to breach the overlying loose sands, which slumped into the underlying pipe within the two days of observed sinkhole enlargement. Initially, the sinkhole was almost dry, but it soon filled to the piezometric level of the Ocala Limestone aquifer, indicating a restricted water flow through the debris-filled pipe beneath the sinkhole. The

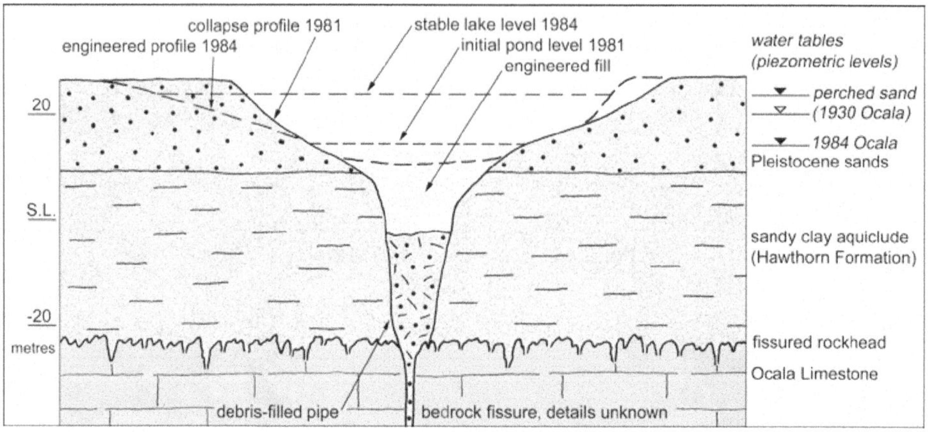

Figure 4.10. Profile through the large subsidence sinkhole that developed in Winter Park, Florida, showing the initial dropout through the lower clay soils, the flared bowl in the upper sands and the subsequently engineered lake.
After Jammal (1984).

remedial works on the slopes included pushing soil down into the sinkhole's throat, thereby choking it to prevent further drainage down to the limestone. Seepage from the sand then fed a lake that almost filled the sinkhole, with its surface at the level of the water table in the perched sand aquifer (Figure 4.10). Some years later the lake level fell close to that of the limestone water table, and this may indicate renewed drainage down the debris pipe, with implications for possible reactivation of the sinkhole. This non-instantaneous event is probably representative of most very large dropout sinkholes, and though infrastructure damage ran to millions of dollars, no lives were lost.

4.3.1 Growth and failure of soil cavities

The critical process in the formation of dropout sinkholes is the growth, migration and failure of ephemeral soil cavities; these are widely known as regolith arches in North America. They are generally initiated at the rockhead, as soil is carried down into bedrock fissures (Figure 4.11). As more soil is lost, the cavity grows beneath an arched roof, until this becomes unstable, causing roof stoping and upward cavity migration until surface failure is induced. Sediment transport is critical to continued cavity growth, and is normally dependant on water flow rates.

Soil cavities are relatively stable when they are small and at depth, so that compression arches may develop within the soil above them. As the cavities increase in size, either by concentric sloughing and ravelling failure of their roof material or by coalescence of adjacent voids, their growth and upward migration accelerate (Drumm *et al.*, 1990). Ultimately the soil arch is likely to fail in shear, but the progressive sloughing is related to variations of pore water pressure within the soil. Critical are situations where the pore water pressure is greater within the mass of the soil than it is immediately adjacent to the cavity wall (Tharp, 1999, 2001, 2003). This situation is created when water table decline creates a falling pore water pressure that lags behind drainage of the cavity, when a new wetting front of rain-fed percolation water advances through a dry soil after a drought, or by exsolution of groundwater air creating bubbles close to the cavity wall. The dominant soil failure process is that activated by pore pressure decrease due to water table decline. When this occurs, the rate of sloughing failure from the void roof is a function of the soil's compressibility, permeability and tensile strength (Tharp, 2001), and where these are known it could be possible to predict cavity migration rates for any planned drawdown.

The processes related to water table decline in open voids are described as vacuum effects in China, whereby the soil cavity fails by suction forces (Xu and Zhao, 1988). The role of suction is demonstrated by Chinese success with inserting open pipes into the ground to allow air recharge and thereby prevent further sinkhole failures in soils impacted by mine drainage. A quantification of soil-arch stability in incipient sinkholes largely relates the cover thickness and arch radius to the soil's angle of friction, but also relies on estimations of groundwater flow velocity, pore water pressure and cave atmosphere pressure (He *et al.*, 2003). Some field evaluations have been successful, but data on the variable parameters may not be available until collapses have occurred and can be observed.

Failure of a soil arch was monitored in South Africa's Pulik Cave, which has an arched soil roof spanning 3 m between walls of dolomite in a passage developed along the rockhead (Jennings, 1966). Processes were accelerated by experimental injection of water through boreholes into the soil profile over the cave. The roof arch was seen to fail progressively as concentric slices of soil about 700 mm thick fell into the cave, while surface subsidence at the same time was very small and considered to be due to elastic deformation.

A soil arch may also lose its integrity by classic piping failure that works headward along flow paths into the cavity, probably defined by soil microfractures. This is not activated by pore water pressures but by seepage flows that are capable of washing the fines out of a heterogeneous soil. Their significance may be observed where soil cavities are made accessible by partial failure of the roof. Most cavities do propagate vertically upwards, ultimately to form dropout sinkholes, but piping may open obliquely to the surface where it is directed by water inflow (Figure 4.12). Piping failures respond directly to flow increases, notably those induced by rainfall events. These are widely recorded as triggering multiple sinkhole events, including the Florida sinkhole maximum during summer thunderstorms (Currin and Barfus, 1989) and the large number triggered by a single storm in Georgia, U.S.A. (Hyatt *et al.*, 1999). In Florida, it appears that sinkhole events do not correlate with long-term rainfall data (Beggs and Ruth, 1984), but do correlate with localised rainfall events (Benson and La Fountain, 1984). Similarly, sinkhole development may be induced by artificially re-directed drainage input (see Chapter 8).

Other factors have been interpreted as contributing to soil cavity failure. Loss of buoyant support during water table decline may be significant (Newton, 1987), but its effects may be confused with those of increased seepage. Soil cavity failure may be

Figure 4.11. A soil arch beneath about 150 mm of cover over a cavity developed over a fissure in limestone rockhead, exposed in the side of a small collapse in Kentucky. Photo: Art Palmer.

Figure 4.12. A thick soil arched over a piping cavity in the edge of a fresh dropout sinkhole in the Xingwen karst in Sichuan, China; rockhead is deeply pinnacled in this mature karst, and the edge of a limestone pinnacle is exposed just to the right of the boy's feet. TW.

due to vibration from artificial sources (see Chapter 8), from earthquakes and from water hammer in collapsing ground (Chen, 1988). Soil weakening by growth of desiccation fractures appears to account for a number of sinkhole failures in drought conditions, and may be more important in sandy soils that gain much of their cohesion from surface tension (Tharp, 1999). Laboratory modelling of sinkholes demonstrates the main failure processes, and also shows that increased stress from slab loading (as opposed to point loading under small foundations) is irrelevant (Chen and Beck, 1989).

Timescales for void migration through soils are rarely recorded, because the sites are usually only known after failure reaches the surface. Rapid development is indicated by the records of new sinkholes developing within the first few hours or days after pumped water table decline in Florida (Bengtsson, 1987), in Alabama (Newton and Hyde, 1971), in China (Waltham, 1989) and elsewhere. These apparently rapid failures could be the final collapses of soil cavities that had developed previously over longer periods of time, and sites in Pennsylvania (Foose, 1968), Alabama, China and elsewhere do record new sinkholes appearing over months and years after the initial drawdown. Numerical modelling of a clay soil 2 m thick, with an unconfirmed compressive strength (UCS) of 0.15 MPa, water on its surface and a drained cave below, indicates that a void 20 mm across on the rockhead will grow to 1 m across in 12.9 years, but will break through to the surface within only another 0.7 years (Tharp, 1999). Combining this evidence with the many records of sinkholes very rapidly induced by drainage modification (Chapter 8), it is clear that dropout sinkholes can form within days or weeks wherever water has access to a soil overlying a fissured limestone, and they will always be a major geohazard in karst.

4.3.2 Evolution of subsidence sinkholes

Perfectly formed suffosion sinkholes are as rare as completely non-cohesive soil profiles. The Ebro basin, in north-east Spain, has hundreds of subsidence sinkholes in its alluvial terraces of gravel, sand and silt that overlie gypsum karst (Soriano and Simon, 2001). Active sites include both suffosion and dropout sinkholes. Four years of monitoring of one suffosion sinkhole, 60 m across, revealed steady subsidence of its bowl at a mean rate of 64.5 mm/y, and two other sinkholes were subsiding at rates of 39 and 21 mm/y. But the same area also recorded a number of dropout events and repeating soil collapses. The great majority of sinkholes in soil-covered karst occupy a spectrum of morphologies and development processes that lie between end-members that are the rare suffosion sinkholes and the usually small dropout events; they are nearly all best described just as subsidence sinkholes. Perhaps the most common type starts with some slow surface settlement, followed by an intermittent sequence of short but rapid soil failures and collapses. Some subsidence sinkholes mature and grow over hundreds or thousands of years, but only one of a series of collapses may impact a construction project within its lifetime. Patterns of repeat activity do demonstrate the folly of simply filling old sinkholes prior to land redevelopment.

Slow surface subsidence may be a precursor to imminent failure on a larger scale, either where this is due to particulate erosion in a suffosion sinkhole, or where a soil arch is settling prior to failure over a soil cavity as it develops into a dropout sinkhole. These movements can provide valuable warning signals, and are especially noticeable as hairline cracks in buildings and bridges prior to more extensive damage. In Kentucky, small ground movements have been ascribed to settlement of shear-bounded soil wedges over a growing soil void (Figure 4.13) and are interpreted as precursors of more serious ground failure as a dropout event (Crawford, 2001). This pattern of ground movements could also be explained by the growth of numerous small soil voids over a network of rockhead fissures (Cooley, 2001), creating a wide zone of soil deformation and settlement before the cavities start to coalesce and fail in larger dropout events.

Many large subsidence sinkholes expand by repeated failures of parts of their soil cover. Marble Pot in the English Pennines (Box 4.1) was reactivated when the boulders, jammed across the top of the rockhead fissure in a previous dropout, were washed out to initiate a second phase of rapid soil loss. The sinkhole profile was then maintained as a new slice of soil slumped into the open fissure. Fresh clearance of the bedrock outlet rejuvenates the sinkhole and helps to maintain its steep profile. In

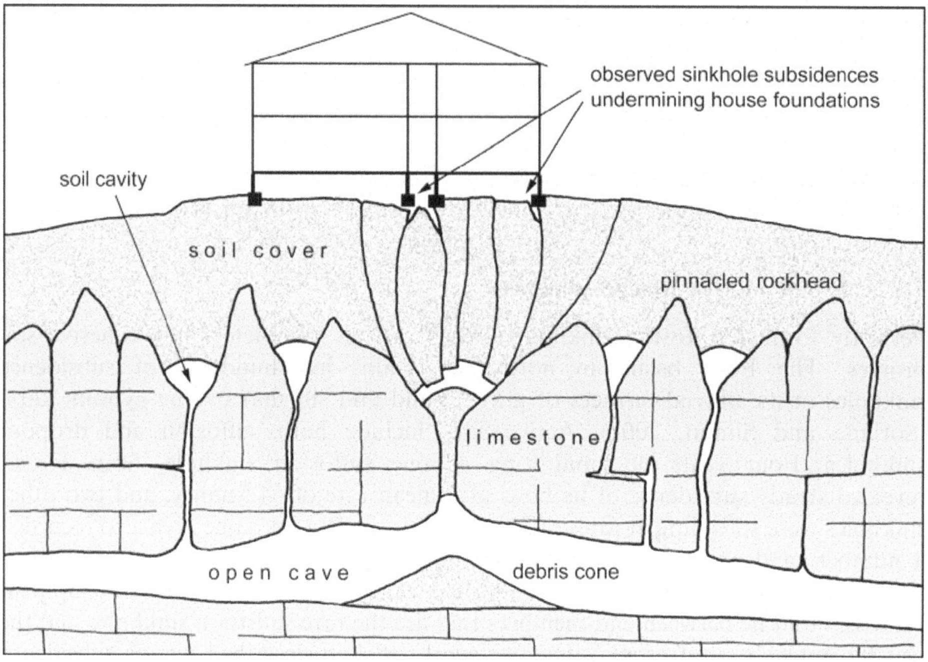

Figure 4.13. The concept of sinkhole growth by shear surfaces that allow soil wedges to fail into a cavity opened by ravelling of the soil into a bedrock fissure; based on a site in Kentucky that was investigated by soil borings and geophysical profiling.
After Crawford (2001).

contrast, a sinkhole with its outlet permanently choked by large boulders matures into old age as its sides degrade, even though water continues to drain through the coarse choke so that a lake does not form within it. The sinkhole's conical profile becomes wider and flatter, and ultimately evolves into a wide bowl where there is little or no scope for sediment transport towards the central outlet into bedrock. This sequence is well observed in Florida (Beck and Sinclair, 1986) where the thick cover of unconsolidated sediment allows the subsidence sinkholes to widen to diameters of hundreds of metres and reach levels below the regional water table, so that they hold many of the circular lakes that are common on parts of the Florida karst (Figure 4.14). Others among Florida's sinkhole lakes are perched in the soil aquifer, and suffer occasional rapid drainage events when small new subsidence sinkholes develop through their floors.

An alternative mechanism for sediment transport is indicated by striated columns of very viscous mud that have been extruded into some caves in Georgia, U.S.A., in limestone beneath a cap of Paleogene clay sediments (Jancin and Clark, 1993). These suggest a creeping plastic flow of the clay mass, in place of granular suffosion. Though sinkholes do not overlie most of these structures, land above their positions within the caves must be regarded as sites of potential subsidence.

Figure 4.14. Lakes stand in many of the very old subsidence sinkholes that have degraded to wide bowls in the thick sediment cover over the limestone in the Florida lowland karst. TW.

Though some of the very large and old subsidence sinkholes become almost permanent features of the karst landscape, they always have the potential for renewed activity, especially when disturbed by construction works (see Chapter 8). It is a sad fact that a new subsidence sinkhole can develop almost anywhere, and without any warning, in a soil-covered karst, but it is still advisable that even the degraded and apparently stable sinkholes should be avoided if possible when planning new development. The case from Pennsylvania of a site with a large dry subsidence sinkhole in the middle of it being offered for sale with the suggestion that "the hole would save on excavation costs for the house basement" showed either a frightening lack of knowledge or an indefensible excess of salesmanship by the vendor.

4.4 SPATIAL DISTRIBUTION OF SUBSIDENCE SINKHOLES

Any recognition of patterns in sinkhole distribution has to be welcomed as a possible tool for predictions of future events (Chapter 10), but sadly the concept is fraught with complexity and difficulty. While it is frequently possible to identify zones or areas that are more susceptible to subsidence sinkhole events, it is unreal to hope to predict where individual sinkholes will next occur. The few available data sets on old and new sinkhole distributions tend to confirm the very limited prospects for predicting the locations of future sinkhole events.

A karst terrain of 71 km^2 in Georgia, U.S.A., with a soil cover averaging 15 m thick, had 329 recorded sinkholes of various types; rainfall from a tropical storm then triggered 311 new subsidence sinkholes in 1994. The new sinkholes were not clustered near the old sinkholes, with the implication that locations of old sinkholes "have limited predictive utility to identifying sites for new sinkholes" (Hyatt *et al.*, 1999). It is possible that the old data was distorted by both the unrecorded filling of small features and also the inclusion of some large solution sinkholes, while the storm-induced events were all subsidence sinkholes; some new sinkholes were clustered within the larger older features. Locations of 2,303 old sinkholes were compared with those of 179 new sites in a karst in eastern Florida (Upchurch and Littlefield, 1987). In areas of bare karst with very thin soils and few sinkholes, there was a spatial correlation between old and new events, but there was no correlation in areas with more than 5 m of soil cover. A second area of Florida karst had 385 sinkholes within about 40 km^2, where 30 new sinkholes were distributed almost everywhere except in the areas of high density of old sinkholes (Bahtijarevic, 1989), and this is a karst with more than 30 m of soil cover. Predictions may be possible in almost bare karst where subsidence sinkholes are a minimal threat, but karst with thicker cover provides the main geohazard and also offers little hope for useful predictions.

Networks of fractures that include open fissures only a few metres apart in each direction are typical in a karst limestone. Beneath a soil cover, these provide an infinite number of locations for soil suffosion and collapse that could create new subsidence sinkholes. The difficulty for the engineer lies in the near-impossible task of recognising where the open fissures lie when they are obscured beneath the blanket of soil. Many attempts at recognising fracture traces, in order to predict new subsidence sinkhole events, notably at their intersections, have been based on analyses of topographic maps or air photographs (LaValle, 1967; Kemmerly, 1976; Littlefield *et al.*, 1984; Brook and Allison, 1986; Ogden *et al.*, 1989). Many of the results are less than convincing where "fracture traces" are interpreted from scattered sinkholes within the soil cover, and such studies have declined in numbers in the more recent literature. A comparable analysis in Minnesota found that sinkhole locations were not controlled by rock structure (Magdalene and Alexander, 1995).

Where the major joints and faults can be mapped in underlying cave systems, it may become clear that there is very little correlation with the distribution of subsidence sinkholes. This can be seen in parts of the Yorkshire Dales karst in northern England (Waltham and Hatherley, 1983). Within the area mapped in Figure 4.15, there is a clear line of large sinkholes along the Death's Head fault zone, but the fracture traces are not recognisable elsewhere from the sinkhole distribution. Fracture recognition by sinkhole elongation may also be unreliable. Along the Death's Head faults, only one sinkhole is elongated within the till, and this lies over an almost cylindrical shaft in the limestone, while the Ashtree sinkhole (Figure 4.16) is conspicuously circular and lies over a major joint. Even where fracture locations are known, sinkhole development is still highly variable. As a means of predicting future subsidence sinkhole events, the value of fracture trace interpretations must be seriously questioned, except where individual and very conspicuous linear features can be seen.

Certain other geological factors may influence the distribution of subsidence sinkholes. They tend to be concentrated close to boundaries where allogenic drainage is supplied from impermeable rock outcrops, and in areas of banded and

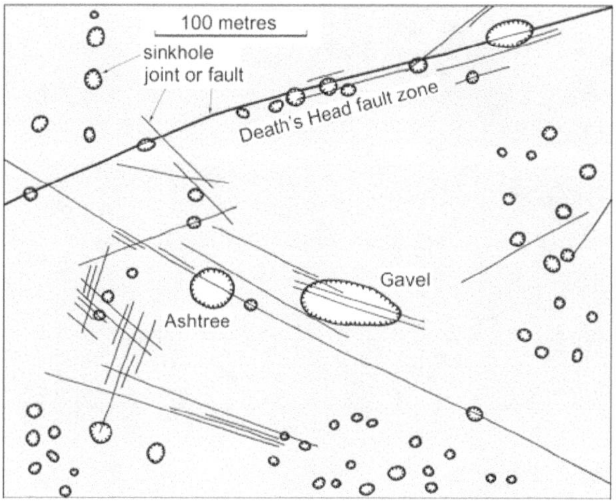

Figure 4.15. Sinkholes within a till cover correlated to fractures in the underlying limestone bedrock on Leck Fell in the English Pennines. The sinkholes were mapped on the ground and from aerial photographs, and the fractures are all recorded in underlying cave systems. Gavel Pot is a large collapse sinkhole, but all the other features are subsidence sinkholes within the till.

Figure 4.16. The deep and almost circular subsidence sinkhole known as Ashtree Hole that lies over a major joint in the Leck Fell limestone (see Figure 4.15).

steeply dipping bedrock they are more numerous over the outcrops of purer and more cavernous limestone. In one part of Georgia, U.S.A., new sinkholes occur more frequently in more permeable sandy soil cover (Hyatt *et al.*, 1999), while in another part of the same state, sinkhole sizes are related less to the permeability of the cover than to its type and age (Hyatt *et al.*, 2001); this may be influenced by more

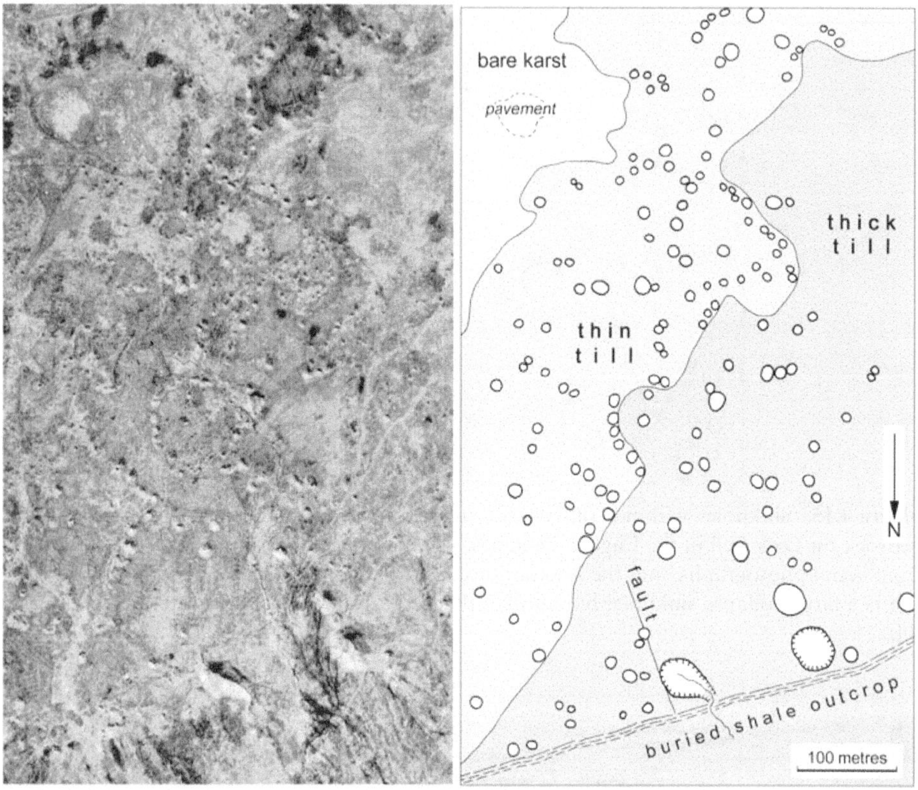

Figure 4.17. The distribution of subsidence sinkholes on part of the till-covered limestone benches that surround Ingleborough in the Yorkshire Dales karst, U.K., with the aerial photograph and the annotated map covering exactly the same area. Bare karst has a thin and incomplete cover of organic and loessic soil. The till is very variable and patchy in thickness, and the drawn boundary between thin and thick till is at an approximated thickness of 3–4 m. The marked fault is mapped inside a cave but has no surface expression. Many smaller sinkhole features are masked by a cover of sphagnum bog that is still expanding.
Aerial photo: Meridian.

of the older sinkholes that have degraded to larger diameters. Sinkhole development can also be inhibited by impermeable clays (Soriano and Simon, 2001). Contrasting lithologies within till soils appear to influence sinkhole distribution in the Pennine karst of England, but patternless variations within these soils generally produce chaotic distributions of the subsidence sinkholes. There is little or no recognisable pattern of the sinkholes on the Hurnel Moss slopes of Ingleborough in the Yorkshire Dales, U.K. (Figure 4.17); within this area, thicker areas of till restrict sinkhole development, mask effects from the one fault that can be mapped underground and almost obscure the buried shale margin. Though valley floors and dry

Figure 4.18. A new dropout sinkhole in the middle of a field on Kentucky's Sinkhole Plain; the top of a limestone pinnacle is just visible about 1 m down, but the open hole is 12 m deep within a fissure between the rockhead pinnacles.
TW.

thalwegs are significant sites for solution sinkholes, they have less influence on the distribution of subsidence sinkholes.

4.4.1 Subsidence sinkholes related to cover thickness

Maximum depths, and therefore maximum diameters, of subsidence sinkholes are broadly limited by the thickness of soil cover in which they form. While larger sinkholes can develop in them, thicker soils still have a preponderance of smaller sinkholes, with floors or apices that do not reach down to rockhead. Also, narrow and deep dropouts can form down previously soil-choked shafts, where the lower part descends between pinnacles of bedrock capped by only a thin soil cover (Figure 4.18). Overall, cover thickness is not a good predictor of sinkhole size

(Hyatt *et al.*, 2001), and models studies suggest that development of subsidence sinkholes is related more to soil type and structure than to soil thickness (Lei *et al.*, 2001). Cover thickness related to the depth to water table is not significant, as subsidence sinkholes can form where rockhead is either above or below the local water table. However, a declining water table, and particularly one that declines past rockhead, is one of the major causes of sinkhole development (Chapter 8).

In general, thicker soil covers develop fewer subsidence sinkholes. China has many thousands of recorded subsidence sinkholes, and data from different areas within the karst indicate that over 60% of sinkholes occur in cover soils less than 5 m thick and over 85% in cover less than 10 m thick (Yuan, 1987; Chen, 1988); a similar picture appears in karsts of many other countries. It is much more difficult to recognise any maximum soil thickness for sinkhole development. For the well-documented karst of Florida, it has been stated that most sinkholes occur in cover less than 20 m thick (Beggs and Ruth, 1984), less than 25 m thick (Upchurch and Littlefield, 1987) and less than 30 m thick (Currin and Barfus, 1989). This trend may reflect improvements in the database, but is probably due to more conservative interpretation. Only where soil cover is more than 60 m thick are sinkholes described as very rare (Sinclair and Stewart, 1985).

Very thick soil covers tend to preclude sinkhole development by preventing seepage drainage through to the underlying limestone, but it is very difficult to recognise any upper limit to cover thickness. Soil cover was 45 m thick where the 1981 dropout sinkhole developed in Florida's Winter Park, and is up to 50 m deep where some of the sinkholes have formed in the Rand area of South Africa, though the latter were in areas of exceptional water table drawdown (Box 8.1). Piping processes operating through cover up to 150 m thick have been indicated in Florida (Arrington and Lindquist, 1987), but no details are available. Tuscany, Italy, has two exceptional dropout sinkholes formed in Pliocene and Quaternary clastic sediments. A dropout 30 m across and 13 m deep destroyed two houses in Camaiore, when it formed where rockhead was at a depth of 100 m. The other was 100 m across, but only shallow, and formed in farmland over a period of two days where the limestone rockhead is 200 m deep. However, these sinkholes over such great cover thicknesses appear to be atypical as they are associated with rising water from geothermal sources (Beck, 2000). Sinkholes are reported as extremely rare in cover > 60 m thick in Florida (Wilson, 1995) and > 70 m thick in China (Xiang *et al.*, 1988), and these figures appear to represent a guideline limit applicable across many karst areas.

4.5 THE SUBSIDENCE SINKHOLE GEOHAZARD

In lowland karst where bedrock of limestone or gypsum is mantled by an unconsolidated soil cover, and therefore accounts for the majority of construction sites on karst, the major geohazard is presented by the occurrence of new subsidence sinkholes. Any small structures that are founded on the soil cover run the risk of losing integrity when the soil is washed down into underlying karstic voids. On complex or extreme karst (of classes kIV or kV), rates of new sinkhole development (NSH) may exceed 1 per km^2 per year – which implies the probability that a new sinkhole will develop inside a one-hectare site within a 100-year lifetime of its structure. Most new sinkholes are small, but can damage or destroy parts of structural foundations. On thick soils, larger sinkholes are less common, but can swallow entire houses. Risks are also lower on less mature karst and on some less permeable soils, but widely scattered individual sinkhole events present an intractable geohazard in any soil-mantled karst.

It is virtually impossible to predict new sinkhole locations in previously undisturbed soil, and there are serious difficulties in economically identifying soil voids that could propagate to create new surface dropouts (Chapter 9). Consequently the engineering response to the subsidence sinkhole hazard is to reduce the risk. The sinkholes are caused by water washing down through the soil, and the vast majority of new sinkholes on construction sites are induced by engineering works that are either unfortunate in their impact or are simply inappropriate to the karstic ground conditions (Chapter 8). Control of drainage is therefore fundamental to good practice in engineering on soil-mantled karst (Chapter 10). This concept does provide special difficulties in the retention of either reservoir water or landfill contaminants in areas of sinkhole karst (Chapter 12). Unfortunately individual subsidence sinkholes present superficially attractive sites for waste disposal to create level ground, but attempts at retaining either landfill or water within subsidence sinkholes usually end in failure. Even where drainage is well managed within new construction projects, natural water flows from rainfall onto any exposed soil still create some potential for the development or reactivation of subsidence sinkholes. It is therefore commonly appropriate either to found structures on stable underlying bedrock, or to create structures on soil that can survive a new sinkhole event under a part of their foundations (Chapter 12).

5

Buried sinkholes and rockhead features

Within the contractual tangle of construction projects, many of the numerous claims of "unforeseen ground conditions" relate to rockhead variations, especially where sound rock is found to lie deeper than had been anticipated. Some of the most extreme rockhead relief is found in mature karst terrains, and the largest individual features are buried sinkholes that have been filled with soil during natural evolution of the site, and now have little or no surface expression. These features may be known as either buried sinkholes or filled sinkholes; both terms are accurately descriptive, but the former is better as it invokes appropriate associations with buried valleys, with which many engineers are more familiar as a rockhead hazard. Very old buried sinkholes may be referred to as paleokarst features (James and Choquette, 1988); the fills in these are generally lithified, so that they have less influence on structural loading capacity except at very sensitive sites (though they may be critical as potential leakage paths from reservoirs). Younger features, with unconsolidated fills that may be the cause of ground subsidence, are sometimes also described as paleokarst as there is no absolute definition of the term.

A significant geohazard in areas of mature karst (of class kIV or kV) is a pinnacled rockhead, where deep, soil-filled, fissures and shafts intervene between bedrock pinnacles. Typically, many of the negative features in such a rockhead are only one or a few metres across, but the larger and deeper rockhead depressions may be described as filled sinkholes. Pinnacled rockheads in the very well-developed karsts of equatorial regions can provide extremely difficult ground conditions with a wildly convoluted interface between soft soil and hard rock. Kuala Lumpur in Malaysia has been described as "the worst ground in the world" with rockhead varying from 4 to 48 m deep beneath a single building site (Bennett, 1997). On one construction site in the same city, bedrock was reached at depths of 5 m and 80 m in adjacent boreholes just 5 m apart. The 80 m deep borehole was probably down a narrow soil-filled fissure, but this would not have been an isolated feature in such a karst. In these terrains, buried sinkholes are certainly very deep between the tall

Figure 5.1. Buried sinkhole within the bare limestone pavements of Ingleborough in the English Pennines; the patch of soil overlies an unconsolidated soil fill of a depth that is unknown but likely exceeds 20 m.
TW.

buried pinnacles that form the deeply dissected rockhead, but widths and profiles of these sinkholes vary hugely and unpredictably.

Filled and buried sinkholes can occur as isolated features in any type of rockhead morphology; they can break the profile of glaciated rockheads that may be otherwise very uniform (in karst of class kI or kII), and can also occur within terrains of exposed limestone where a patch of soil may either rest on the limestone surface or may be the top of a deep fill in a buried sinkhole (Figure 5.1).

5.1 BURIED SINKHOLE MORPHOLOGY

A rockhead depression that constitutes a buried sinkhole may have formed by dissolution with or without collapse when it was an active surface feature. It may therefore have any of the great variety of profiles and dimensions that are found in the solution or collapse sinkholes described in Chapters 2 and 3. Buried solution sinkholes that appear as broad rockhead bowls are readily recognised by a competent ground investigation, and therefore offer minimal hazard to engineering works, though they may constitute considerable inconvenience and perhaps additional costs in establishing foundation. Smaller and deeper dissolutional features, and steep-sided buried sinkholes that had collapse origins, provide the greater geohazard, as they are statistically less easy to find during an exploration programme and may constitute small zones of very unstable ground.

Figure 5.2. Various profiles within some of the buried sinkholes exposed in a single road cut, about 7 m deep, through a limestone hill outside Huntsville, Alabama.
TW.

Buried solutional potholes appear as vertical cylinders of soft soil perforating areas of limestone that otherwise offer solid ground to the engineer. Climatic changes, during and at the end of the Pleistocene, were frequent causes of process changes – whereby soil could be newly introduced to terrains previously dominated by dissolution. Sinkholes buried by thin or thick soils are therefore common throughout the world's karst regions. A cutting for a new freeway into Huntsville, Alabama, exposed a profile with a series of buried sinkholes beneath the soil mantle on a low limestone hill (Figure 5.2); some sinkhole walls were vertical, but others were flared out or overhanging. The narrowest of the buried and filled solution shafts within karst are often known as pipes, or soil pipes, and are a particularly common feature in chalks and softer limestones (Section 5.3).

Immense variety in the profiles of buried sinkholes means that individual features can only be fully comprehended after extensive ground investigations. Rockhead encountered at great depth in a single borehole may be at floor level in the wide bowl of a buried sinkhole, or it may be within the narrow confines of an isolated deep fissure. Investigations for a factory site on mantled karst in Pennsylvania found rockhead at depths of 15–22 m, except for one borehole that did not reach bedrock at 70 m. It is more than likely that this borehole happened to pass straight down a narrow sub-vertical fissure that was largely filled with in-washed clay. Such a feature presents a minimal geohazard to construction, but it would have to be checked with further boreholes nearby to ensure that it was not a wider soil-filled sinkhole that could induce significant ground subsidence. A borehole at a Malaysian construction site only reached rock at a depth of 55 m, and was interpreted as the location of a conical buried sinkhole 30 m across between adjacent boreholes that found rockhead at about 18 m depth (Bergado and Selvanayagam, 1987). A more plausible interpretation for this site in the class kV karst of Kuala Lumpur is that of deep narrow fissures within a terrain of pinnacled rockhead, which may or may not include a buried sinkhole within its very erratic rockhead profile (Figure 5.3).

The karst in Florida contains buried sinkholes of all sizes. Buried solution sinkholes hundreds of metres across and up to 60 m deep are distinguished by the disturbed clastic sediment sequences within them that indicate multiple and ongoing

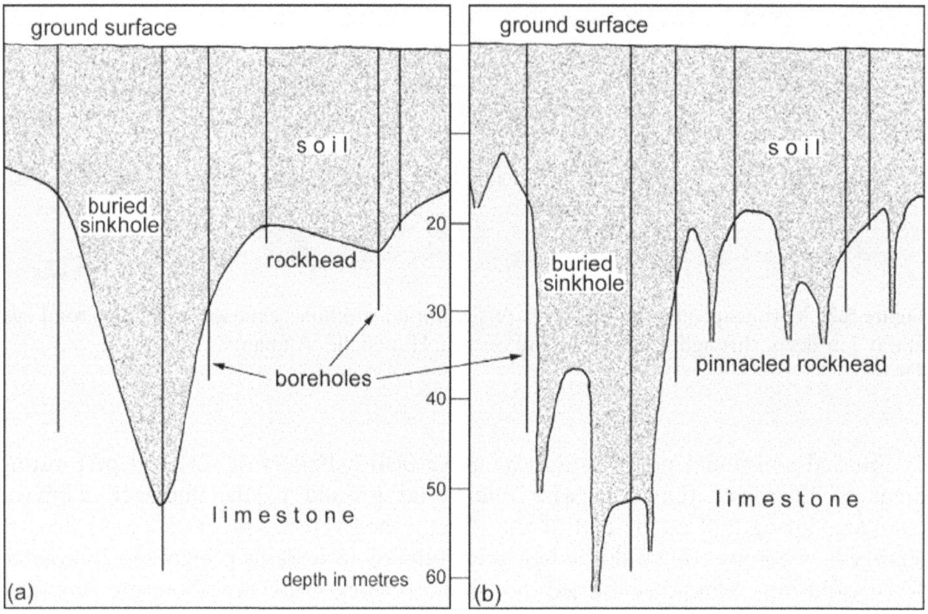

Figure 5.3. Two interpretations of the rockhead profile under a tower block in Kuala Lumpur, Malaysia. (a) is the basic interpretation of the borehole data; (b) is an interpretation of the same data, incorporating the concept of the pinnacled rockheads that are common in the area; limestone in the left-hand borehole was recorded as weathered, suggesting the proximity of the wall of the buried sinkhole.
(a) After Bergado and Selvanayagam (1987).

phases of subsidence instigated by suffosion into the underlying limestone (Horwitz and Smith, 2003). All karst terrains with a soil cover in north-west Florida have buried sinkholes filled with sand, and densities of buried sinkholes can be 50–3,000 per km^2 (Wilson, 1995), but more than half of these are small solution shafts and pipes mapped by ground radar – which could be described as the normal negative features within a pinnacled rockhead on the soft Tertiary limestones. There are smaller numbers of buried sinkholes that are both more than 10 m across and 5 m deep, but these still constitute a significant geohazard for construction in soil-mantled karst terrains (Figure 5.4); as with nearly all sinkholes, there is no recognisable pattern to their distribution that can allow prediction of unseen features.

Unroofed caves are a significant feature of the karst terrains of Slovenia and probably elsewhere (Knez and Slabe, 1999), and are effectively extremely elongated buried sinkholes. Formed where surface lowering has breached an old cave, these appear in the modern landscape as lines of depressions or sinkholes, some of which are collapses separated by surviving segments of intact roofed cave (Case study #3). These unroofed caves are mostly filled with sediment (much of which is remaining old cave sediment), and when this is removed for construction works they are seen as sinuous channels entrenched in rockhead (Figure 13.3.2). They are therefore

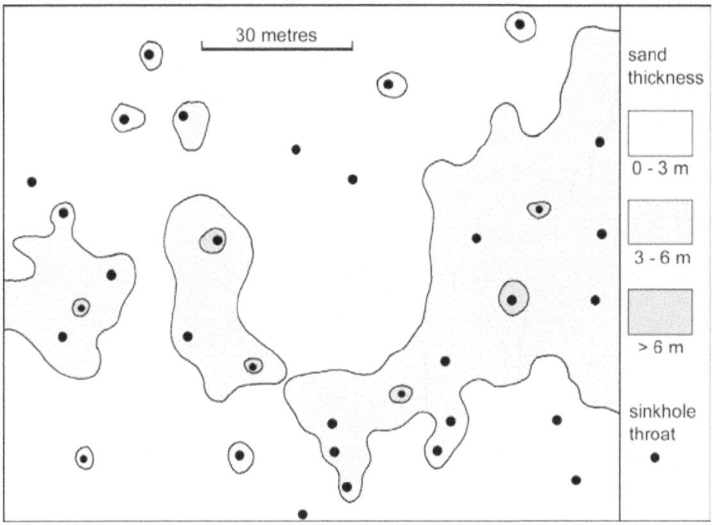

Figure 5.4. Distribution of buried sinkholes in a small part of the Florida karst, interpreted from ground radar measurements of the thicknesses of surficial sand that overlie a clay whose profile closely follows that of the underlying karstic rockhead on limestone.
After Wilson (1995).

comparable to buried valleys, except that their profiles are even steeper than subglacial tunnel valleys. There is also the additional geohazard of fissures or other cave passages in their floors that may act as suffosion outlets unless treated prior to construction over them.

The sediments that fill buried sinkholes may consist partly of locally derived insoluble residues from the limestone. These include wad, a silty clay, rich in manganese and iron oxides, derived from dissolution of dolomites that originally contained 2–3% of each oxide; it may be easily eroded or may develop into a hard pan within the fills of buried sinkholes, notably in the karst of the South African Rand. However, most sinkhole fills are dominated by allogenic materials carried into the karst after environmental change. The relative abundance of clay and sand largely accounts for the contrasts in subsidence movements over different sinkholes, especially when their fills are either loaded or de-watered. Dolomitic limestones in England's southern Pennine karst are punctured by more than 60 buried sinkholes filled with leached, Tertiary, fluvial sands and clays that are extracted as refractory materials (Ford, 1984). Quarrying of the fills reveals widths up to several hundred metres and depths of over 50 m between the vertical walls of these large buried collapse sinkholes (Figure 5.5). Synclinal sags in the sediment fills indicate that the sinkholes subsequently evolved by dissolution of their limestone floors and suffosion of some of the fill into bedrock caves that are now choked with sand (Walsh *et al.*, 1972). Karsts in Jamaica, China and Hungary contain large buried sinkholes filled with extractable bauxite deposits, while others contain valuable placer deposits of ore minerals in either sand or clay matrices.

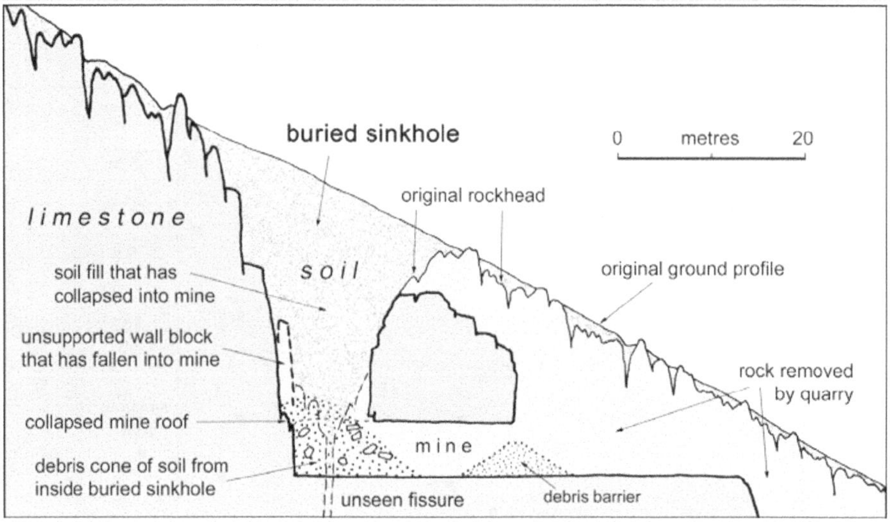

Figure 5.1.1. Profile through the buried sinkholes exposed by an abandoned limestone quarry and mine in the Chunglim Valley, Korea.

Figure 5.1.2. The upper bench of the Chunglim quarry, with the subsoil outcrop of the now-empty buried sinkhole picked out by the dotted line. Entrances to the mine workings are almost hidden behind the debris barriers piled up on the main bench at the foot of the picture.
TW.

BOX 5.1. BURIED SINKHOLE

An example – Chunglim Valley, Jecheon, Korea

The mountains of eastern South Korea contain steeply folded bands of limestone and dolomite that are locally eroded into moderately mature karst of classes kII and kIII. Some outcrops are pitted with large solution sinkholes. Rockhead relief is generally 3–5 m, with many small, buried and filled sinkholes. The buried sinkhole in the Chunglim valley was intersected at depth by a large heading in a limestone mine, and consequently its morphology is well exposed. The hillside profile was transformed by the benches of an open quarry, but working continued underground to follow the material of highest quality. The upper hillside is thickly wooded, with scattered small pinnacles of limestone rising through the thin cover of clay and loess soil. Mine tunnels, each 7 m high and wide, were driven into solid rock, but one heading collapsed when it breached the buried sinkhole nearly 25 m below the steep hillside (Figure 5.1.1). The soil fill within the sinkhole then ran into the mine, blocking the heading with a cone of loose debris, and the slope above is now scored by the gaping hole of the empty sinkhole (Figure 5.1.2).

The buried feature appears to have been about 20 m in diameter at the surface, and tapered to about 7 m across at the depth of the mine roof. Its soil fill was a mixture of red–brown clay, silt, rounded pebbles and angular limestone blocks. Details of the soil-filled fissures and any open cavities below this depth were obscured by the collapse and its consequent debris pile, though a solutionally opened cave extends for a few metres to one side (Figure 5.1.3, right of the person). When the soil was emptied from the sinkhole into the mine beneath, slices and blocks of rock came away from the unsupported walls. This left the nearly planar backwall (Figure 5.1.3, left of the person), that has since been washed clean by rainfall, and is revealed as a major joint that had been previously opened by dissolution, and may or may not have once been filled with in-washed soil. The original sinkhole appears to have developed in a zone of joints that strike roughly parallel to the hillside contours, but this degree of fracturing is normal in the well-folded limestones of Korea.

Purely by chance, this deep buried sinkhole had not been exposed by any soil slumping from the hillside above the advancing quarry face. It appeared only as a patch of soil between the widely spaced rock pinnacles that breach the surface of the thickly wooded slope. In this respect, it is typical of so many buried sinkholes, and its existence was not known until the mine headed into it from below.

Figure 5.1.3. Looking up the debris pile, from the Chunglim mine tunnel into the open daylight shaft that is the emptied buried sinkhole; the shadowed roof of the mine is scored by narrow soil-choked fissures around the perimeter of the old filled sinkhole.
Photo: Hyeong-Dong Park.

Figure 5.5. A large buried sinkhole in England's southern Pennines, exposed after its fill was extracted as a valuable source of refractory minerals; scale is indicated by the old excavator. TW.

A feature of many borehole explorations in karst is the revelation of sections of sound limestone core followed by soil, unconsolidated sediment or open voids at greater depths. Where not matched by comparable core records in adjacent boreholes, these records are commonly described as "floaters" and interpreted as isolated blocks of limestone within the soil profile (Figure 5.6). But these apparent floaters may represent a variety of features. Possibilities include transported limestone boulders, limestone blocks in a collapse zone, limestone bedrock with caves at multiple levels or a single large cave with ledges and notches in its walls (Tan, 1987). Fallen or detached pinnacles should be added to this list. These floaters are sometimes referred to as corestones, but this is misleading as they are not residual blocks surrounded by their own weathering products. Transported boulders are very unlikely at most sites, but collapse blocks are likely to occur in many buried sinkholes. All floaters must be regarded as potentially unstable. However, many of them eventually prove to be extensions of solid bedrock. These may offer stable load-bearing rock if they are thick enough (Chapter 7) and are continuous to bedrock on three or four sides, but are unstable if they are rock projections cantilevered out from one side only. Further investigations are normally required to elucidate valid 3-D ground models (Figure 5.7), on which sound engineering decisions can be based.

5.1.1 Compaction and suffosion in buried sinkholes

Ground subsidence over a buried sinkhole may occur either by compaction of the sediment fill, or by suffosion of the fill into the underlying karst, or by a combination of both. Compaction causes slow and generally modest amounts of subsidence (that

Figure 5.6. Two small limestone "floaters" within a red, clay-rich soil above a rockhead of only modest relief exposed in a new road cut in Korea.
TW.

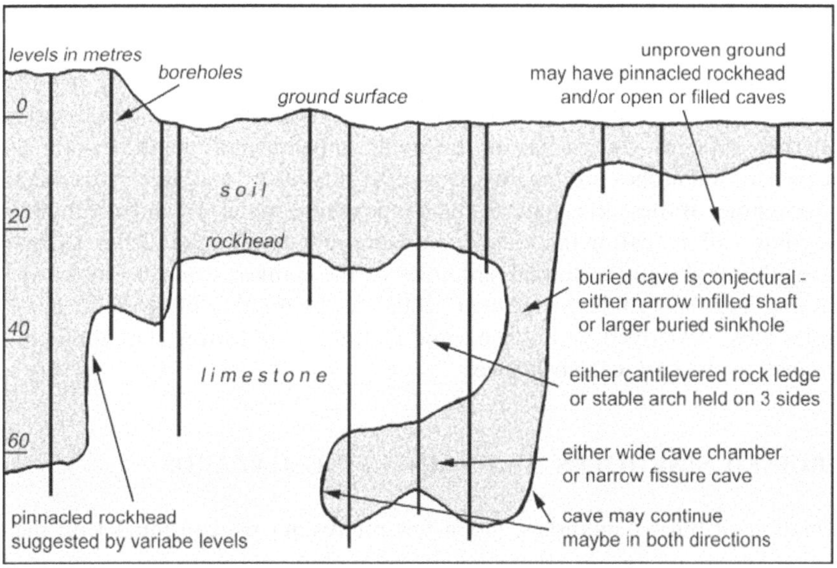

Figure 5.7. Ground conditions at a site in Kuala Lumpur, Malaysia, with a 2-D bedrock profile drawn by the original authors. The annotations have been added, and the stability of the central overhanging area can only be assessed through further borehole data and construction of a 3-D model (as was subsequently completed on this site).

can however destroy brick structures very effectively), while renewed suffosion can lead to more rapid or even catastrophic subsidence if the fill is washed down into underlying fissures and caves. Soil cavities that are found by boreholes into sinkhole fills indicate that suffosion is active and that ground subsidence is likely to occur over the buried sinkhole when the soil arching migrates to the surface. At one site in Florida, test borings into 46 buried sinkholes revealed soil cavities and active suffosion within half of them, while only two out of 32 buried sinkholes at another site in the same karst revealed signs of current activity (Wilson, 1995). The geohazard posed by buried sinkholes is therefore greatly variable. However, surcharge by redirected surface drainage may instigate both suffosion and compaction, and has certainly been relevant at many sites where buildings have been damaged by subsidence shortly after their construction with inadequate drainage control.

Many of the fills in the large buried sinkholes within the South African Rand have compacted in response to the mine de-watering (Section 5.2. and Box 8.1). The resultant shallow surface depressions are known specifically as compaction sinkholes where they form over the deep buried sinkholes (only within South Africa are these known as dolines to distinguish them from the subsidence sinkholes, but this is not international practice). They develop slowly over a period of a few years, and can reach up to 200 m across (Jennings, 1966). Usually only a few metres deep, the largest compaction sinkholes have reached depths of 8 m and are distinguished by concentric cracks and steps within the soil. Bedrock is not exposed. There were so many very large buried sinkholes in the Rand karst, that during the main period of mine de-watering, subsidence by compaction destroyed more houses than did the isolated but more spectacular subsidence sinkholes.

Within the Rand karst, the compaction sinkholes are identified by the installation of telescopic benchmarks (Jennings, 1966) that record greater subsidence at the surface than at depth – in contrast to the greater movement at depth beneath a static surface where soil arches are sagging over soil voids prior to a large dropout failure. Many boreholes at these sites have found compressible materials and a complete lack of soil voids – all indicating the role of compaction in causing the surface subsidence. However, boreholes in the buried sinkholes of the Bank district of the Rand karst have found large soil voids (Swart *et al.*, 2003), and a number of dropout subsidence sinkholes have developed within the same features – indicating that suffosion also contributes to the surface subsidence.

5.2 BURIED SINKHOLES AS ENGINEERING HAZARDS

The smallest of buried sinkholes, just a few metres across, can impact on the construction of column bases or other small foundation points, especially where footings were intended to stand on bedrock. In immature karsts the rockhead depressions may be only a few metres deep, and therefore pose only a minor inconvenience when exposed in the construction stage (Statham and Baker, 1986). Deeper foundations may be incorporated with little delay and the minimal extra costs are justifiably

Figure 5.8. An ancient buried sinkhole bisected by a road cutting in the limestone of Egypt's Kharga Plateau; much of the cohesive fill remains in place in the modern dry environment. TW.

claimed against unforeseen ground conditions. Only a slight increase in size and scale of the buried features, in a more mature karst, can have far greater impact. Investigation drilling for construction of a water tank in Pennsylvania found rockhead at 5 m, but excavation for the foundations revealed buried sinkholes and soil-filled fissures up to 15 m deep and new boreholes found large caves beneath rockhead (Knight, 1971). The original boreholes had given a false picture by ending on floaters and pinnacles. The water tank was relocated.

Buried sinkholes are a widespread geohazard for foundations and may also diminish integrity of shallow cut faces (Figure 5.8). Most close downwards into narrow fissures, but some larger filled caves (often known as pipes) continue to great depth. At a depth of nearly 100 m, the heading of the Dodoni Tunnel in northern Greece breached a filled sinkhole 1.5 m in diameter (Marinos, 2001). Its saturated fill ran into the heading, and a sinkhole 3 m across opened on the surface above. The fact that 1,200 m^3 of fill entered the heading indicates that it was emptied from an unknown and more extensive system of caves and fissures, as its volume far exceeded that of a single narrow pipe.

Larger buried sinkholes may demand major redesign of foundations when they are only revealed during the construction phase of a project. Construction of a large

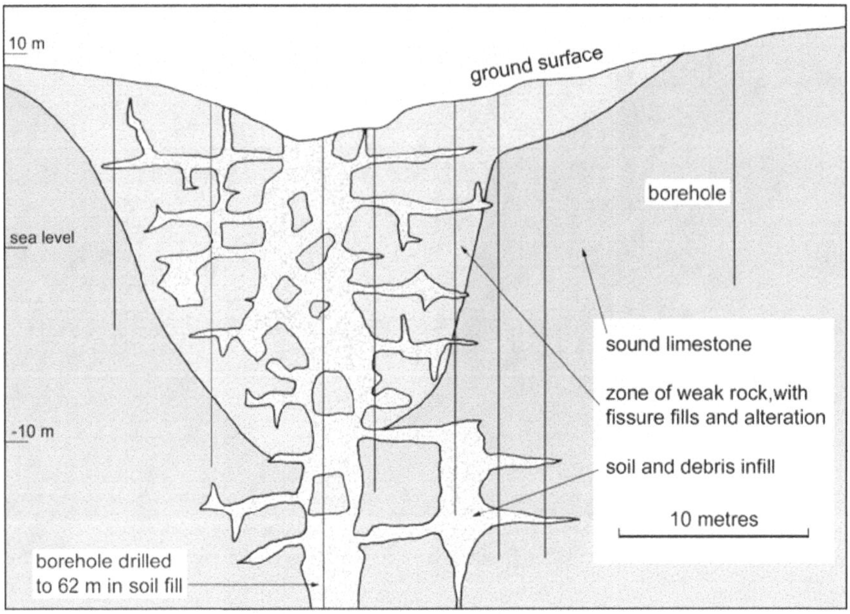

Figure 5.9. Schematic section through a large buried sinkhole found in limestone under a construction site near Limerick, Ireland.
After Clark et al. (1981).

industrial plant on limestone glaciokarst near Limerick, Ireland, revealed a number of buried sinkholes (Clark *et al.*, 1981). The largest was a classic example, underlying an area of about 60 by 20 m. It was explored by drilling to reveal a tapering zone of fissured, weathered and weakened rock down to about 20 m deep, with soil filled fissures extending below and around this, and a central pipe that was drilled to 62 m without reaching rock (Figure 5.9). The size of this feature meant that structures were moved to avoid it, but other smaller features either had their fill cleaned out before backfilling with concrete, or were spanned by reinforced rafts.

The EPCOT site is typical of the Walt Disney developments in central Florida in that it stands on about 30 m of clastic soils overlying karst limestone (Handfelt and Attwooll, 1988). Boreholes on a 30-m grid found a buried sinkhole 120 m across and about 70 m deep, together with a thick cone of buried peat (Figure 5.10). The off-centre location of the peat indicates a past phase of ground subsidence, probably due to underlying sediment suffosion, on the flank of the buried sinkhole. Layout design of the site avoided both the peat zone and the sinkhole centre where potential compaction would have required expensive foundations. In Pennsylvania, the Hershey Medical Centre stands on 5–8 m of soil over a karst rockhead, at a site 100 m from its original design position, where a buried sinkhole, 80 m across and 40 m deep, created prohibitive difficulties in achieving sound footings (Foose and Humphreville, 1979).

The very large buried sinkholes in the Rand karst of South Africa are well

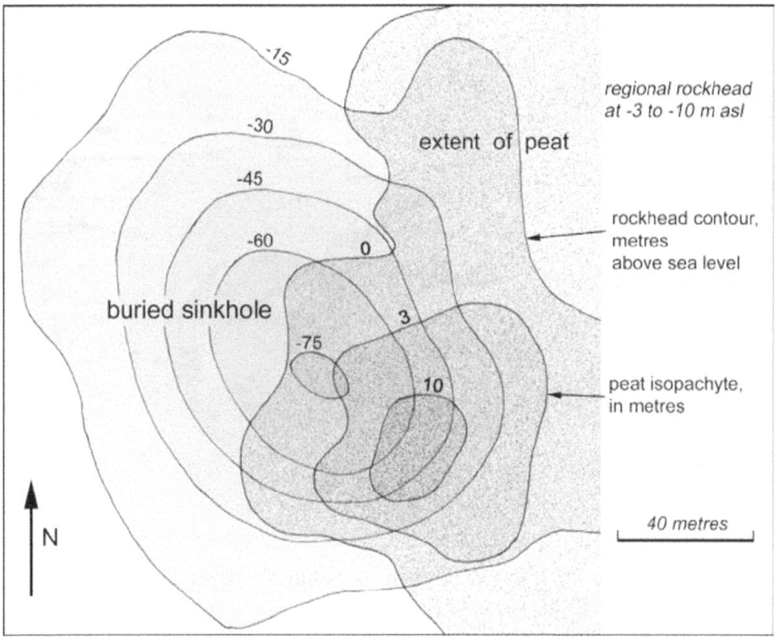

Figure 5.10. Outline map of part of the EPCOT site in central Florida, with rockhead contours picking out a deep buried sinkhole, and isopachytes on buried peat identifying the site of a past phase of ground subsidence offset to the south-east.
After Handfelt and Attwooll (1988).

known for their subsidence events triggered by large scale de-watering (Box 8.1). Most then expressed themselves as shallow compaction sinkholes, which eventually contributed more damage to the local infrastructure than did the dramatic but isolated dropout sinkholes (De Bruyn and Bell, 2001). At Carletonville, Schutte's depression subsided by over 8 m with ground cracks reaching over a diameter of 180 m (Figure 5.11). A second compaction sinkhole, at nearby Lupin Place, was wider even though it sank just 7 m, and caused 24 houses to be demolished. Subsidence damage was worst over the buried sinkhole margins, while some houses over the centres of the sinkholes subsided almost undamaged (Brink, 1979). Widespread subsidence in compaction sinkholes caused abandonment of the entire township of Bank, when activated by the mine de-watering (Box 8.1). Boreholes and gravity surveys indicated that the buried sinkholes under Bank are up to 500 m across and 70 m deep (Kleywegt and Enslin, 1973). Compaction within these caused up to 3 m of surface subsidence, which slowly destroyed all buildings founded within the surface soils (Figure 5.12). Smaller dropout sinkholes developed at the same time, mainly over the steep rockhead margins of the buried sinkholes – a preferred location that has also been observed over Florida's buried sinkholes (Wilson, 1995).

Activation of compaction sinkholes by de-watering is amply demonstrated by the multiple events within the Rand karst, but also occurs where water table decline

Figure 5.11. The compaction sinkhole known as Schutte's depression, when it was active during de-watering of the Rand karst of South Africa.
Photo: Consolidated Goldfields.

Figure 5.12. A destroyed house within a large compaction sinkhole induced by de-watering the fill in a large and deep buried sinkhole; in the abandoned Bank township in South Africa's Rand karst.
Photo: Wal Gamble.

is much smaller in any compressible clays or peats within buried sinkholes. Surface subsidence over buried features may also be induced by input of uncontrolled drainage that can initiate renewed suffosion through the floor of the buried sinkhole. The formation and collapse of soil voids makes suffosion-induced subsidence more erratic and potentially dangerous than the slow and uniform compaction subsidence. Less widespread than the effects of de-watering, localised drainage surcharge onto or into the fill of a buried sinkhole is equally effective at inducing ground subsidence. Such events have been recorded in the South African karst (Case study #12) and elsewhere, including the chalk karst of Europe (Section 5.3). Proper drainage management is therefore essential on any karstic sites where buried sinkholes are known or suspected.

5.3 BURIED SINKHOLES AND SOIL PIPES IN CHALK

The soft white chalk of England and northern Europe contains buried sinkholes comparable to those in any other limestone, but is also distinguished by its abundance of smaller and narrower, sub-vertical, filled solution features. These are commonly known as pipes (either chalk pipes or soil pipes) and were clearly formed as potholes, shafts or swallow holes before being filled with clastic sediment. The typical pipe is circular or slightly elliptical, 1–3 m in diameter, and extends to depths of 10–30 m before tapering to a narrow fissure (Figure 5.13). Its fill may be sand or clay that is loosely packed which may develop unstable bridges within the pipe.

Many pipes formed beneath a cover of soil or sediment, and their distribution across the chalk outcrop is heavily weighted towards a zone adjacent to the feather edge of the overlying Paleogene clastic rocks (Edmonds, 2001). Within this outcrop zone, pipes are commonly clustered so that only narrow pinnacles of chalk survive between the sand-filled pipes and buried sinkholes (Figure 5.14). Columns of highly weathered chalk in solution features at a site in Berkshire, U.K., appear to be remnants amid a number of coalesced pipes (Rhodes and Marychurch, 1998). In some large solution features only small pinnacles of chalk remain between sand-filled pipes, and dissolution has enlarged some bedding planes into fissures now also choked with clastic sediment. Wider, bowl-shaped "pipes" are buried solution sinkholes, generally 5–20 m across and only a few metres deep.

Beneath thin covers of either soil or disturbed Tertiary sediments, the chalk rockhead is commonly highly convoluted within a local relief of 2–5 m. Not only is it broken by the numerous narrow pipes and the wider buried sinkholes, but the surface layers of frost-shattered chalk have been distorted by cryoturbation and solifluction. At depths greater than 5–10 m below rockhead, the chalk is generally more intact, except that conspicuous fissure zones may underlie the larger buried sinkholes, and a proportion of the pipes continue to greater depths. A soil-filled pipe 1 m in diameter was breached in a tunnel heading under the English North Downs at a depth of 80 m below ground surface (Warren and Mortimore, 2003). Cave passages

Figure 5.13. A single clay-filled pipe in chalk, exposed in both plan and profile on a building site in southern England.
Photo: Lawrance Donnelly.

Figure 5.14. A cluster of clay-filled pipes in chalk exposed in a quarry in the English Chilterns.
TW.

Figure 5.15. Subsidence of about 300 mm over a chalk pipe whose fill was disturbed by construction activity on a site in the English Chiltern Hills.
Photo: Lawrance Donnelly.

are not common in chalk, but they do exist, and this was merely one that happened to be filled with sediment; it was described as chimney karst, but this is not a widely used term.

Pipes in chalk commonly develop beneath a shallow cover of sand or gravel, which subsides as the pipes grow. These areas of sands and gravels are typically marked by heathland vegetation that promotes the formation of humic acid, which may aid the dissolution process. The sand and gravel bulks as it subsides into a pipe, giving rise to what Edmonds (1988) termed loose zone ground conditions. Such conditions are distributed widely in areas where permeable, granular soils and sediments overlie piped chalk bedrock. Subsequent consolidation of the loose infill material in larger pipes can give rise to shallow surface depressions, but not all pipes are accompanied by surface depressions. Suffosion of the clastic fills, aided by dissolutional enlargement of the pipes, can lead to their sudden collapse. The collapsed material occupying a pipe may possess a metastable structure due to it being more loosely packed than the *in situ* material from which it was derived. Voids may occur in the infill beneath unstable soil arches – which can collapse for a number of reasons, causing the void to migrate towards the surface.

Pipes in chalk karst constitute a significant geohazard in that they occur as scattered, hidden, unstable ground features. A building site of about 2 ha, in England's Thames Valley, recorded 27 subsidence features and 15 zones of loose soil undermined by suffosion (Edmonds, 1988). Each feature was 2–10 m across, and lay over either a buried sinkhole or a pipe. They were all revealed during construction activity on the site, when loss of soil cover, imposed loading and drainage disturbance are prime triggers that renew either compaction or suffosion of the sinkhole and pipe fills. Most such subsidence events are little more than inconveniences on construction sites (Figure 5.15), and most are small enough that subse-

Figure 5.16. Closely spaced pipes within a soft raised limestone on Efate Island, Vanuatu; some of the soil fills have been washed out since the face was exposed on a building site. TW.

quent subsidence events can be safely spanned by appropriate structures. New pipes often appear at the surface after periods of heavy rainfall. The suggestion that a pipe a metre in depth could form within ten years where infiltration of rainfall into the ground is concentrated (West and Dumbleton, 1972) is not compatible with the very low dissolution rates of chalk, but it is reasonable to expect suffosion of the fill inside an existing pipe within that timescale, resulting in comparable surface subsidence.

The use of raft and reinforced foundations is commonly appropriate for sites in sinkhole-prone zones of the chalk lands (Rigby-Jones *et al.*, 1993). It is also essential to control drainage and avoid the risk of pipe enlargement or disturbance due to any concentration of water into chalk as run-off from hard surfaces such as major roads, or from leaking water mains and sewers. Critically, any soakaway drains must be placed well away from structures. Renewed subsidence over chalk pipes can be instigated by very minor pipeline leakages, or even by excessive watering of domestic gardens, in that any surcharge of water can reactivate dormant sinkhole processes (Case study #8).

Though pipes are a characteristic feature of chalk, comparable features occur in other soft limestones, notably the young reef rocks that form the raised terraces of many tropical islands (Figure 5.16). Rapid dissolution in these environments creates narrow vertical shafts that take rainwater, until they are filled by in-washed soil to form pipes in prematurely mature karst. They then become just one more variety of buried sinkhole that can provide geohazards for engineering works in karst terrains.

6

Sinkholes in insoluble rocks

While the vast majority of sinkholes are found in the more soluble rocks, notably limestone, dolomite and gypsum, such features also occur in other rock types. The most common insoluble rock type that hosts sinkholes is basalt that was extruded as lava flows during effusive (rather than explosive) volcanic eruptions. These sinkholes are formed by the collapse of the lava tubes. However, landforms that resemble sinkholes in their processes and morphology also have been described in other rocks, notably sandstone and loess, where drainage has passed through conduit-type voids or pipes. All these features of pseudokarst, and many other more obscure landforms, differ from those in true karst by having been formed by processes other than dissolution (Halliday, 2004b).

6.1 LAVA TUBES

Lava tubes form as natural pathways through which lava travels beneath the surface of a lava flow. They exist on most effusive basaltic volcanoes because a buried tube, insulated from the cooling atmosphere, is the main means by which lava can travel long distances to form a very long lava flow without cooling and solidifying nearer to its vent. This accounts for the widespread distribution of lava tubes; they are recorded on nearly all oceanic basaltic islands and occur in most regions of Quaternary basalt volcanism (Figure 6.1). The greatest influence of lava tubes, with respect to their collapses and ground subsidence, has been on the more densely populated outcrops of young basalts, mainly in parts of western U.S.A., Hawaii, Japan and Korea. Not all basalt lavas contain tubes; the huge areas of flood basalts, including those of the Columbia Plateau, U.S.A., and the Deccan Plateau, India, accumulated by repeated phases of sheet injection deep inside the lava piles (Self *et al.*, 1996) and have never contained open tubes, except at a few isolated and atypical

Figure 6.1. Breakdown pile beneath the sinkhole entrance to Indian Tunnel, a lava tube 20 m wide beneath a roof less than 5 m thick in the Craters of the Moon lavafields, Idaho; people on the right indicate the scale.
TW.

sites. Lavas other than basalts (and closely related types) do not normally contain tubes.

Lava tubes originate as open channels down which the molten lava drains. When the supply of lava reduces or ceases, the tubes drain under gravity and are left wholly or partially empty. These channels may start either on or in the lava flow (Kauahikaua *et al.*, 1998). The former start as lava streams open to daylight on the surface of flows with gradients typically of 2–5°. The open channels then roof over, either by coalescence of accreting marginal levees, by growth of a crust across a cooling flow or by welding of rafts of cooled lava (Dragoni *et al.*, 1995). Once the tube is formed, its encased lava ceases to lose heat, so it may evolve into either a rounded tube or an entrenched canyon passage. The second type develops entirely underground by injection of lava inside expanding pahoehoe flows, typically on gradients of 1° or less. These conduits generally start with low elliptical profiles, but may then entrench themselves by thermal erosion while cooling lava accumulates along their walls to create the more common tubular profile. The flowing lava in tubes of either type may ultimately solidify and fill the original conduit, or it may drain away to leave an open cave.

It is normally difficult or impossible to determine how an old lava cave originated, but more important to the civil engineer is the structure of the roof that spans the open void. Most variations in roof morphology are the result of six alternative processes, of which the first four form the initial roof (Waltham and Park, 2002):

Figure 6.2. A skylight into an active lava tube on the Kilauea volcano, Hawaii; the open hole is just a few metres across, and the failed basalt crust is less than a metre thick. TW.

- accretion onto open banks and levees, until they coalesce over the flow, leaving vertical fractures in the arch centre;
- accretion and wedging of floating slabs, leaving an irregular blocky structure;
- crust formation in toes of pahoehoe lava, leaving fractures around sub-horizontal lava lenses;
- tube formation by injection and inflation beneath a thin or thick roof of solid lava;
- addition of shells inside the tube, formed when lava pulses fill the tube and leave chilled linings against the cooler roof; and
- addition of subsequent lava layers to thicken the roof.

Many active lava tubes have only very thin roofs of chilled basalt, and the flowing lava may be exposed in its tube by small roof collapses that form skylights (Peterson *et al.*, 1994). These are essentially collapse sinkholes, but are distinguished by the absence of fallen roof debris – which has been carried away by the lava or has melted into it. Though some active tube roofs are dangerously fragile (Figure 6.2), they constitute no engineering hazard, as construction works are neither required nor advised on an active lava flow.

When the flow rate decreases the level of the flowing lava may fall within the tube, and stabilise for long enough to develop a solid crust as a further roof inside the conduit. Flows can also erode their channel floors, sometimes breaking into lower, older tubes. Multi-storey lava tubes can therefore develop, and many are recorded in the Mauna Ulu lava fields in Hawaii (Wood, 1981), but the common situation is a single level of tubes lying at shallow depths. Once the tubes have drained of their lava they are left with floors that are generally flat, and may have stalactites of solidified lava hanging from the roof.

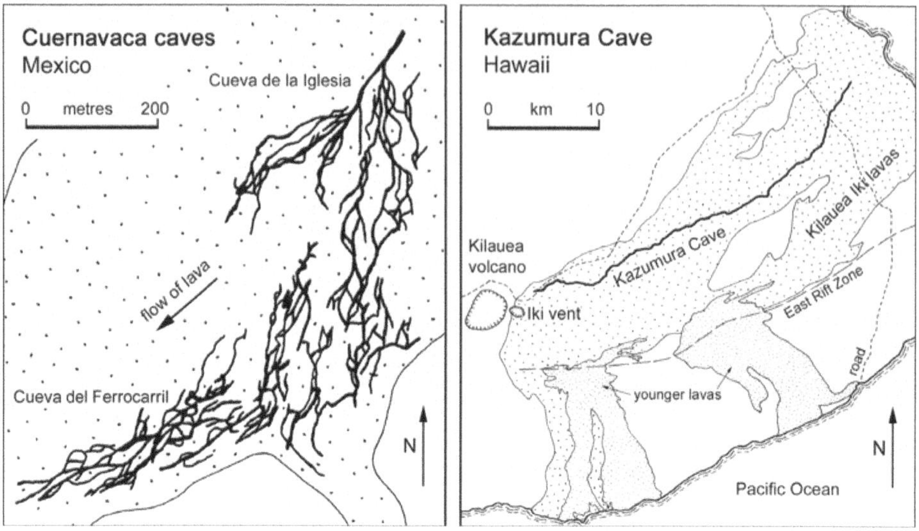

Figure 6.3. Complex and simple lava tube patterns at Cuernavaca, Mexico, and Kazumura Cave, Hawaii.
After surveys by Ramon Espinosa and Hawaii Speleological Survey.

On gradients much steeper than 1°, lava tubes tend to develop as single conduits that may reach great lengths with few braids and very few distributaries (Figure 6.3). Kazumura Cave has a mean gradient of 1.9° and reaches 32 km down the flank of the Kilauea volcano on Hawaii (Allred and Allred, 1997). In contrast, pahoehoe sheet flows develop where very fluid lava spreads across ground that is almost level, and a stationary crust forms while the molten lava continues to move outward and upward to produce small distributary tubes at the front of the main feeder tube (Hon *et al.*, 1994). In plan the tube system resembles a classic fluvial delta, and the same pattern may develop on a larger scale where lava has been temporarily ponded before breaking out across nearly level ground or has been produced at very high effusion rates (Figure 6.3).

6.1.1 Lava tube stability

The stability of the rock forming the roof of a lava tube depends on the rock thickness, the unsupported width over the tube and on how the roof was formed (Figure 6.4). Tubes formed by injection might be at a depth of tens of metres, but most open tubes lie only 1–5 m below the ground surface. The width of a lava tube is commonly anything up to about 10 m. A minority of wider caves are recorded; Manjang Gul, on Korea's Cheju Island, has a single tunnel that is 15–25 m wide along some kilometres of its length, but this is exceptional, and most of it lies beneath solid rock more than 10 m thick.

In unweathered basalt with a low density of fractures, a tube width to roof

Figure 6.4. The entrance to the Flothellir lava tube in Iceland where a very thin roof has collapsed across a very wide span.
Photo: Chris Wood.

thickness ratio of 3 : 1 has been proposed as being adequate to establish a sound rock arch that would support all but the heaviest of structures (Waltham and Park, 2002). A more conservative safe cover ratio of 2 : 1 may be more appropriate as a general guideline, bringing it in line with the high factors of safety applied over limestone caves (Section 7.3). Basalt is an extremely strong rock in the intact state (unconfined compressive strength (UCS) >200 MPa). However, the various processes by which lava tubes are formed can lead to different stability conditions. Roof rock may have been weakened by subsequent melting, by fracturing due to tectonic stresses on an active volcano, by weathering or by inflation pressures within the tube; or it may be intrinsically weak due to highly scoriaceous or frothy horizons or to a structure of poorly welded shells added to the roof from below. Alternatively a roof may be strengthened by rewelding under subsequent flows.

The winter of 1993 saw the collapse of the often photographed basalt arch across the Ofaerufoss waterfall in Iceland (Figure 6.5). Though it was an erosional feature created by river scour of a scoriaceous horizon, this arch had the proportions of a lava tube roof, with a low arch about 2 m thick across a span of about 14 m, giving it a cover ratio of 7 : 1. Though it collapsed without any imposed load, its rock was significantly broken as it was formed of a mass of pahoehoe toes and partly welded blocks, and it was therefore much weaker than many lava tube roofs formed of more homogeneous basalt.

Collapse sinkholes over lava tubes vary in size from small holes where single blocks of basalt roof have dropped out, to major collapses as wide as the tube and tens or even hundreds of metres long. The floor of each is a pile of breakdown blocks, generally sloping down into segments of intact tube on each side (Figure 6.6). Some collapse zones are very localised, while others have left varying depths of fallen roof material along considerable lengths of tube. Most lava tube caves are distinguished by piles of breakdown beneath sections of roof collapse that may have evolved into stable arches, but may be in the process of migration toward the surface and yet further collapse sinkholes (Waltham and Park, 2002; Wood, 1981).

Figure 6.5. The thin basalt arch across the Ofaerufoss waterfall, before its collapse in 1993.
Photo: John Middleton.

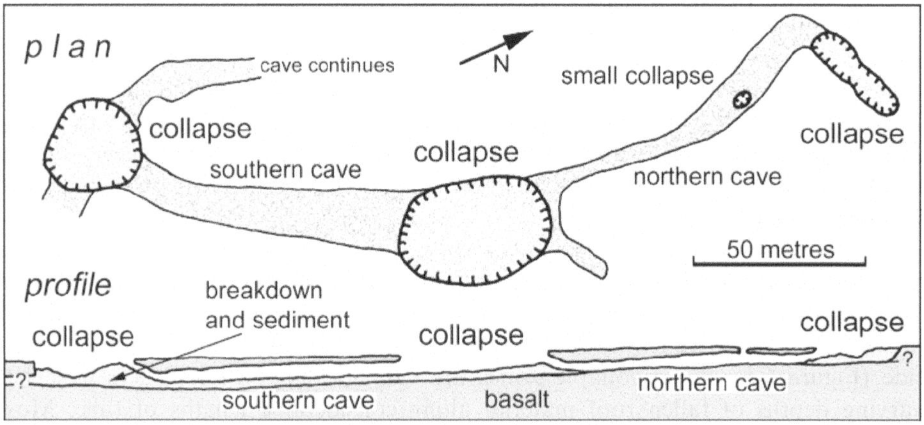

Figure 6.6. Plan and profile of the Flothellir lava tube within the basalts formed during the 1783 eruption from the Laki fissure, Iceland; the photograph in Figure 6.4 is of the entrance to the southern cave from the middle collapse.
After Wood *et al.* (2002).

Figure 6.7. Newly collapsed sinkhole entrance to the lava tube of Apua Cave only 3 m below the basalt surface in Hawaii.
TW.

Skylights that form in active lava tubes, and large areas that deflate and subside as lava drains out from beneath them, are structural features of young basalt lava. Some tubes also collapse as or soon after their lava drains out, to form sharply defined "valleys" across outcrops of very recent basalt. The notable geohazard on basalt is created by the collapse sinkholes that develop where rock collapses into open tubes long after the eruption activity has ceased. A large proportion of tubes have thin roof spans that must be regarded as only quasi-stable, even prior to any weakening of the rock by subsequent weathering. It is significant that the entrances to most of the thousands of lava tube caves known around the world are through natural collapses, where their roof rocks have failed without any imposed load. Most collapses pre-date recorded observations, but the two entrances to Apua Cave, in lava flows formed on Hawaii in 1973, were only found after they had collapsed during a minor earthquake in 1975 (Figure 6.7). Apua Cave is about 15 m wide and lies about 3 m below the surface, indicating a cover ratio of 5 : 1, but it is not known if roof stoping had reduced this ratio prior to the collapse. There would appear to be significant numbers of sites on recent basalt lavas where ground is waiting to collapse over open tubes, and will readily collapse under any loading imposed by engineering activity.

The different processes by which lava tubes originate, and then have their roofs modified, make it difficult to predict where sinkholes will occur by the collapse of tube roofs. Similarly, the difficulty of predicting the location of lava tubes means that any building and construction on basalt lavas that are susceptible to tube formation carries an element of potential risk.

6.1.2 Engineering failures related to lava sinkholes

Though construction is unlikely in active lava fields, there are many basalt terrains where volcanicity is extinct or dormant and where sinkhole collapses into old lava tubes pose a significant hazard. Recorded events are rare largely because of the minimal scale of engineered development in most of these regions. The Hawaiian islands are almost entirely formed of basalt, and tube roof collapses have occurred on construction sites, where heavy equipment has fallen into unsuspected lava tubes. Similar problems have been encountered with expanding development on Cheju Island, off the south coast of South Korea (Case study #4). However, many roads and houses, on both Cheju (Waltham and Park, 2002) and Hawaii, stand on very thin arches of basalt over large lava tubes, in some cases intentionally so that the tube can be used as a basement. Some lava tubes have been deliberately collapsed to allow road construction across the broken rock pile instead of over an unstable cave, but such action has been deemed inappropriate where caves with scientific or recreational values have had to be left intact under some sites on Hawaii's Big Island.

Foundation design for an extension to a hospital near Portland, Oregon, had to be modified following the discovery of collapsed ground along the lines of lava tubes in deeply weathered Plio-Pleistocene basalt beneath up to 9 m of silt cover (Bekey and Rinne, 1975). During construction of the main hospital building in 1968 two collapsed lava tubes were found, each about 12 m wide, and isolated collapse sinkholes were noted both upflow and downflow. Prior to building an extension to the hospital, a ground investigation in 1974 used probe, auger and cored boreholes to characterise the ground, and revealed another major collapsed lava tube (Figure 6.8). A zone of broken basalt is up to 50 m wide, containing breakdown blocks up to 12 m long with the spaces between either open or filled with loose silt. Considered to be a single, very wide lava tube, this could equally be a series of smaller braided tubes, but in either case the collapse was almost total. The margin beneath the hospital extension has a sudden drop-off from solid rock into at least 10 m of basalt rubble. A composite foundation was therefore used, with spread footings on competent bedrock to the south, while the northern half of the extension was placed on steel H-piles driven onto the collapsed basalt, and meeting resistance at depths varying by 10 m where only 1 m apart. Design pile capacities were 1,000 kN, except that they were only 500 kN at the very northern end and only 300 kN along the margin of the tube (Figure 6.8). The design was conservative, as it was anticipated that any individual pile could rest on an unstable breakdown block and could therefore fail without threatening the integrity of the structure.

Not far away, on the south side of Mount St. Helens in Washington, a large canal 9 m deep has for 50 years fed water from the Swift Reservoir to a power station. Part of its embankment failed in 2002 where it had been built over basalt. Nearly 75 m of embankment were lost, and the resulting escape of 30 M m^3 of water caused damage with a bill of $9 million for remediation works. About half the

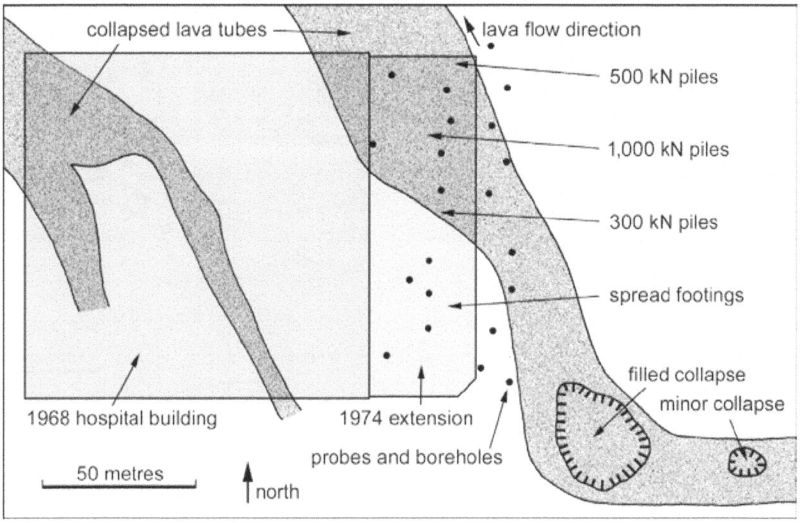

Figure 6.8. Outline map of the hospital at Portland, Oregon, build over a series of large collapsed lava tubes.
After Beekey and Rinne (1975).

canal's flow was lost when a sinkhole developed in its floor. A 10 m thick soil fill had been piped into a previously unknown lava tube, at least 3 m in diameter, to produce a subsidence sinkhole (CH2MHill, 2003). The water flowed through the tube, and then broke through the floor of the basalt into an underlying alluvial sand, from where it emerged under the canal's embankment about 200 m downstream of the sinkhole. This caused the normal progression of piping, undermining, slope failure, overtopping and scour to completely destroy the embankment about four hours after the initial sinkhole collapse. The lava tube, in the Cave Basalt flow dated to AD85, had not been detected during construction of the canal, when jetting was used to clean out fissures in the basalt that was exposed prior to placing the clay fill for the canal's floor. Following some leakage losses in 1973, the canal had been emptied and its floor was found to have been eroded and washed into caverns within the basalt; two of these were exposed, each some metres across. Though the 2002 sinkhole was largely a suffosion feature within the fill, there appears to have been significant collapse of basalt rock into an open tube; the rockhead, intact at the time of construction, was breached by a hole large enough to swallow a massive flow of water.

6.2 SINKHOLES IN OTHER FORMS OF PSEUDOKARST

Some of the world's largest collapse sinkholes, hundreds of metres deep and wide, are formed in quartzite. Over timescales of 1–10 million years, weathering by rainwater dissolves quartz along grain boundaries, turning the quartzite into sand

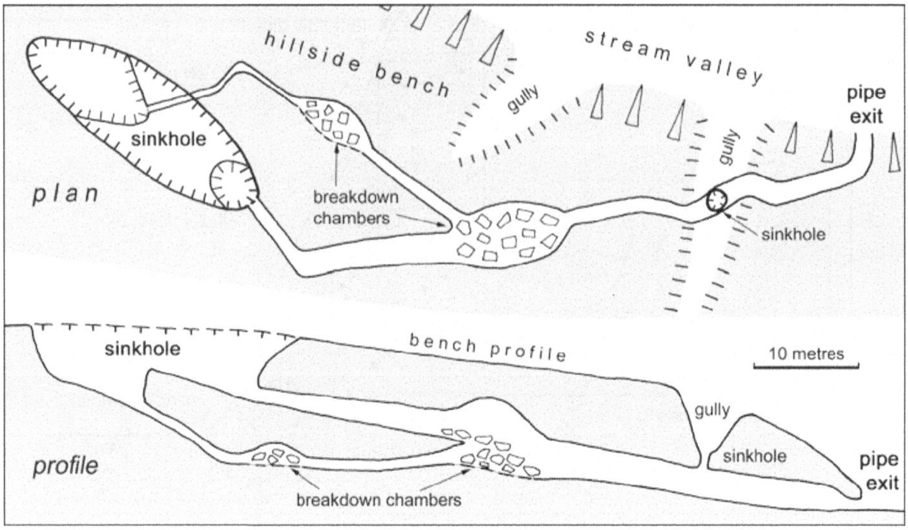

Figure 6.9. Plan and profile of a piping system and its sinkholes within the soil cover in Big Muddy Valley, Canada; the breakdown material is shown diagrammatically.
After Drew (1972).

along fracture lines, where piping then develops open caves and sinkholes (Martini, 2004). The process is only significant on some stable continental plateaus, notably in Venezuela, where there is no engineering activity and no collapse events have been recorded. Comparable pseudokarst on granite, phyllite and other strong and insoluble rocks is only on a much smaller scale.

Piping failures in unconsolidated, weathered or poorly consolidated soils are widespread, and develop a form of pseudokarst where sinkhole failures over unstable soil cavities may constitute a significant geohazard in some terrains. Their common feature is the washing of fines from granular soils in a process that increases in scale as progressively coarser material is removed until an open pipe is created; this extends headward from its outlet, and may ultimately lead to sinkhole development in the headwater ground surface. Pipes develop only where hydraulic gradients are steep enough to initiate flow in uncemented soils that are either transported or residual, and reach significant sizes almost exclusively in areas of semi-arid climate. In Saskatchewan's Big Muddy Valley, Canada, piping was initiated along dipping bedding planes in the Paleogene sands (Drew, 1972); pipe caves have formed under slopes into incised gullies, and sinkholes have then developed on the hillside benches (Figure 6.9).

Piping failures are recorded in many unconsolidated materials (Halliday, 2004a). Some of the largest pipes have formed in materials that are partially soluble, including salt-rich lake sediments in Kenya and landslide debris containing limestone blocks in Nepal (Figure 6.10). Construction in Brasília, the capital of Brazil, has encountered problems with piping failures and sinkholes in the cover

Figure 6.10. A sinkhole 30 m across, formed by a combination of piping and dissolution, in calcareous landslide deposits in the Pokhara valley, Nepal.
TW.

of lateritic soils on Precambrian metamorphic rocks (Mendonca *et al.*, 1993). Sinkholes 5–25 m across occur where water drains through the highly porous laterite into deep gullies in the transition zone between plateau areas and river valleys. The distribution of sinkholes appears to be related to fracturing in the bedrock, probably because of the control that these have on drainage. More than 50 buildings were damaged in 1986 following a severe storm that caused renewed gully incision and the development of related sinkholes, all of which were in the areas of laterite.

Loess is generally susceptible to piping, notably in the vast loess terrains of central China (Derbyshire *et al.*, 2000). On uncultivated slopes, water penetrates fissures in the loess, enlarges them and ultimately develops systems of pipes, generally a few metres in diameter, that lie sub-parallel to the slope surface. Much of the land has been terraced for cultivation, and smaller sinkholes are more common over short pipes through the terrace rim, but some of the larger sinkholes are set well back from the terrace edges. Sinkholes over some of the pipes are as deep as 30 m, and others develop as collapses up to 25 m across when the loess fails over individual piping chambers that approach that diameter (Figure 6.11). The hazard from piping and sinkholes is widespread in the loess, but is minor in comparison to the failures by landslides or hydrocompaction.

In any soil terrain where piping is observed, there is the possibility of sudden collapse of the ground at the upper end of unseen pipe systems. The soil arches are so

Figure 6.11. Sinkholes in the loess terrain near Lanzhou, China – a recent small collapse over a youthful pipe, and a larger old collapse further from the terrace edge.
Photos: Tom Dijkstra.

weak that they are likely to collapse naturally due to continuing piping and erosion. They have negligible bearing capacity, and any that already exist are likely to collapse under the disturbance and loading of construction traffic. New pipes and sinkholes may be formed during single or multiple rainstorm events, within the lifetime of an engineered structure. It is therefore essential to control surface drainage and ensure that opportunities for rainwater infiltration are eliminated from development sites.

7

Rock failure under imposed load over caves

An ever-present geohazard in karst terrains is the collapse of bedrock into open caves when engineering works inadvertently impose new loadings on the unsupported spans over unknown caves. However, rock collapse is a rare event. A scatter of collapse and caprock sinkholes exist across most karst terrains (Chapter 3), but their small numbers have developed through geological time, albeit without imposed loading. Nearly all collapses induced by engineering activity in karst are subsidence sinkholes that develop entirely within the soil profile (Chapter 8). Induced rock collapse may be rare, but events can be catastrophic and should be avoided by appropriate engineering design.

A widely used guideline figure for safe (or allowable) bearing on limestone is 4 MPa (British Standards, 1986), but this assumes sound rock and takes no account of large, unseen voids. Column loads in large structures may be 5–10 MN, which therefore bear on pads 2–3 m across and generally reduce imposed stresses to no more than 2 MPa. Small pads over potentially large caves are effectively point loads on the cave roof. Raft and strip foundations impose less critical loads, especially where they are reinforced to span potential voids. Pile tips generally impose stresses no greater than about 1 MPa, especially where loads are partially carried by skin friction through the soil and rock cover, but these can impose high stress concentrations over small footprints that are also effectively point loads. Safe bearing pressures are lower on weak limestones, and are generally taken as around 750 kPa on sound chalk, but caves tend to be smaller in such rock types. Potential settlement may have to be considered where heavy loads are imposed on some of the chalks, weaker limestones and evaporite rocks, even where cave hazards have been accommodated. Highways impose much smaller loads. In British highway design, overall imposed stresses at ground level are assumed to be 17 kPa, with an additional 33 kPa along a metre-wide strip at the worst-case site. However such stresses are generally distributed by the underlying soil, and by the roadbed, before imposition on the bedrock.

Structural design in cavernous karst has to allow for loading imposed directly over the largest cave likely to exist in the area. Maximum safe loads are therefore a function of the width of the cave and the thickness of its intact rock roof (beside the rock mass strength, which normally dictates the safe bearing pressures, cited above, for sound rock). This may be expressed as thickness/width, either as a percentage or as a cover ratio. A broad rule-of-thumb has been that integrity is ensured where roof thickness exceeds cave width. Though this may be applicable in weak limestones and chalk, it appears to be excessively conservative for typical cavernous karst in strong limestones – where safe roof thicknesses that are 50% or 70% of cave width appear to be more appropriate. There is then the problem of assessing the maximum likely size of a cave beneath any construction site – and this can be based only on local knowledge, or broadly by reference to the engineering classification of karst (Waltham and Fookes, 2003; Waltham, 2002b).

Any justification or refinement of the thickness/width cover ratios for construction over caves is limited by the difficulties of defining rock mass strength, especially within a thin cave roof that will be subjected to point loading and consequent flexural distortion. This is especially problematic where dissolutionally opened fissures have to be characterised within karst ground. The available literature, both in textbooks and academic papers, is minimalist on this subject and refers to "calculations by established principles of rock mechanics", while studiously avoiding any presentation of useful numbers. Even where ground conditions are well explored and tightly defined, data is scarce, but the question of "how thick is a stable rock roof?" requires attention.

7.1 RECORDED COLLAPSE SINKHOLES INDUCED BY LOADING

Though true rock collapses over caves are rare, some have occurred during construction works. At Dhahran, Saudi Arabia, a cave roof 2.5 m thick collapsed under the 50 tonnes load of a Caterpillar D9 bulldozer (Grosch *et al.*, 1987). The cave was then filled with 1,200 m^3 of concrete, but its width was not stated, so the cover ratio, in the weak limestone of the region, is unknown. Twice in 1995, heavy bulldozers dropped through cave roofs in basalt during construction projects on Hawaii. On these young volcanic islands, lava tubes about 2–10 m wide lie beneath rock roofs 1–8 m thick, so many are inherently unsafe, but no great damage has been done, and collapse details have not been recorded.

Four spans of a concrete bridge near Tarpon Springs, Florida, failed in 1969 when three supports dropped out of sight – in a rare case of infrastructure loss into a new, self-induced, collapse sinkhole. The piers were H-piles driven into the porous bedrock limestone, and their simultaneous failure suggest that this was due to the total collapse of a zone of bedrock as it fell into an underlying, unknown, flooded, karstic cavern (Sowers, 1975). Corrosion could not have accounted for such a total failure, but may have contributed by reducing skin friction on the piles and thereby increasing their end loads.

Highway loading is normally so low that roads are unlikely to induce failure in

Figure 7.1. Fractured rock around the margin of the new collapse sinkhole that destroyed Dishman Lane in Bowling Green, Kentucky, in 2002.
Photo: Hilary Lambert, KEEP.

bedrock, but an induced collapse sinkhole did drop one Kentucky road into an underlying cave chamber (Case study #2). The cave ceiling lay about 7 m below the road, but 2 m of this was soil cover, and the upper half of the rock was heavily fissured and pinnacled. The effective rock slab was therefore only 2 or 3 m thick, and it spanned a cave chamber over 25 m across. This represented a cover ratio of about 10% – at which roof failure was almost inevitable (Figure 7.1). It appears that the road itself did not provide the critical loading (and there was no heavy vehicle there at the time), but failure of the rock slab probably occurred under the impact loading of the road as it dropped when voids collapsed within the intervening soil profile.

There are numerous reports from around the world of "limestone collapses" into sinkholes, but most of these appear to be "ground collapses in limestone regions". Reports and accounts emanate from construction sites, from highways and from just a few completed structures, but when the sites are examined or reported in detail, they are nearly all found to be failures of soil over fissures and cavities within stable bedrock. Induced collapses of limestone, gypsum or basalt that span natural cavities may be rare, but they do offer a significant geohazard to the unwary engineer who might impose inappropriate loads on ground regarded as solid and safe.

7.2 BEARING CAPACITY OF CAVE ROOFS

Observation of cave ceilings reveals that very few roofs are formed by single unbroken slabs or beds of intact rock. They must therefore be regarded as fractured rock masses. However, rock mass strength is notoriously difficult to assess; it may be safely estimated where it can be seen in the roof of an accessible cave, but it is very difficult to quantify for an unseen rock mass within the ground that may be straddling an unseen cave. In typical cavernous karst, the limestone is strong (intact unconfined compressive strength (UCS) of 50–100 MPa), is massively bedded (with beds 0.5–5.0 m thick) and has a mean fracture spacing of 1 m or more, typically across three or more fracture sets. Many of the fractures are open and locally enlarged by dissolution, tectonic folding may mean the dominant bedding plane fractures are steeply dipping, and lateral confining stresses are normally low at shallow depths. These parameters dictate that typical karst limestone can generally be regarded as a rock mass at the weaker end of Class III, with Q of around 4 on the classification scheme of Barton *et al.* (1974), and RMR of about 40 on the rock mass rating scheme of Bieniawski (1973). Strong cavernous limestones in Tennessee have been assigned RMR of 50 to 65 (Siegel and McCracken, 2001). Of the other rocks that may contain caves, gypsum, chalk and the weaker limestones generally constitute rock masses of Class IV or V, while many basalts are of Class II (caution should be exercised when these classifications of rock mass, though very useful, are used for purposes other than as originally intended with respect to tunnel excavation and support). All these estimates of rock mass strength and class are open to adjustment, up or down, on inspection of the rock conditions at any specific site.

Structural analysis of a cave roof can treat the rock mass as either a beam under flexural stress or an arch in compression. Both cases must be examined with or without imposed loads that are either spread or applied over small areas. Strengths of rock masses are negligible in tension and flexure, but remain high in compression. Where a cave roof profile is analysed, calculated bearing capacities for rock beams are smaller than those calculated for compression arches within the same profile. This is at least partly due to the rounded profiles of most caves that leave marginal buttresses to support a beam that only thins over the crown of the cave. Some beam analyses indicate factors of safety of less than unity for caves that are still standing, and are therefore clearly wrong, probably due to inherent difficulties in estimating rock mass properties. Loaded cave roofs appear to derive their integrity from arch development within their profiles, and beam analysis appears to be unhelpful in their assessment. There are no known beam failures under imposed loading in the Miami Limestone of southern Florida. Even though this rock is typically weak and thinly bedded, and is structurally loaded *in situ* over loose, deformable sands (and not normally over open caves), full-scale foundation load tests have not achieved beam failures of the rock (Kaderabek and Reynolds, 1981).

7.2.1 Integrity of loaded rock arches

A rock mass is stable over a void where an arch, capable of carrying its own load and any imposed load, can develop within its profile (Figure 7.2). This structure may be known as a voussoir arch (derived from the French for a stone arch) where the entire load is carried in compression. A bridge of this form is built of an arch of stone blocks that carry all loads normal to their interfaces, and therefore requires no cementing material; the analogy to a fractured rock mass depends on the existing fracture pattern. Structural analysis indicates that an arch within rock normally fails in compression (Sofianos, 1996); shear failure at the buttress occurs only in very short spans, and buckling failure is limited to arches in very thinly bedded rock. Punching failure is not a function of arch structure, and is a hazard only in thin rock slabs (Section 7.2.3).

The bearing capacity of a voussoir arch increases with its thickness, as it is ultimately limited by the unconfined compressive strength of its material. It also decreases as the rise of the arch is reduced from an optimum profile toward a flat arch that can fail as a beam. Stable arches in unsupported mine roofs develop a rise that commonly approaches half the span width (Franklin, 1989), which is the profile inside an optimum circular voussoir arch. Any material that does remain beneath the compression arch within a rock mass is in tension, and its incidental loss increases arch stability by reducing the load on it. Buttress stability is critical to voussoir arch integrity, but normally offers no problem in a rock mass over a cave. Load on the arch is imposed by any material above the compression zone that is not self-supporting by its own voussoir arch. Distributed loads, from either a soil cover, a road base or a raft foundation require only a modest thickening of the arch to carry the increased stress in compression. Point loads, from column bases or other small foundation pads, threaten to fail an arch by its distortion and ultimate buckling, and are only safely carried where their stresses are distributed through an adequate

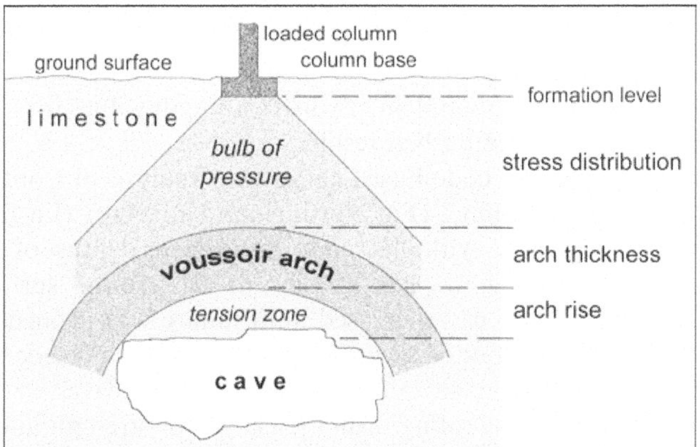

Figure 7.2. Elements of a voussoir arch developed in a fractured rock mass that spans a natural cave and carries additional load imposed by engineering works.

thickness of passive cover across the compression arch. If load distribution is assumed to spread beneath an angle of 45°, a zone with a depth that is about 0.7× the cave width is adequate to ensure no buckling of the voussoir arch, and less depth is sufficient where the foundation pad occupies a large proportion of the cave width.

The thickness required for stability in a rock roof over a cave is therefore the sum of the three components – arch thickness, arch rise and stress distribution zone (Figure 7.2). Structural analysis may determine safe values of these components for any cave width and for any applied load. Cave height is irrelevant as wall failure is not a threat in strong karstic limestone, and there are none of the thin pillars that are critical to the evaluation of mine stability. A cave with a steeply arched roof has no rock within the tension zone inside the arch rise, and also offers efficient load transfer to its wall rock; it may therefore be stable with a thinner roof over its centre-line.

However, such analysis is only with respect to the compressive strength of the rock mass, for which an overall value may be assigned. The additional hazards in a natural rock mass derive from the more variable factors. Fractures may fail in shear where they are orientated so that stress is imposed at highly oblique angles (unlike those imposed normal to the joints that are radial within an engineered dry stone arch). Wedge-shaped blocks can destroy arch integrity, though they are commonly held in place by confining stress and high friction angles on rock fractures. Fissures that have been partially and irregularly opened by dissolution, may carry no shear stress across their voids, but are likely to be locked by compression across areas of block contact. Thin beds may be overstressed to the point of buckling, but roof loads are generally carried by the thicker beds (where these are lacking, natural collapse should be widespread and recognisable). There is commonly no answer to these variations except to raise the factor of safety yet further, though inspection of accessible caves may reveal rock structures that should be treated as if they are either more or less safe than those in the "typical" rock mass.

7.2.2 Modelling the failure of cave roofs under load

Numerical analyses of loaded cave roofs suffer from the twin difficulties of modelling fracture patterns reliably and of applying realistic strength values for a complex rock mass. There is a shortage of published results.

Using the finite difference code Fast Lagrangian Analysis of Continua (FLAC), cave roofs have been modelled in 2-D at Nottingham University (Lu Zhengxin, pers. comm.). Caves, 3–50 m wide with flat roof profiles at depths of 2–10 m, were modelled under loads applied to pads of 1 m^2 at the ground surface above the centre-line of the caves. Load was increased until failure was defined by settlement of 25.4 mm, which indicates loss of integrity and is likely to precede total collapse, besides causing significant damage to built structures. The caves were modelled in materials of various rock mass rating, using packages of strength and deformation values that have evolved through research and appear to provide realistic results (Asef *et al.*, 2000). Results for values of RMR of 20–50 effectively show ultimate bearing pressures in terms of cave width, roof thickness and rock mass strength

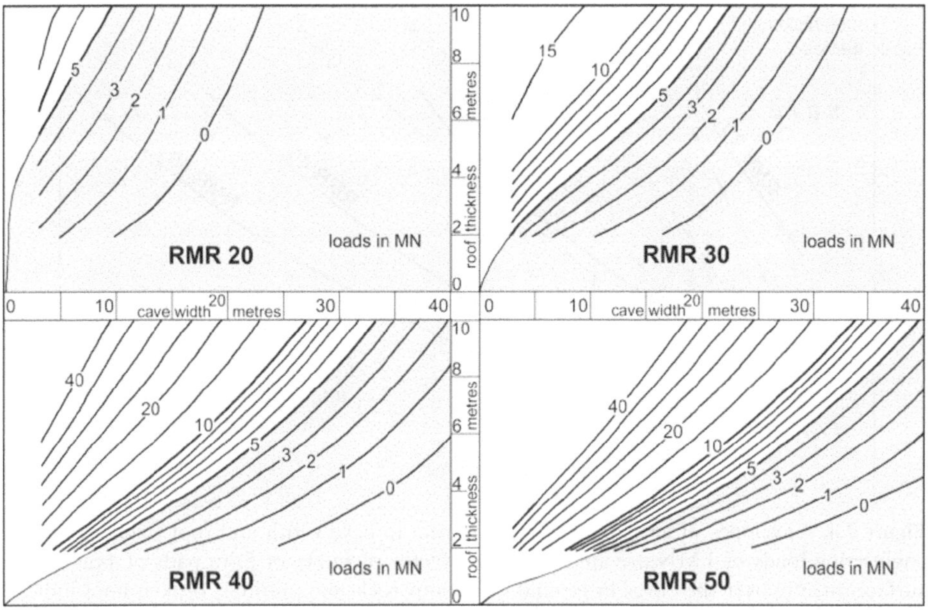

Figure 7.3. Nomograms that relate failure loads to cave width and roof thickness in ground of various rock mass ratings; loads are applied to foundation pads 1 m^2 on the surface directly above the caves; the shaded areas represent situations with respect to cave width and roof thickness where loading of 1 MN on the pad leaves a factor of safety <5, and are therefore considered unsafe.
After Lu Zhengxin, pers. comm.

(Figure 7.3). For any design load, selected factor of safety, estimated RMR and known or inferred cave width, a safe roof thickness can be determined from Figure 7.3. Continuation of the modelling showed that, in typical karst ground of RMR = 40, caves about 50 m wide collapse with no imposed load even at depths of 10 m, while caves 30 m wide collapse naturally when their roof is reduced to about 2.5 m thick. Where no cave exists, the same rock mass exhibits settlement of 25 mm at loads of 45 MN, indicating a safe bearing pressure of about 9 MPa.

Cave dimensions at failure loads of 5 MN can be extracted from these FLAC models to define safe conditions for the single case of 1 MN loading, on the 1-m^2 pad, with a factor of safety of 5 in any given rating of rock mass (Figure 7.4). If RMR for typical cavernous karst in strong limestone is taken conservatively as between 30 and 40, a cover ratio of the roof thickness being half the cave width ($t = w/2$) appears to be adequate for most engineering practice. In karst terrains on chalk and some other weak limestones, RMR may be estimated as nearer 20, and a cover ratio whereby roof thickness equals cave width ($t = w$) may be required for safe construction. Implications from this numerical modelling, with respect to cave roof integrity under imposed load (Figure 7.4), are only based on generalised estimates of the strength parameters for rock masses of the various rating values.

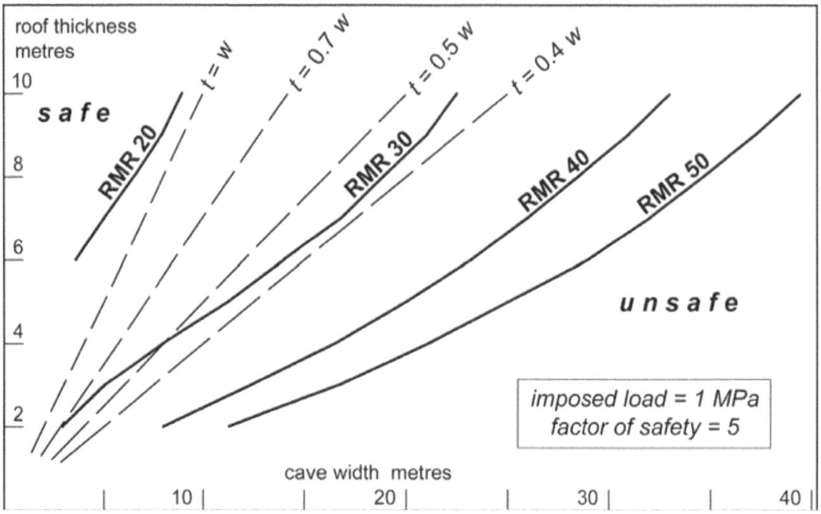

Figure 7.4. Envelopes of acceptability, with respect to cave width and roof thickness, where engineering loads of 1 MN are applied, with a factor of safety of 5, to pads of 1 m^2 on the surface directly over the caves in ground of various rock mass ratings; broken lines indicate various ratios of roof thickness (t) to cave width (w).

These estimates can only approximate the variable conditions in karst, where strong intact rock is broken by open or soil-filled fissures, in styles very different from those in insoluble rock masses. The data on caves could be improved by more specific modelling of fissured karst, with models designed to address the block mechanics of individual sites, and research is continuing with this aim. The current data is also derived from purely 2-D modelling, and failure loads are likely to be higher where some roof support is provided in the third dimension.

Finite difference analysis, also 2-D, of a cave roof in strong limestone in Tennessee did not incorporate structural loading, but did assess the effects of comparable seismic loading with a vertical peak acceleration of 0.14 g (Siegel *et al.*, 2003). Results showed that a roof thickness equal to that of the cave width provided a factor of safety of 2.5–3.0, which reduced to 1.5–2.0 where the thickness was half that of the width (Figure 7.5). This stability was under a dynamic seismic loading that was equivalent to little more than normal highway loading. When compared to practice from elsewhere, these factors of safety appear to be low, but probing to a depth of 3 m was regarded as adequate for the local site conditions where caves are typically up to 6 m across.

Numerical and physical modelling of artificial caves beneath loaded column bases has provided data on the stability of weak, massive sandstone that has a rock mass rating of 30–40, rather lower than that of typically stronger, but more fissured, karst limestone (Waltham and Swift, 2004). In the homogeneous sandstone, plug failure was the main cause of structural collapse over the artificial caves. A roof thickness that was half the cave width ($t = w/2$) was shown to be stable under the

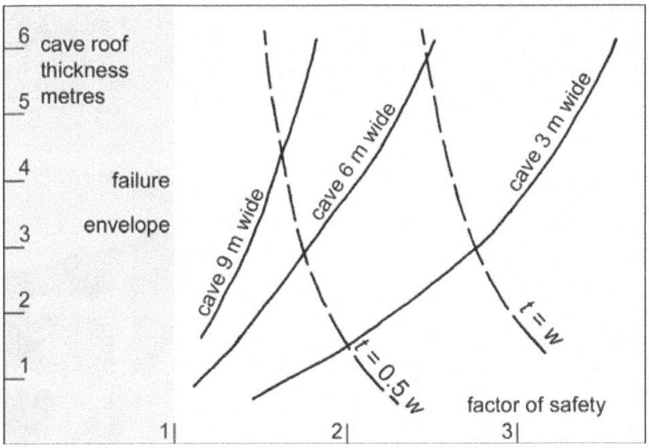

Figure 7.5. Numerical modelling of roof failure in strong limestone over caves of various widths under modest seismic loading in Tennessee; ratios of roof thickness (t) to cave width (w) are also indicated.
After Siegel *et al.* (2003).

maximum loads permitted by local building regulations. This implied that such a ratio was appropriate and perhaps conservative over typical limestone caves. Perhaps more significant than the absolute results, comparative data showed the importance of load position over the cave. Even where only a proportion of the column base footprint is over solid ground beside the cave, the bearing capacity is greatly increased. This shows that caves obliquely below structural foundations do not present a hazard; probes that are splayed out from the corners of a structural footprint appear to be unwarranted in the proving of solid ground. A full-scale loading test of a cave roof over the same sandstone was used to validate the modelling data. Failure of a roof 0.5 m thick, over a cave 4 m wide, occurred under a load of 340 kN applied to a pad 400 mm square, creating a plug through a zone of oblique stress fractures (Figure 7.6). The results from that rare opportunity of a real loading test do conform with the interpreted and calculated values for cave roof stability elsewhere.

7.2.3 Punching failure of cave roofs

A punching failure may develop where a thin slab of rock fails in shear round the perimeter of a small loading area – such as a pile tip or a small column base – so that a plug of rock is displaced into an underlying void (Figure 7.7(a)). Resistance to failure is therefore a function of the shear strength of the rock mass and the wall area of the plug. Shear strength is immensely variable for a rock mass, as it depends on the immediate disposition of fractures and fissures; a very approximate figure may be taken as about one-tenth of the shear strength of the intact rock, therefore about 3 MPa for strong karstic limestones. Plug wall area increases with rock depth (roof thickness), but part of the depth is lost in the development of a flare on the lower part

Figure 7.6. Broken rock in a cave roof loaded to failure; a full-scale test (by loading upwards on hydraulic jacks) on homogeneous sandstone in Nottingham, U.K.
TW.

of the plug, where it therefore fails in tension at lower unit values; this is a normal failure mechanism in homogeneous rock, but is distorted by pre-existing fractures within a rock mass (Figure 7.7(b)). Plug wall area also increases in proportion to the perimeter of the loaded pad. A square foundation pad quadrupled in area only has its perimeter doubled, so can accept only half the loading stress where plug failure may occur; guideline values for safe bearing pressures are less relevant than total loads in this situation.

Under these conditions, safe loading of a cave roof may be defined theoretically by roof thickness and pad size (Figure 7.7(c)); these calculated bearing capacities incorporate a safety factor of 5, and any lower factor would be inappropriate where unknowns remain with respect to the behaviour of a fractured and fissured rock mass. The implication is that rock proven to conventional limits of around 3 m will not develop a plug failure over a cave. Risk is eliminated by using larger foundation pads, but a small inherent risk can remain with respect to pile tips.

A plug failure could conceivably develop under lower loads where a roof block, bounded by pre-existing fractures through the entire roof thickness, is punched through by a point load imposed exactly over it. The chance of this happening is remote. In reality, structural loading over a cave generally closes the rock fissures and thereby develops a stronger voussoir arch within the roof mass. Punching failures have been recorded in the weak Florida limestones (Sowers, 1975) but only where excessive loads were placed on crusts of limestone just 1–2 m thick over unconsolidated sand. The sand's presence was known, but had been thought to be stronger; conventional proving of the ground should reveal any caves at such shallow depths beneath planned foundations.

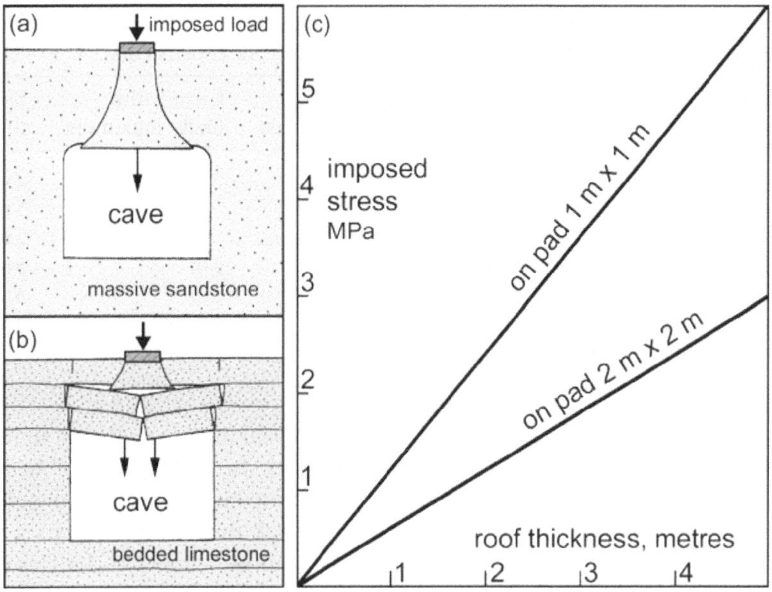

Figure 7.7. Plug failures over caves. (a) Typical plug profile in homogeneous rock. (b) Typical complex failure in bedded and fractured rock. (c) Bearing capacity on limestone with respect to plug failure over a cavity, in terms of the size of foundation pad and the thickness of sound roof rock; the values have a safety factor of 5, and the rock mass has a shear strength of 3 MN/m^2.

7.3 SAFE COVER THICKNESS OVER CAVES

All the available information suggests that the "rule-of-thumb" that cover thickness should exceed cave width is excessively over-conservative in most of the strong limestones that form cavernous karst. Evidence from the various available sources suggests that a roof thickness of about half the cave width is stable and safe under most conditions of loading. In view of the extreme variability of karstic ground conditions, a guideline that roof thickness should exceed 70% of cave width (i.e., thickness/width = $t/w = 0.7$) is probably more appropriate in most karst terrains in strong limestone. This value is still conservative under normal structural loading, and is very conservative under highway loading. Guideline values of safe cover thickness must be modified to accommodate cavernous rocks of varying strength, notably increasing in gypsum, chalk and weak limestones, while decreasing in strong basalt (Table 7.1). Within this table, imposed loads are maxima that are 50% of the Safe Bearing Pressures applied to sound rock; quoted safe roof thicknesses are of sound rock, ignoring any soil cover, or any epikarst zone of weathered, fissured or pinnacled bedrock.

At individual sites, specific site details, local knowledge and local experience may dictate safe roof thicknesses (and therefore probing requirements) that are higher or

Table 7.1. Safe roof thicknesses for various cave situations – and therefore a guideline to appropriate depths for probing to prove sound rock prior to construction.

Rock	Imposed load	Karst class	Cave width – likely maximum	Safe roof thickness
Strong karstic limestone	2,000 kPa	kI–kIII	5 m	3 m
		kIV	5–10 m	5 m
		kV	>10 m	7 m
Weak limestone and chalk	750 kPa		5 m	5 m
Gypsum	500 kPa		5 m	5 m
Basalt lava	2,000 kPa		5–10 m	3 m

lower than the tabled values. Site inspection may reveal limestone that is significantly more or less fractured and fissured than is typical. In local zones of heavy rock fracturing, accurate figures for safety factors cannot be achieved, and only broad guideline concepts can be applied. Experience in rock tunnelling suggests a variable cover ratio (Sowers, 1996), where sound rock requires a cover of half cave width while more fractured rock requires cover that exceeds cave width (though apparent typing errors inverted this relationship on page 157 of that excellent book). Long-term decline in rock strength, due to fracture propagation where it is under stress (Tharp, 1995), has implications for the stability of individual rock beams; however its impact on a compression arch within fractured rock would appear to be minimal, and is adequately covered by the very high safety factors incorporated into concepts of rock mass strength.

An alternative approach to the safe cover thickness is based on the decline of imposed stress at increasing depths beneath a loaded foundation structure. It has been suggested that induced collapse of a cave roof is unlikely where the loading stress is less than 5–10% of the existing overburden stress (Sowers, 1996). Reference to the undistorted bulbs of pressure perceived by foundation engineers suggest that this stress ratio is reached at a depth of about 4 m beneath a foundation pad 1 m^2 carrying a load of 1 MN, where overburden stress increases by 25 kPa per metre depth. This takes no account of cave width, and assumes there is no cave roof at a critical state of imminent collapse. It is however slightly conservative because it does not account for stress redistribution around an open cave, where wall failure is unlikely. A safe thickness of 4 m is commensurate with guideline figures derived from other considerations. Where a foundation pad 2 m^2 carries a load of 4 MN, still with an applied stress of 1 MPa, the imposed stress exceeds 10% of overburden stress at a depth of about 6 m. This implies that greater thicknesses of sound rock should be proven where heavy structural loads are placed on karstic rock that may contain large caves. There are multiple benefits in using larger footings that impose lower stresses on cavernous ground.

Any accessible cave may be inspected so that its roof stability is assessed at least semi-quantitatively, and reasonable precautions are then be taken with respect to either remedial works or reduction of imposed loads. The obvious requirements for

Table 7.2. Some of the various guidelines that are recommended or applied during ground investigation to prove that limestone is free of voids by probing or drilling beneath planned foundations.

Source	Location	Probe depth	For loading	Cave width	Rock type
City of Rochester	New York	1.5 m	< 4.8 MPa	n.s.	Strong dolomite
City of Allentown	Pennsylvania	1.5 m, or 2 × pd	implied	n.s.	Strong limestone
Foose *et al.* (1979)	Pennsylvania	2.5 m	n.s.	n.s.	Strong limestone
Cooley (2002)	Alabama	2.0–3.0 m	0.9-m piles	n.s.	Dolomite, 70 MPa
Sowers (1996)	Florida	1.5–3.0 × pd (1.5–6.0 m)	implied	n.s.	Limestone
Knott *et al.* (1993)	Pennsylvania	3.0 m	1.0 MPa	n.s.	Dolomite
Wyllie (1999)	n.s.	3.0 m	n.s.	n.s.	Limestone
Raghu (1987)	N.E. U.S.A.	3.0 m	n.s.	< total pile cap	Strong limestone
Erwin and Brown (1988)	North Carolina	3.0 m*	< 1.2 MPa	n.s.	Weak limestone
Waltham *et al.* (2003)	Karst kI–kIII	3.5 m	< 2.0 MPa	< 5 m	Stong limestone
Higginbottom (1966)	U.K.	3.6 m	high-rise	n.s.	Weak chalk
Sotiropoulos *et al.* (1979)	Greece	4.0 m	heavy pile	n.s.	Limestone
Wagener and Day (1986)	South Africa	4.0 m	n.s.	n.s.	Strong dolomite
Tan and Batchelor (1981)	Malaysia	4.5 m	high-rise	may be large	Strong limestone
Garlanger (1991)	Florida	4.9 m	n.s.	n.s.	Limestone, 30 MPa
Tan (1987)	Malaysia	5.0 m	high-rise	may be large	Strong limestone

* Specified as 1.5 m of sound rock and the next 1.5 m below with no void >150 mm.
n.s. = not specified. kIII = karst class. pd = pile diameter.
Strong limestone = typical karstic limestone with UCS >70 MPa.

safety, and natural desires to err on the side of caution, mean that most caves with rock cover less than their width will be filled with concrete where they lie beneath a construction site. The problem lies not with the known caves, but with the unknown caves. Many of these can only be found by protracted and expensive programmes of probing, with or without geophysical surveys. A pragmatic approach on cavernous ground can be to omit the costs of potentially inconclusive ground investigations and spend the savings on improving any new construction so that it will not cause or suffer from ground collapse. Buoyant foundations and spread footings on rafts or beams can reduce imposed loads so that rock collapse is not induced in ground that has been stable when undisturbed through geological time. Piled foundations can support structures designed so that any single pile failure does not destroy integrity. In the words of one practical engineer, "you can buy a lot of reinforcing steel for the cost of a geophysical survey, and then not worry about the chance of caves being there".

Without taking the pragmatic approach in its entirety, some guideline value of safe cover is required to plan any ground investigation by probing, and a variety of numbers is available from published sources (Table 7.2). Some of these variations are in respect of local available data on the maximum and typical widths of caves that can be anticipated. It would appear that probing depths of <2 m are inadequate except where there is a substantial databank confirming that only very small caves are likely to occur. It is also significant that the only two probing depths greater than 4 m are one on a weak limestone and one in a karst terrain notable for its large caves. Heavy-duty, cast-in-place, end-bearing piles may require 4–5 m of proven sound rock in any environment. However, large-diameter concrete piles carrying loads up to 10 MN have been placed on the weak limestones of Florida simply on the basis of load testing of each pile, without any specification for probing beneath the pile tips.

It is notable that guideline values of safe cover, cited above or in Table 7.2, bear no relationship to the figures widely used in assessing stability over old mine workings. A safe cover of 10× mine height relates only to the hazard of void migration (to form crown holes) in thinly bedded Coal Measure rocks; it is meaningless with respect to cave roofs in karst limestone.

7.4 EXISTING STRUCTURES OVER CAVES

Numerous built structures have been placed over open caves that lie at shallow depths, which were either known or unknown at the time of construction. Some were built after due remedial works had been completed. Others were built on natural ground. A few of these have failed (Section 7.1), but the remainder survive – some in conditions that may be described as marginal.

Natural Bridge, in Virginia, is effectively a very short cave that is about 27 m wide with a roof thickness of 14 m ($t/w = 0.5$) of thickly bedded strong limestone (Figure 7.8). It has carried a two-lane highway for a hundred years and shows no sign of distress. A main road in Iceland crosses over the lava tube of Raufarholshellir where it is about 10 m wide under a basalt roof just 4 m thick ($t/w = 0.4$). When the cave was found (from a nearby entrance) long after the road was built, concerns led to checking it for movement and roof breakdown, but none was found over some years of monitoring, and the cave and road are now regarded as stable. In Hungary, the urban sprawl of Budapest includes roads and buildings that stand over some 30 km of known cave, much of which lies at shallow depths and where no collapse has ever been recorded.

In the karst of western Ireland, the railway from Dublin to Sligo stood for many years on limestone only 2.5 m thick over a passage more than 6 m wide ($t/w = 0.4$) in the St. Augustine's Cave. When the single track railway was replaced by a main road around 1960, a ground-slab of reinforced concrete effectively carried the road over the cave in order to minimise any risk of collapse, and a length of masonry walling was built inside the cave to reduce its widest section of unsupported span (Figure 12.13). A major building in Huntsville, Alabama, lies over a cave 12 m wide under about 5 m of rock ($t/w = 0.4$), but is supported on piles that pass beside or through the cave (Case study #6).

Bowling Green, Kentucky, has many roads and buildings that stand over the large cave passages of the Lost River System. Many stand, in apparent stability, on rock spans with t/w ratios that are around 0.5, though the Dishman Lane collapse was inevitable where it stood on a cave roof with $t/w = 0.1$ (Case study #2). On the other side of town, a four-lane highway passes directly over a chamber in Bypass Cave, where a flat roof nearly 10 m wide has only about 2.5 m of rock and 2.5 m of soil cover between itself and the road. With a ratio of $t/w = 0.25$, this lies outside the perceived wisdom for safe engineering, even with the low imposed load from only a

Figure 7.8. Natural Bridge, Virginia, with a road over its thin but stable arch. TW.

highway. The continuing survival of the road suggests that guidelines on safe cover ratios are comfortably conservative (unless the road is in fact dangerously close to collapse).

An open cave was discovered under the main runway of Palermo airport, in Sicily, Italy, during adjacent routine maintenance works (Jappelli and Liguori, 1979). The critical part of the cave was a single chamber 40 m long and 30 m wide with an uneven but roughly horizontal roof draped with stalactites (Figure 7.9). This was separated from the base-course of the runway by just 2–4 m of limestone, and the upper part of that was described as brecciated due to weathering. With a cover ratio (t/w) of 0.1 or less, the risk of collapse was deemed unacceptable, even under only the distributed load of the reinforced concrete runway; the entire cave was filled with concrete (injected through a grid of boreholes on 5 m centres). This cave could be regarded as having been a close call on the runway's safety. On the other hand, it

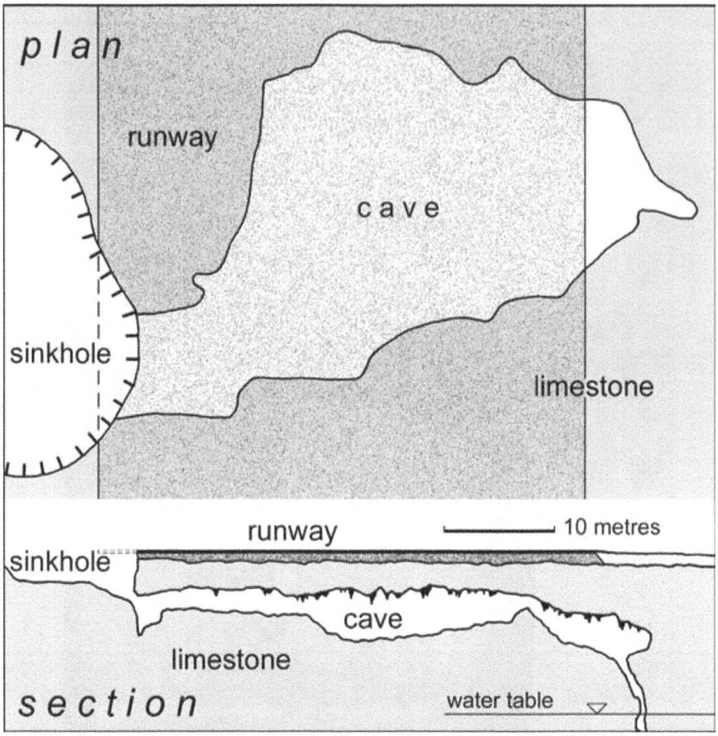

Figure 7.9. Plan and section of the cave found beneath the runway at Palermo airport in Sicily; on the section, part of the limestone immediately beneath the main cave and the sinkhole floor is a layer of breakdown blocks of unknown depth.
After Jappelli and Liguori (1979).

may be viewed as an indicator of the stability of a very thin span of fissured rock over a natural cave. It may indicate the importance of spread loading, by the runway structure, and it was perhaps surprising that the cave did not collapse under the disturbance of construction traffic. Overall, it does appear that guidelines demanding rock cover at least half as thick as a likely cave span ($t/w = > 0.5$) appear to be adequate and appropriate in engineering works on most cavernous ground.

8

Sinkholes induced by engineering works

Subsidence sinkholes that develop within a soil cover are the major geohazard in karst. They are created rapidly by water transport of the soil, and any disturbance of the water regime that increases through-flow is more than likely to promote new sinkhole development. Such disturbance, either designed or unintended, is all too commonly due to engineering works, and in the developed modern world the great majority of subsidence sinkholes are induced by man's activities.

The proportion of new sinkholes that are induced in populated areas is generally accepted as being around 90%. Records in Alabama from 1900 to 1980 mark over 4,000 new sinkholes, mostly occurring since 1950, while there were less than 50 reported new sinkholes that were natural events (Newton, 1981). The sinkhole database from China (Table 8.1) is so large that its proportion of sinkholes that are induced (between 87% and 93%) should be representative of the overall picture in karst. Clustering of multiple sinkholes around single causes does have an impact on the statistics, and natural causes are assigned to 23% of the Chinese sinkhole sites (as opposed to numbers of individual sinkholes). Artificial water table decline and increase in imposed drainage are the two main factors that induce subsidence sinkholes, and which one is dominant depends on the local situation. The urbanised karst areas of Pennsylvania are characterised by deep water tables, where induced decline has little impact on sinkholes in the soil profile; its effects are therefore not recorded in Table 8.1, which partly accounts for the state's atypically low proportion of sinkholes that are induced as opposed to natural.

Sinkholes may be induced by engineering disturbance that comes in a variety of forms (Table 8.2). The major critical factor is any increase in water flow down through the soil and into cavernous rock beneath. This is most commonly due to drawdown in response to water table decline; whether this is due to abstraction or by de-watering, it may have to be regarded as an environmental factor for any particular site. Alternatively, water may be added at the surface to achieve the same effect.

Table 8.1. Proportions of new sinkholes that are either natural or induced by various means in China and parts of the eastern U.S.A.

After Gao *et al.* (2001) and Myers and Perlow (1984).

Location	China	Alabama	Missouri	Pennsylvania
Water table decline (including mine drainage)	80.0%	87%	8%	n.a.
Water impoundment	3.4%	12%	21%	
Construction works and drainage modifications	3.6%	1%	71%	58%
Total induced	87.0%			58%
Natural	6.2%	n.a.	n.a.	42%
Unknown	6.8%			
Total number of sinkholes	44,904	1,324	87	1,574

n.a. refers to statistics that are not available.

Table 8.2. Listing of the main processes by which sinkholes are induced through engineering activity.

Partly after Newton (1987), Waltham (1989) and Cai (1991).

Increased water input to soil cover
Uncontrolled run-off drainage from a site or structure Installed drainage ditches that are unlined Broken pipelines Soakaway drains (dry wells) within the soil profile Unsealed boreholes Irrigation for agriculture Impoundment of reservoirs and floodwater retention basins
Increased through drainage due to water table decline
Over-pumping for groundwater abstraction Well pump-testing De-watering to maintain dry quarry operations De-watering to maintain dry mine workings
Physical disturbance to the ground
Partial removal and consequent thinning of the soil profile Partial or total removal of vegetation on the soil cover Vibrations from blasting or plant movements Structural loads on foundations within the soil profile Water table fluctuations.

In the short term this is commonly related to rain-storm events, but in the long term is a common result of drainage diversion within construction or built works, and these should be avoidable by best-practice engineering. The third group of inducement processes covers any physical disturbance of the soil conditions. This may include the effects of foundation loading on the soil profile, but excludes the rare collapses of rock due to imposed load over caves (see Chapter 7).

8.1 SINKHOLES INDUCED BY INCREASED WATER INPUT

Any disturbance of the natural patterns of run-off and infiltration of rainwater can lead to localised increases of drainage down through the soil, which can lead to renewed or accelerated soil suffosion at those points, and the ultimate development of new subsidence sinkholes. About 70% of new sinkholes in the Bowling Green area of Kentucky are induced by drainage disturbance either by farming practices or by urban development (Crawford, 1999). The disturbances may be unintentional diversions of drainage, either during construction works or where housing development places a quarter of the land under concrete and diverts its rainwater run-off elsewhere. Industrial or commercial development, that can totally seal large areas of land, demands installation of extensive drainage structures. At Bowling Green's airport in Kentucky, 25 sinkhole collapses have been recorded in the soil cover, but all are where run-off from the runways has been directed into channels, ditches and sinkhole basins, while none has occurred under the paved areas (Crawford, 2003). A large proportion of new sinkhole events are triggered by new drainage works that are inappropriate or inadequate for installation in or on soils that cover cavernous karst.

Unlined drainage ditches are a major source of problems in karst regions, and many develop sinkholes soon after their construction. By their very existence, they concentrate water flow at sites that previously took only direct rainfall infiltration. Run-off from an airport runway in Pennsylvania was fed into an unlined gully that is now broken into a series of large sinkholes (White *et al.*, 1986). Ditches on low gradients are traditionally left unlined along transport corridors. Seepage losses from unlined ditches accounted for 14 out of 54 severe sinkhole collapses along China's railways (many of the others were induced by water table decline; Guo, 1991), and also for 49 out of 72 sinkhole events along Tennessee's highways, where only 4 sinkholes developed along the lined ditches (Moore, 1988). Seepage into any unseen rockhead fissure can carry soil from beneath a ditch floor; by the time surface subsidence is apparent (Figure 8.1) it will require remediation to prevent potentially destructive expansion, when it may then be turned into a stable outlet to bedrock (Chapter 11). Though many unlined ditches do keep sinkhole initiation away from the actual road or railway, concrete-paved ditches can prove cost-effective in sinkhole-prone karst (Moore, 1988).

A leaking water main is a very effective means of supplying water to a soil and thereby initiating a subsidence sinkhole as both water and soil are lost into bedrock fissures. A mains failure under a rural road on the South Downs, in England, created about 80 dropout sinkholes in adjacent fields within a few hours (McDowell and Poulsom, 1996). There was no surface flow, as the water spread through the cover of loose sands and gravels until it found outlets in pipes and fissures within the underlying chalk. Some of the dropout failures occurred within minutes, suggesting that

Figure 8.1. Early stages of a sinkhole developing in the untreated soil floor of a ditch beside a Pennsylvanian highway; it is temporarily choked with clay, while the riprap in the foreground was placed to prevent erosion at a culvert outlet.
TW.

these may have formed as caps of more cohesive topsoil collapsed into soil voids already formed by suffosion of the sand and gravel into chalk fissures (Figure 8.2).

Pipeline fractures in cover soils are more destructive in urban areas, especially where large buildings are not founded on bedrock. Minor soil settlements induced by a building's loading can fracture utility lines, whose leaking water then accelerates the soil loss in a feedback loop that is both dramatic and inevitable. In the urban karst of Pennsylvania, utility failures account for nearly 30% of sinkhole events (Myers and Perlow, 1984), including some major structural failures (Case study #7). A single break in a 360 mm water main in Phillipsburg, New Jersey, created a cluster of sinkholes under a suburban street, the first of which completely destroyed a house within 4 hours of ground movement first being noted, even though the water was shut off within 3 hours (Figure 8.3). The houses, street and pipeline all stood on sandy soils that lay 3–15 m thick over a pinnacled rockhead of Cambrian dolomite. Subsequent drilling found a dozen cavities within the soil profile beneath the street, with the largest all around the water main break; these were grouted to prevent future failure (Canace and Dalton, 1984).

Early response to a water main leak can minimise expanding damage. However, a loss of pressure in a Pennsylvanian water main prompted a water company engineer to go out looking for the site of the escaping water, which he found when his van fell into a sinkhole that opened in the street in front of him (White *et al.*, 1986). At other locations, sinkholes are generated beneath towns due to widespread small-scale leakages from utility lines that are either poorly built or poorly maintained (Case study #12). The desert city-state of Kuwait has neither significant

Figure 8.2. One of the cluster of subsidence sinkholes induced by a broken water main on the English chalk downs; there was no surface flow, but the dropouts developed above suffosion points in the soil over bedrock fissures.
TW.

karst nor significant rainfall, yet four large sinkholes opened in the Al-Dhahar suburb in 1988/9. These all formed in sandy soils 40–60 m thick over the cavernous Dammam Limestone, and appear to have been triggered by water from leaking pipelines and also from garden irrigation within the zone of new housing (Abdullah and Mollah, 1999).

While sinkholes caused by pipeline fractures and ditch leakages may be counted as unintentional, those caused by dry wells (or soakaway drains as they may be known) can only be ascribed to engineering practice that is unacceptable on karst. Rainwater drainage from a new factory and its car park in Pennsylvania was directed entirely into dry wells sunk into the soil cover over a pinnacled limestone rockhead across the floor of a large and shallow closed depression. Within a few months, over twenty sinkholes had developed, mostly adjacent to the dry wells (Knight, 1971). Prior to development, rainwater had sunk into the soil with minimal temporary ponding in small hollows, but the new plant had placed half the site under concrete and had concentrated all its run-off at the dry wells dug entirely in the soil profile; sinkholes failures were inevitable. Soakaway drains are commonly installed in England's chalk, which is appropriate for their use because of its very high diffuse permeability. However, it is well established that they are not placed where there is any possibility of ground cavities (either natural or mined) that could permit suffosional loss of the soil cover. Such a ruling could well be applied more widely, to exclude their use in any soil-covered karst – except where appropriate measures were utilised.

It is possible to use dry wells to dispose of storm-water in soil-mantled karst as long as they drain directly into bedrock fissures so that there are no pathways for suffosion of the soil (Figure 8.4). The critical factor is to ensure that the well casing is sealed through to bedrock, with no routes for water to escape into the soil and through to other fissures, or for influent flow to wash soil into the dry well.

Figure 8.3. Terminal damage to a timber-frame house in Phillipsburg, New Jersey, caused when sinkholes developed underneath it and under the sidewalk after a water main burst beneath the street.
Photo: Rick Rader.

Hundreds of dry wells operate successfully in and around Bowling Green, which stands entirely on Kentucky's Sinkhole Plain. Many were drilled directly into open caves that had been previously located at minimal depths, and they have various grilles and traps that ensure they are not choked by sediment (Crawford and Groves, 1995). However, among 80 recent sinkhole collapses in the town, 24 were on poorly constructed drainage wells, while another 28 were beneath ponded storm-water in broad sinkhole depressions that lacked engineered conduits into the limestone below (Crawford, 2001). Such retention basins within sinkholes can be acceptable, but only with appropriate design and management (Chapter 12). The effect of diverting water into an existing sinkhole depression with no engineered outlet was demonstrated by the overnight collapse of the Boxhead sinkhole in England (Figure 4.9). A shallow subsidence sinkhole had developed by years of percolation suffosion, but its floor dropped out immediately after a small moorland stream was diverted into it.

There are many recorded cases of well drill rigs collapsing into sinkholes triggered by their own disturbance to the groundwater patterns. In 1959, a driller lost his life in Florida when he was buried 10 m deep in a dropout collapse. His well boring was approaching rockhead at a depth of 24 m where it breached a clay aquiclude and allowed rapid drainage and suffosion from the perched aquifer (in

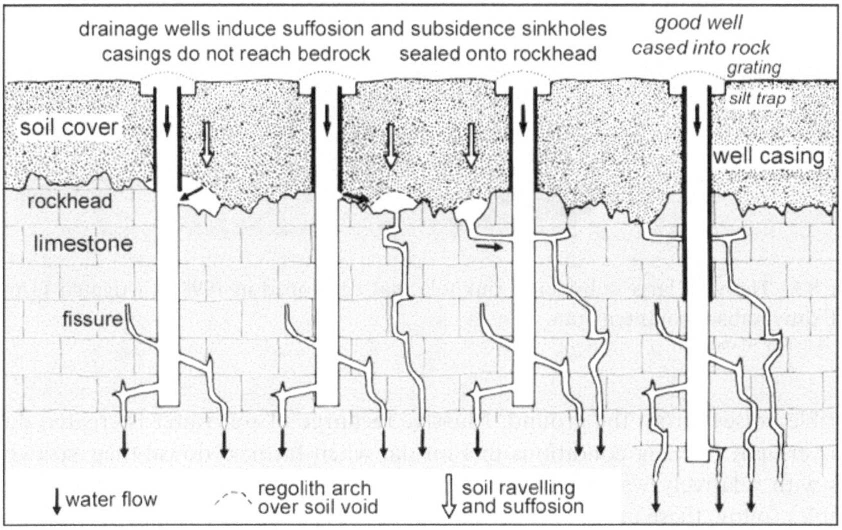

Figure 8.4. Dry wells in soil-covered karst. The three wells on the left provide pathways for suffosion of the soil and are likely to promote future sinkhole development, while the well on the right is sealed deep into bedrock fissures and should provide a stable drainage outlet. After Crawford and Groves (1995).

the upper sandy soil) down into the fissured limestone (Newton, 1987). Sudden downward drainage of a soil aquifer is the most common sinkhole trigger induced by drilling. The reverse was the trigger at a drilling site in South Dakota, where an artesian aquifer in cavernous gypsum was breached and the upward flow of water collapsed the well and created a large sinkhole, from which water continues to be discharged (Rahn and Davis, 1996). Investigative drilling at sites of ground subsidence always run some risk of triggering more catastrophic collapses, and water-flush drilling is certainly inappropriate for investigating potential sinkhole collapses. There have been cases where exploratory drilling of a site with modest prior subsidence provided the input of water, that drove the suffosion, that induced a sinkhole, that destroyed the structure under investigation.

Land irrigation may be a necessary and planned recharge of water to the soil, but it can have disastrous side-effects where the soil overlies cavernous rockhead. Small subsidence sinkholes are common in ploughed and irrigated farmland in karst regions all round the world, but their economic impacts are generally small, and could not justify stopping the irrigation except in the smallest of sinkhole-prone areas. An unusually large sinkhole opened in a pistachio field that was kept well irrigated in the desert lands just west of Kerman in Iran (Figure 8.5). It formed in alluvial soils more than 40 m thick, and though the groundwater pumping was causing steady water table decline, the irrigation recharge onto the surface appears to have been the critical factor in the sinkhole's very rapid collapse (Atapour and Aftabi, 2002). Dissolution of gypsum horizons within the alluvium sequence may have aided growth of large soil cavities, whose dramatic collapse was then marked by

Figure 8.5. The very large subsidence sinkhole that developed in 1998 in irrigated farmland near Ekhtiyarabad, southern Iran.
Photo: Habibeh Atapour.

"horrible noises" from the ground. Massive recharge of soil water is created during rare overnight freezing conditions in Florida, when farmers douse their strawberry crops with relatively warm groundwater to keep the frost off them, and rashes of sinkholes follow these events (Bengtsson, 1987). Though water tables decline due to the pumping, the speed of events suggests that the sinkholes are at least in part triggered by the surface recharge (Section 8.2.1).

8.1.1 Sinkholes caused by reservoir impoundment

The most massive recharge of water to any soil cover on karst is created by the impoundment of a reservoir. New sinkholes commonly develop in the floor of any reservoir on karst directly after it is first filled, causing rapid drainage of the impounded water, though the impact of the sinkholes themselves is usually minor in comparison to the loss of the reservoir's integrity (Chapter 12). Seven complete failures of reservoirs or sewage lagoons were recorded as due to sinkhole development and rapid drainage into the limestones of Missouri's Ozarks karst (Aley *et al.*, 1972), and six more are recorded in the eastern U.S.A. (Newton, 1987).

Nearly all these reservoir losses on karst create subsidence sinkholes entirely within the soil profile, but the massive through-flows of escaping water can scour out buried features and expose open solutional voids that reach deep within the bedrock. Shortly after the Al Marj dam was built on karst in the Jabal Akhdar (Green Mountains) of northeastern Libya, its reservoir was filled during a major rain storm. Local people gathered in celebration, only to see a huge whirlpool develop in the middle of their new lake, which was soon completely empty. Sinkholes left on the ex-reservoir floor included a vertical open shaft, down through 1 m of soil and 26 m of limestone to a rubble choke (Figure 8.6). This sinkhole appears to have developed when a previously filled shaft within the bedrock was washed out from below, where the main water flow (from the whirlpool) had entered through a side fissure 20 m down. Collapse of limestone bedrock is rare beneath a reservoir, but some measure of dissolution and subsequent rock collapse must occur during a catastrophic reservoir drainage event that occurs on gypsum.

Figure 8.6. The Karrubah sinkhole, in Libya, that formed beneath the waters of the short-lived reservoir impounded by the Marj dam, visible in the distance.
Photo: Attila Kosa.

Repeated sinkhole activity may be induced where reservoir levels fluctuate. In western Turkey, the May reservoir lost most of its water through 33 new sinkholes when it was first filled in 1960. These were all subsidence sinkholes that developed in the alluvial sediments forming an unstable reservoir floor over cavernous limestone (Dogan and Cicek, 2002). The sinkholes were repaired and sealed with clay, but the reservoir continued to suffer from massive leakage. When the reservoir again reached its maximum level, in 2002, three new large sinkholes developed under water, and drained most of the reservoir. These formed in valley-floor alluvium that is 20–50 m thick, when suffosional soil removal down into the limestone was driven by the increased head difference across the soil that was self-imposed by the reservoir. It appears that these sinkholes did not have time to develop in the few months of 1960 before the reservoir was drained through a sinkhole elsewhere, indicating a minimum timescale for suffosional piping through the thick soil.

8.2 SINKHOLES INDUCED BY WATER TABLE DECLINE

The most widespread cause of induced sinkholes is any form of water table decline, where the vertical drainage is quite literally drawn down to greater depths, in many cases carrying more soil with it. The effect is most powerful where the water table declines past the rockhead. In these cases, a regime of slow, multi-directional, phreatic flow beneath the water table is replaced by rapid, downward, vadose drainage above the water table within the critical zone where soil arches over rockhead fissures are ready to fail. Lesser impacts are felt through other regimes of water table decline, but may still be enough to induce sinkhole failures.

Natural water table decline is normally very slow, and its impact on sinkhole occurrences is lost within the slow evolution of landforms during ongoing rejuvenation. Hundreds of sinkholes, each up to 20 m wide and deep, have developed around the shores of the declining Dead Sea, in Jordan and Israel (Yechieli *et al.*, 2003). Sinkhole events started in the 1980s, and noticeably accelerated about 10 years later, after the Dead Sea had fallen by another 8 m. They have been caused by sediment collapse over cavities in salt beds at shallow depth, where the accelerated dissolution of the salt was due to an invasion of freshwater drawn in from the mountains in response to the falling level of the Dead Sea. Among induced sinkholes, this process of rock dissolution is unusual, but its relationship to water table decline compares to that of accelerated soil suffosion forming subsidence sinkholes.

Shrinkage of the Dead Sea is at least in part due to the over-use of the limited water resources in its basin, so those sinkholes may also be viewed as induced by man's activities. Similarly, sinkholes may be described as natural when induced by water table decline during a prolonged drought, but in any populated area that decline is likely to be increased by extra groundwater pumping (Newton, 1987). Most induced sinkholes are related to artificial water table decline that is either an accidental consequence of groundwater abstraction or the intentioned result of de-watering mines or quarries.

8.2.1 Sinkholes induced by groundwater abstraction

Though karstic limestones can be very productive aquifers, their water tables can decline in the face of intensive or long-term abstraction. Sinkholes are therefore an almost inevitable consequence of urbanisation, industrial expansion or over-thirsty agricultural development on many areas of soil-mantled karst. Water table decline may be limited to the cone of depression of a single well that is over-pumped for an industrial plant. In Brazil, a sinkhole collapsed in a karst that was not previously recognised as such because the cavernous rock was hidden beneath a deep soil cover (Figure 8.7); the new sinkhole was induced by water table decline around an over-abstracted well on an industrial site near Sao Paulo (Prandini *et al.*, 1990). On a regional scale, the water table decline is commonly due to abstraction for municipal supplies, as is all too common in the densely populated lowland karst regions of China and eastern U.S.A.

Sinkhole collapses have occurred in all China's many cities built on karst, and groundwater pumping is the main cause (Chen, 1988; Guo, 1991). More than 1,450 subsidence sinkholes have developed in the alluviated and soil-covered plains around Guilin, that lie between the limestone towers in Guangxi's famous fenglin karst. They have caused extensive damage, mainly in farmland but also destroying roads and buildings, and 90% of the new sinkholes were induced by water table decline due to groundwater abstraction. The alluviated Shuicheng basin is similarly underlain by

Figure 8.7. Collapse of a house in Cajamar, Brazil, in 1986 over a sinkhole induced in a thick soil cover by excessive groundwater abstraction at a nearby industrial site.
Photo: Jose Labegalini.

limestone within Guizhou's fengcong karst, but was the most convenient site for abstraction wells to supply the expanding new town of Liupanshui (Waltham and Smart, 1988). Wells were placed out in the rice fields, where clay soils are 3–10 m thick over a pinnacled rockhead, and the original water table was only 1–2 m below ground. Within about 8 years, 1,023 new subsidence sinkholes were formed over cones of depression 8–15 m deep and reaching up to 400 m from 14 of the 17 wells (Figure 8.8). Sinkholes were mainly in the open fields, but 89 buildings were damaged, roads were broken, sewers were reversed, and at least two of the pump houses were damaged in classic cases of self-induced destruction (Figure 8.9). The broad conclusions from experience in China's karst suggests that sinkholes are induced mostly where the water table declines past the rockhead and/or where it declines more than 3 m, while surface leakages are more important than water table decline where its original level is below rockhead (Lei *et al.*, 2001).

Decline of a water table that lies deep below rockhead is observed to have minimal influence on subsidence sinkholes in the dissected and deeply drained karst of Pennsylvania (White *et al.*, 1986), matching the data from China. The major sinkhole problems due to water table decline in America are in the lowland karst of Florida. Across most of the state, the Floridan aquifer is formed by the Ocala Limestone with its rockhead 30–50 m below ground level. The soil profile has a perched aquifer in sandy materials overlying an aquiclude formed by the clay-rich Hawthorn Formation that rests on rockhead. The piezometric surface of the

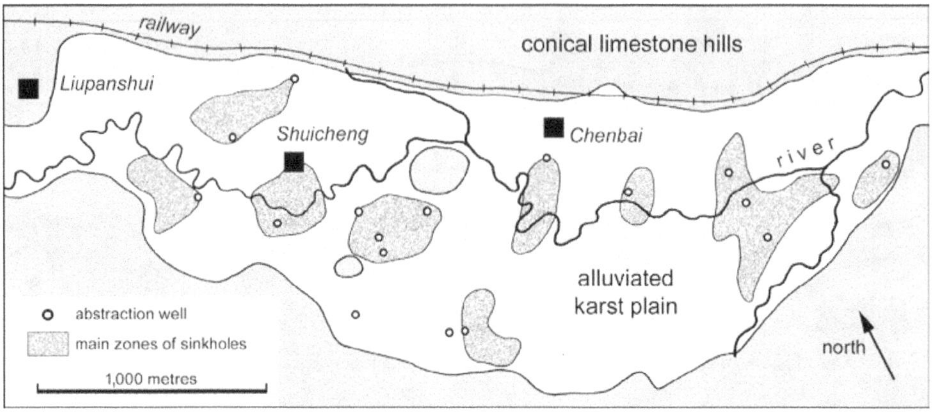

Figure 8.8. Distribution of abstraction wells and their induced sinkholes in the alluviated karst basin of Shuicheng, China.

Figure 8.9. A subsidence sinkhole induced by water table decline as groundwater was pumped through the well house visible beyond, in the Shuicheng karst basin, China. The crest of a limestone pinnacle is exposed in the sinkhole only a metre below the original ground level. TW.

Table 8.3. Sinkholes induced in three areas of western Florida.

Site	Date	Water table decline	Period of activity	Number of new sinkholes	Diameter of area of activity	Source of data, reference
North Tampa	1964	3.5 m	3 months	64	3 km	Sinclair (1982)
West Dover	1977	2.9 m	5 days	22	3 km	Metcalfe and Hall (1984)
East Dover	1985	5.2 m	3 days	27	6 km	Bengtsson (1987)

Floridan aquifer normally lies below the perched water table. The head difference drives seepage from the upper sands down through the lower clays, causing soil suffosion into open fissures in the limestone rockhead. Groundwater abstraction is nearly all from the Floridan aquifer, as the shallow aquifer is prone to pollution, and water table decline therefore increases the head difference across the Hawthorn, and increases sinkhole development. The Winter Park sinkhole in Orlando occurred when the water table in the Floridan aquifer had declined by 6 m, increasing the head difference from 4 m to 10 m (Figure 4.10).

Clusters of sinkholes have developed in the Tampa area of West Florida during at least three events of major water table decline due to pumping from the underlying Floridan aquifer (Table 8.3). These sinkholes are 1–23 m across and only 1–7 m deep, even though local depths to rockhead are 20–30 m. The two events in adjacent areas near Dover each occurred during short periods of sub-zero weather when massive quantities of warm groundwater were pumped from field wells and sprayed onto the strawberry crops to prevent them freezing. Both sites were marked by the very rapid development of sinkholes, on the third day of pumping in the 1985 event. The earlier event north of Tampa was induced by expansion of a new wellfield for municipal supplies, so no water was immediately recharged to the soil, and sinkholes developed over a period of about three months. Irrigation water added to the soil joins the perched aquifer above the Hawthorn Formation, while suffosion processes that form sinkholes are normally initiated at rockhead where soil cavities grow over bedrock fissures. However, it appears that the substantial water recharge to the soil directly over the pumped cones of depression at the two Dover sites had its own impact on sinkhole development in addition to that of the water table decline, and accounted for the new sinkholes appearing more rapidly than at the un-irrigated Tampa site.

Pump-testing and the development of new wells commonly induces massive water table decline, albeit under small areas, in terrains that have not previously been disturbed. They are therefore very effective at inducing new subsidence sinkholes in any available soil cover. Nine over-pumped wells in the eastern U.S.A. all developed sinkholes within their immediate vicinity and within a few days of the start of pumping (Newton, 1987). At five of these sites, the pumping was extended or intensified in attempts to clear muddy water, which was a sign of active suffosion and a warning sign of the sinkhole collapses that soon followed. A staged pumping test in the Guizhou karst, China, followed years of pumping from older wells that had induced over 70 subsidence sinkholes in cover soils 2–8 m thick (He *et al.*, 2003). In the first stage of pumping, drawdown was 6.8 m and no new sinkholes were formed, but subsequent stages that took drawdown eventually to 24 m induced 38 new sinkholes within two months. Sinkholes are most conspicuously induced when water tables first decline to levels not previously reached.

Well development by cyclic flushing is singularly effective at moving sediment through cavernous ground. In 1998, a well drilled 48 m into the Florida limestone aquifer triggered hundreds of sinkholes within just six hours of development flushing. These were all in a woodland area where 6 m of sand covered the limestone. Though most of the new sinkholes were only about a metre across, some expanded or coalesced into depressions nearly 50 m wide. The only more disastrous impact of well drilling is where a well is left uncased or unsealed through salt beds, so that it creates a new route for freshwater to reach the buried salt and there initiate large-scale dissolution. The end result can be large new sinkholes at the surface, as occurred at Wink, Texas (Case study #11).

8.2.2 Sinkholes induced by de-watering

Whether by surface quarries or underground mines, the extraction of limestone or any minerals within it invariably encounters serious water problems in karst terrains. Deep workings require massive pumping operations, and the effect of these is to create deep cones of depression. In any soil-covered karst, subsidence sinkholes are inevitably induced over the surrounding areas of water table decline, which extend well beyond the quarry or mine limits and continue to extend further as workings reach to greater depths while having to be kept dry. Numerous deep mines in limestone have encountered massive water problems underground, but internal drainage measures have minimal surface impact where the mines lie deep beneath mountainous karst terrains. Sinkholes are mostly induced in soil-mantled karst lowlands, and the major events have occurred in South Africa (Box 8.1), eastern U.S.A. and eastern China.

America's classic case of quarry de-watering involved the Pennsylvania factory that produces Hershey chocolate (Foose, 1953, 1968). The main quarry extracted limestone from a valley side 3 km upstream of the chocolate factory, which stands on valley alluvium over the same limestone. At the quarry, the water table was about 20 m down, and massive pumping was required as workings went deeper and then continued down as a mine following the steep dip of the good limestone. De-watering removed 0.45 m^3/s from the quarry and mine, dropping the water table by 75 m. The cone of depression extended across 30 km^2 (in which head decline was >3 m), and the valley floor springs ceased to flow. A pinnacled rockhead underlies valley soils that are about 20 m deep, and the original water table was about 8 m down. As the cone of depression grew, over 100 subsidence sinkholes developed within about 5 months. They were 2–7 m across and up to 8 m deep, and most were where water table decline exceeded 15 m – which is about where it declined past rockhead (Figure 8.10). The chocolate factory stood on the 15-m contour of head decline, with the area of sinkhole development expanding across the fields

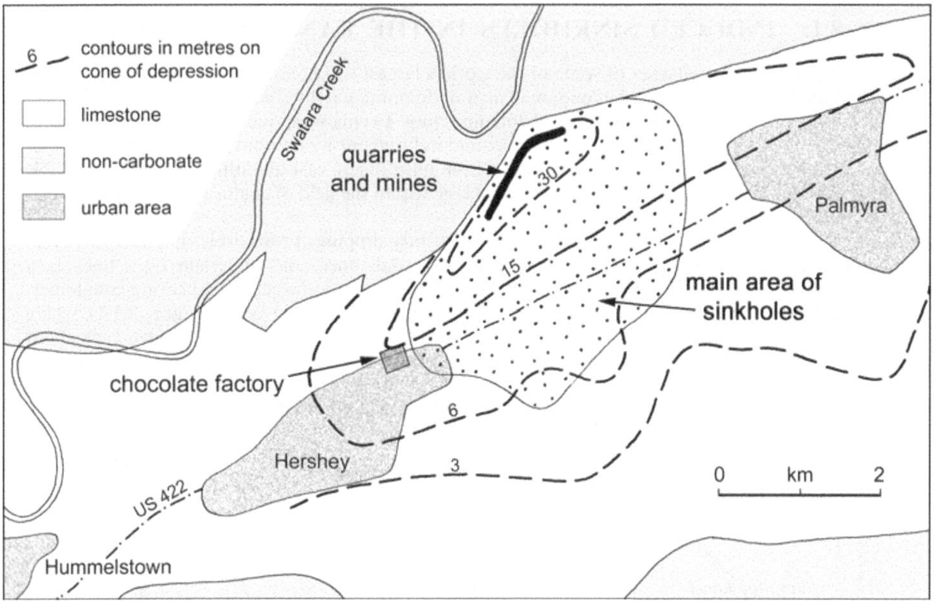

Figure 8.10. Map of the Hershey valley with the area of new sinkhole development related to the cone of depression around the de-watered quarry and mine.
After Foose (1968).

toward it, and the company resorted to groundwater recharge to inhibit the local water table decline. Legal actions ensued, and the mining company then sealed its workings behind a grout wall to isolate its zone of de-watering. This stopped sinkhole development. Mining and pumping have since ceased, so that the quarries are now flooded to the original water table, and chocolate production continues undisturbed.

The Hershey events were almost repeated 30 years later in the Dry Valley area of Shelby County, Alabama. Soils across the valley floor are only 1–6 m thick, and the original water table generally lay above the rockhead of heavily fissured limestone. Deep quarries and even deeper mining of the limestone have required de-watering at rates up to 0.9 m^3/s, which has dropped the water table by 120 m (LaMoreaux and Newton, 1986). Over 2,000 subsidence sinkholes, nearly all small, have formed in the valley soils within the induced cone of depression. Over 600 m from the mine, Highway 16 crosses the valley where water table decline has been 70 m, and sinkholes necessitated more than 30 repairs to the road within six years. The mine continued in use, and sinkholes continued to threaten the road, especially after periods of heavy rain. The advisory road signs (Figure 8.11) appear over-cautious in dry weather, but continue to offer pertinent advice during and soon after major rain events. Higher in the wooded hills of the same karst, a subsidence sinkhole collapsed and flared out to 90 m in diameter in 1972. The water table was 18 m

BOX 8.1. INDUCED SINKHOLES IN THE RAND MINING FIELD

The most catastrophic collapses of some of the world's largest sinkholes, and the greatest loss of life in any sinkhole disaster, occurred in one small area of dolomite karst in the Far West Rand district of Transvaal, South Africa, just south-west of Johannesburg. Events were due to an exceptional combination of a very mature karst, a thick soil cover and inducement by an unusually massive decline of the water table. Though the sinkholes have now been filled in (by vast quantities of readily available mine waste), the events of 30 years ago remain a classic within the field of engineering on karst (Brink, 1979; Swart *et al.*, 2003).

The Rand karst is formed in over 1,000 m of gently dipping, Proterozoic, impure, chert-rich dolomites. These are capped to the south by Karoo sandstones, and underlain by a thick lava sequence below which lie the sedimentary Witwatersrand Group with their gold-bearing conglomerates. The dolomite outcrop is over 10 km wide along the Wonderfontein valley (Figure 8.1.1), where the karst rockhead is buried beneath soils generally 10–100 m thick. Vertical syenite dykes each about 50 m thick break the dolomite aquifer into hydrologically discrete compartments, each originally drained by a valley-floor spring against its downstream dyke (Figure 8.1.1). The very old karst has matured into a complex terrain with a pinnacled rockhead, large buried sinkholes and many shallow

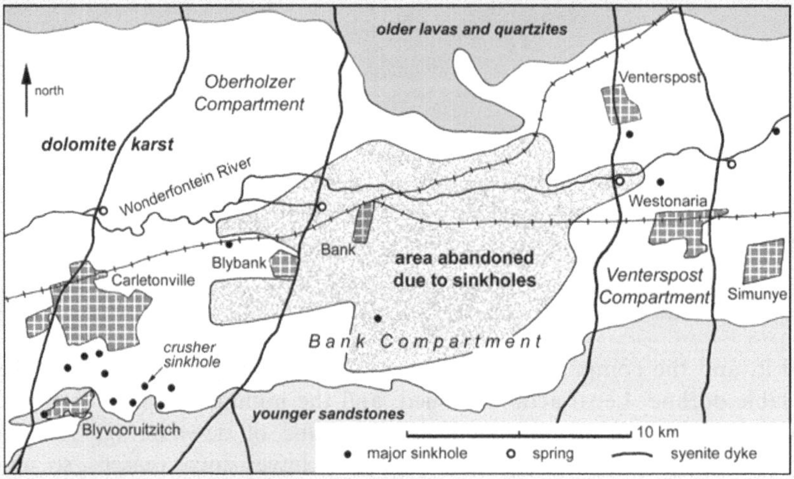

Figure 8.1.1. Map of the Rand sinkhole area.

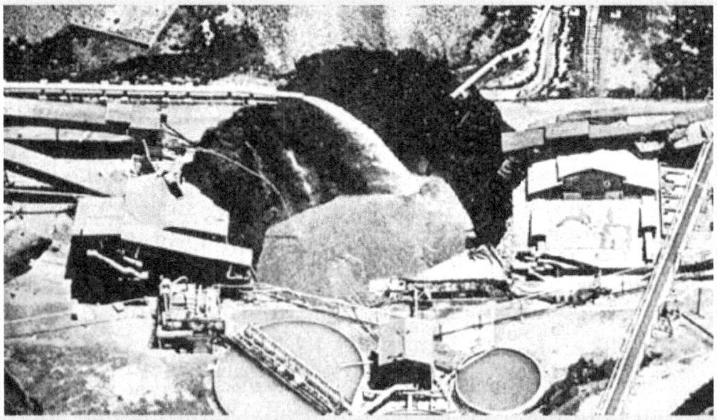

Figure 8.1.2. The subsidence sinkhole that engulfed the mine crusher plant in 1962.
Photo: Consolidated Goldfields.

phreatic caves; the original water table lay 2–26 m above the average rockhead level. Pumping records indicate that porosity in the 30 m below the original water table is up to 10%, largely in sub-vertical fissure caves, but this declines rapidly with depth to values of about 0.1% more than a few hundred metres down.

In 1934 the first successful deep mine worked the valuable gold deposits in the Venterspost Compartment, but encountered water problems as the workings extended beneath increasing areas of the karst aquifer. Massive de-watering was therefore instigated, and the calendar of events (Table 8.1.1) recorded major sinkhole development following major water table decline in each compartment as the mines extended through the goldfield. The worst single event was the sudden collapse of the sinkhole that destroyed the mine's crusher plant at West Dreifontein and killed 29 people (Figure 8.1.2). The plant had been built on a grouted soil mat where rockhead was 117 m deep in a buried sinkhole elongated along a fault in the dolomite. Over a period of three years, minor precursor movements were on a scale regarded as normal over the buried karst (De Bruyn and Bell, 2001), but suffosion must have developed significant soil voids within the fill of the buried sinkhole to allow the instantaneous collapse large enough to swallow the entire crusher building.

Six years later, the same mine was flooded to a depth of 800 m before an inrush of nearly 5 m^3/sec could be controlled by splendid emergency operations (Cousens and Garrett, 1970). A heading had reached through the Bank dyke, where stopes were opened with just 30 m of quartzite separating them from the dolomite karst aquifer, then undrained within the Bank Compartment. Fractures opened up in the de-stressed hanging wall to let the water into the mine. Stemming the flood saved the higher workings in the Oberholzer Compartment, but left those in the Bank Compartment totally flooded, and the only way to resume mining was by total de-watering of the Bank Compartment. It was appreciated that this would induce sinkholes within the compartment, but the foreseeable costs of the damage were outweighed by the value of the mineable gold. In the event, new subsidence and compaction sinkholes were very effectively induced, and a threat of major dropout sinkholes was recognised after drilling revealed large soil voids. Consequently, the Bank township was completely evacuated (Figure 8.1.1). Abandoned houses in the closed area were slowly destroyed (Figure 5.12) by differential compaction over buried sinkholes up to 70 m deep. The greatest compaction subsidence was recorded where the original water table was less than 30 m deep within the soil profile, and the prime site for new dropout sinkholes was over the steepest margins of the large buried sinkholes. The land remains undeveloped today, as the dolomite is still de-watered.

In subsequent years, new sinkholes have continued to develop throughout the de-watered buried karst. Five more people have died in dropout sinkholes around Westonaria; after each event, nearby houses have been demolished and the sinkholes have been filled – and have remained stable. By 1987, a recorded 271 sinkholes had an average volume of 9,000 m^3 – a spectacular testimony to the effects of unusually massive de-watering of a soil-covered karst.

Table 8.1.1. Calendar of events in the Rand karst followed de-watering in the three mined compartments.

Venterspost Compartment	
1957	*De-watering was largely completed*
1958	First sinkhole, 80 m across, collapsed in December
	Over the next 25 years, 165 more sinkholes occurred
Oberholzer Compartment	
1959	*De-watering was completed*
1962	29 people died when sinkhole 55 m across, 30 m deep, swallowed the crusher at West Dreifontein Mine
1963	Schutte's compaction sinkhole developed, reaching 180 m across after 3 years
1964	5 people died when a sinkhole 30 m deep collapsed beneath houses in Blyvooruitzicht
1966	*The water table had declined to 160 m below ground level*
	The largest of 8 sinkholes was 125 m across and 50 m deep in open countryside
	Over the past 4 years, 454 houses had been deomolished or evacuated
Bank Compartment	
1968	*The West Dreifontein mine was flooded where workings had breached the Bank dyke so that they were beneath the undrained karst aquifer in the Bank compartment*
1969	*De-watering was largely completed in the first 6 months*
1970	Bank township was evacuated in January, the road was closed and the railway made freight-only, after large compaction sinkholes developed and soil voids were found
1971	*The water table had declined to 300 m below ground level*
1975	The railway through Bank was reopened to passenger traffic; 8 days later a train was derailed into a sinkhole 20 m wide and 7 m deep

Figure 8.11. Road signs that warn of the ongoing hazard to a road across Alabama's Dry Valley, where sinkholes are induced in the soil by de-watering at a nearby limestone mine. TW.

down and still above rockhead, but the dropout occurred just a few months after a quarry 2 km away commenced working on a new lower level (Sowers, 1996).

Sinkholes may also be induced by de-watering on a very much smaller scale. In southern Ireland, a quarry worked limestone just 6 m deep beneath gravel soils 1–2.5 m thick (Beese and Creed, 1995). While pumping kept the excavation dry in a cone of depression, thirteen sinkholes, each up to 1.6 m across, developed along 200 m of nearby road where suffosion was induced as the water table declined below rockhead. Subsequently the quarry was flooded to form a landscaped lake, and no more sinkholes developed, even while heavy construction traffic was working on the covered karst. Dry working of the shallow quarries in the Ukranian gypsum has only lowered the water table by about 15 m, but the cavernous and nearly horizontal gypsum has been drained over distances of many kilometres (Klimchouk and Adrejchuk, 2003). A consequence has been the development of numerous new sinkholes, both subsidence sinkholes due to accelerated suffosion and also caprock sinkholes due to loss of hydrostatic support when the cave passages were drained. Sinkholes can also be induced by localised drainage into a

Figure 8.12. A dropout sinkhole 5 m deep between the railway tracks at Tai'an, eastern China, one of 24 sinkholes induced over a period of five years by water table decline in response in groundwater pumping.
Photo: Guo Xizhe.

tunnel under construction in karst; when the Nanting railway tunnel was being driven for 6 km in eastern China, 40 new sinkholes developed in the soil cover over the limestone that was draining into the tunnel (Guo, 1991).

China has an extraordinary record of inducing more than 30,000 sinkholes by mine de-watering, largely because its very productive coal seams are interbedded and capped by cavernous limestone and gypsum, notably in the provinces of Hunan and Guangdong (Yuan, 1987; Xu and Zhao, 1988). Numerous records prove beyond all doubt the correlation between de-watering the mines and development of the subsidence sinkholes that are now accepted as inevitable side-effects of safe mining beneath soil-covered karst (Figure 8.12). Induced sinkholes have damaged farmland, houses and infrastructure, and effects reach far beyond the mines. In Hunan, pumping at the Enkou mines induced 6,100 sinkholes across an area of 25 km^2, followed by another 5,800 sinkholes as increased de-watering widened the cone of depression beneath previously unaffected farmland (Li *et al.*, 1993). Also in Hunan, 7,290 new sinkholes were spread across 74 km^2 of farmland around the de-watered Liansao mines. Sinkholes that have captured surface river flows have directly caused disastrous inrushes of mud and water deep underground, so that some mines have had to be abandoned in Guangxi, in Guangdong and in Hubei (Chen, 1988). Remedial works have concentrated on minimising water flow into the mines, as it is recognised that predicting and pre-treating new sinkholes at the surface is next to impossible (Li and Zhou, 1999). Some sites have lent themselves to deep tunnels draining the mines at depth while the shallow karst is left relatively undrained, but others have required major grouting programmes, either sealing entry fissures within the mines or isolating the de-watered mines behind major grout curtains.

Though collapse of a cave's rock roof may be induced when buoyant support is lost due to drainage, there does not appear to be any recorded case where this is considered to have occurred due to artificial decline of the water table in limestone karst. The effects of water table decline have been limited to the movement of soils, and to the compaction of soft and weak limestones at some locations.

8.3 SINKHOLES INDUCED BY GROUND DISTURBANCE

Construction works inevitably involve ground disturbance, but the two principal activities by which sinkholes are induced are thinning of the soil cover – and removal of plant cover both of which increase potential rainwater infiltration and therefore increase potential suffosion. Mere access by construction traffic, which involves the stripping of topsoil and removal or destruction of the original vegetation, can be enough to initiate new sinkholes (Figure 5.15). These are most likely to be subsidence sinkholes, but may form over compacted or suffused fill in buried sinkholes. A large sinkhole developed in bare soil between the foundation trenches for a warehouse in Alabama (Newton and Hyde, 1971), while sinkholes clustered in a part of Shelby County, Alabama, were mostly in a zone where trees had been clear-felled or were close to infrastructure lines (Newton, 1987). Infiltration and suffosion increases in a soil cover that is reduced across karst, so that a road construction site in Ireland was broken by four small subsidence sinkholes after a shallow cut was excavated during a dry summer in soils 7 m thick over limestone (Beese and Creed, 1995).

Sinkhole dropouts can occur when soil arches fail due to vibration, and five new sinkholes added to the destruction wreaked by China's Tangshan earthquake in 1976. During explorations for water resources at Liangwu, in China's Guangxi karst, 157 sinkholes were triggered by blasting during engineering works, and the village had to be abandoned (Yuan, 1987). Vibrations from construction traffic could conceivably induce sinkholes in unstable soil cover, but their effects appear to be masked by the greater influence imposed by drainage modifications. At least three sinkhole collapses in Florida occurred when cars were driven over them (Metcalfe and Hall, 1984), but the events marked only the failures of the road surface under vehicle loading, after the sinkholes had been initiated previously with no influence of vehicle loading or vibration. Similarly, built structures are unlikely to induce subsidence sinkholes by their imposed loads on soil profiles. Instead, structures may cause loading compaction of the soils, so that associated ground deformation fractures buried pipelines, when escaping water may induce suffosional loss of the soil into underlying bedrock fissures. This secondary impact of broken drains under structural loading is the main cause of house damage on soft clay soils, and can be even more disastrous in soils that overlie karst.

Direct filling of an old sinkhole imposes a modest load, but creates a far more serious hazard by encouraging development across it. Within the karst lowlands of Pennsylvania, a sinkhole 60 m across was filled with rubbish and dead vegetation until the ground was level. This appears to have been a caprock sinkhole developed through a weak shale cover and already partially filled with natural soil (Figure 11.9). The site was then lost in the middle of a corn field. About 20 years later, the housing estates of Macungie extended across the farmland, and a road was built directly over the old and forgotten sinkhole. A few years after that, the sinkhole was reactivated, probably by leaking sewer and pipe lines, and the road was lost in a hole 26 m wide and 12 m deep (Dougherty and Perlow, 1988). Though the Macungie sinkhole was casually filled with rubbish, there is a direct need for planning legislation that bans the intentional but uncontrolled infilling of sinkholes by a developer who is long gone by the time that the almost inevitable subsequent collapse occurs. Some states and counties in eastern U.S.A. have such ordinances, but many have no protection.

8.4 THE AVOIDABLE GEOHAZARD OF INDUCED SINKHOLES

A college in Georgia, U.S.A., leased out part of its valley-floor land for a limestone quarry. The bedrock had 3–6 m of soil cover and the water table stood within the soil profile. Deepening and draining of the quarry depressed the water table, and subsidence sinkholes started to develop in an area expanding towards the college that was 750 m away (Sowers, 1996). However the college's royalties from the quarry were considered to outweigh the risk to their buildings. So quarrying continued with the proviso that de-watering was restricted to just the currently active part of the quarry (though this constraint would have minimal impact in a karstic limestone without installation of major grout curtains). At the same time, a small lake in the college grounds was enlarged. It almost immediately drained through new sinkholes, and the new section was subsequently filled in. By imposing both water table decline and also new surface water, this site was doomed to be peppered with sinkholes.

The Georgia site demonstrated everything that was inappropriate in a karst, and its subsequent problems were totally avoidable. However, the march of civilisation and its infrastructure expansion inevitably lead to disturbance of the natural environment, which is in so many ways all that is needed to induce new sinkholes in soil-mantled karst. In these cases, the potential costs of sinkhole damage must feature on the balance sheet for budgeting site development on karst.

Even though mine and quarry de-watering clearly has massive impacts on sinkhole development in soil-covered karst, the lateral extent of the zone of influence may be open to debate at any one site until the complexities of the local karst hydrology are proven by dye-tracing, and are mapped and fully understood (see Box 8.2). Debate may be significantly protracted where new sinkholes can have serious legal and economic implications. At many karst sites, sinkhole damage may

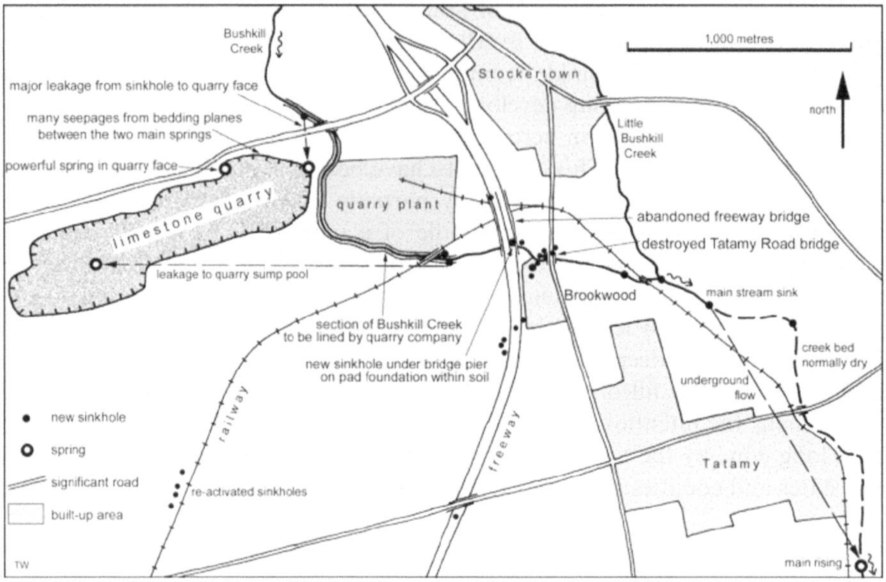

Figure 8.2.1. Outline map of the recent sinkholes and related subsidence features in the Bushkill Creek area, Pennsylvania; there are many other smaller and/or ephemeral sinkholes along the entire length of the creek covered by this map.

Figure 8.2.2. The bridge that carried Tatamy Road over Bushkill Creek, destroyed by an expanding and ever-changing cluster of sinkholes.
TW.

BOX 8.2. DISPUTED SINKHOLE ORIGINS IN PENNSYLVANIA

In the karst of eastern Pennsylvania, north-east of Allentown, a cluster of new sinkholes has destroyed road bridges and threatens a small rural community. Remedial action is delayed because the local karst hydrogeology is not fully understood and consequently the exact processes that induced the new crop of sinkholes are open to debate.

A nearby limestone quarry stays dry by pumping out water at a mean rate of $2\,m^3/s$. The quarry floor is about 50 m below the level of the adjacent small river, Bushkill Creek (Figure 8.2.1), which now loses all its water into its bed until it is replenished by the quarry pumping outfall. Water pours out of at least two open cave passages in the quarry wall, and also seeps out along certain bedding planes. A cone of depression (mapped at $5.6\,km^2$ but probably now larger) created by the pumping has caused a number of sinkholes to appear around the quarry. One sinkhole in the river bed downstream of the quarry diverted a major flow into the far end of the quarry, and was soon sealed, as the quarry pumping was merely re-circulating the same water. The quarry company plans to line about 1,200 m of the river bed to prevent new leakages, but 80 sinkholes have been recorded along 2.5 km of the Bushkill Creek.

Of greater concern are the more distant new sinkholes that started to develop in October 2000, in and around the Brookwood community beside the river a kilometre downstream of the quarry (Figure 8.2.1). Sinkholes opened in the river bed, damaging the Tatamy Road bridge and also removing much of the garden of one of the houses (Perlow, 2003). These initial sinkholes were filled with soil and rock but more have developed since. Three years later, the eastern abutment of the road bridge had been totally destroyed by an ever-changing cluster of four sinkholes adjacent to each other (Figure 8.2.2). The bridge originally stood on pads founded in the alluvial gravel that has since been washed into the limestone. An impermeable liner for more of the downstream river bed is being considered to provide long-term stability for a new bridge. No homes have yet been destroyed, but some residents have moved out because they have justifiably lost confidence as more sinkholes have developed around them.

In 2001, the abutment of the railway bridge over Bushkill Creek close to the quarry, was rebuilt after it was lost into a sinkhole. A freeway, that passes between Tatamy and the quarry, crosses the creek on two large bridges. Early in 2004, the east-side bridge (carrying the north-bound highway) sank 150 mm when a subsidence sinkhole developed under one of its southern piers. Nearly $100\,m^3$ of concrete was poured into the hole with no effect, except that the pier sank further and the bridge was abandoned; it had stood on pad foundations within the soil profile, and was therefore destroyed by a single subsidence sinkhole. At a cost of $6 million, a replacement bridge on deeper foundations was opened later in the year. Just 15 m to the west, the southbound bridge rested on piers that stood partly on bedrock limestone, but one of them was also damaged by subsidence later in 2004. Probing revealed depths to bedrock of 9–42 m, and this bridge was also closed for rebuilding on deeper footings.

It would seem reasonable to ascribe these new sinkholes to induced development within the quarry's cone of depression, but this is currently in dispute. Increased urbanisation since the 1970s, the realignment of stream channels, and even the construction of the freeway, have been proposed as causes of (or contributors to) the rash of new sinkholes. The river flow is lost into various sinkholes along, around and downstream of the road and freeway bridges, so that the channel becomes dry, and stays so as far as risings 2 km further downstream. The lower part of this underground stream channel may lie outside any area of water table decline around the quarry, but the destination of water sinking nearer to Brookwood is currently unknown. More data on levels of the water table are needed to determine the cause of the sinkholes. A dye trace showing that water from the Brookwood sinkholes flows back to the quarry would point a finger at the quarry pumping, though a negative result would still be inconclusive.

have to be budgeted as an essential cost of quarrying or mining. This could compare to the costs of obligatory repairs to subsidence damage that were accepted at 10–15% of the total production costs by longwall mining for coal in Britain. This could preclude quarry or mine development near to high-value sites, in the same way that Britain's deep coal mines could not afford to work beneath urban areas. It is notable that, since the case was taken through the courts, Hershey chocolate is still produced in Pennsylvania, while the local quarries and mines have closed. Widespread damage from induced sinkholes may have been tolerated by the remote officialdom that once made the decisions in China, but values have since changed, where fields once worked by peasant farmers have been replaced by new industrial plants within an expanding enterprise economy. This is reflected in the massive research effort by China's industry, academia and government, which has had considerable success in understanding and, more significantly, containing or reducing sinkhole damage (Li and Zhou, 1999; Lei *et al.*, 2001).

9

Ground investigation in sinkhole terrains

In karst or any other terrain, a thorough site investigation precedes construction to assess the suitability of locations and appropriate designs for buildings and engineered structures; it involves the acquisition of all necessary information on the characteristics of the sites relevant to design, construction and the security of neighbouring land and structures (British Standards, 1999). Each investigation should be designed to meet the requirements of the building or construction to be carried out. A preliminary stage of the investigation involves a desk study and reconnaissance survey; this is followed by the main stage of detailed field exploration and ground investigation; data review then continues during the construction activities when ground excavations expose more details of the ground conditions.

In karst terrains, prior to any development and construction operations, a geohazard assessment of the possibility of sinkholes or subsidences occurring at any specific sites is necessary to determine its overall suitability for development (Chapter 10). Where a site is designated suitable, this assessment should help evaluate the risk of damage occurring to any of the buildings or structures that are erected subsequently. It should also help in design of any precautionary or mitigating measures that are required to reduce or eliminate this risk. However, an accurate assessment of the likelihood of sinkhole development is usually difficult where there is incomplete data relating to the potential sinkhole processes. Karst ground conditions are so highly variable that every site on karst can be regarded as unique. An overall description of karst ground conditions at a given site might prove of value in terms of the scale of anticipated foundation difficulties, but a full description should consider not only the karst class (Figure 2.11) but also the mean sinkhole density, typical cave size and rockhead relief (Waltham and Fookes, 2003).

Particularly important in sinkhole terrains, a feasibility study should be carried out before any development plans are drawn up, and this must evolve into a full ground investigation prior to final layout of a site and the design of its buildings and structures. A ground investigation in karst should not only attempt to determine the locations of any voids or caves in the ground, but should also determine the properties and character of the relevant soil and rock masses, the rockhead configuration and the hydrogeological conditions. Rock structure is important as dissolution voids are normally enhanced along fracture zones and at the intersections of discontinuities, while soil properties can indicate the susceptibility and characteristics of potential subsidence sinkholes (Figure 9.1). As sinkholes frequently make their appearance after periods of heavy seasonal rainfall or prolonged water table decline, long-term information on local meteorological conditions should be gathered, as should data on the location and status of water pipelines and drains. In terms of geotechnical engineering, the depth and relief of the carbonate rockhead may influence excavation and foundation design. The final evaluation also has to identify any restrictions on land use and the type of development that is suitable. Examination of ground conditions should continue during excavation and foundation works, as many of the details and peculiarities of the karst ground are unlikely to be revealed by cost-effective site investigation.

9.1 PRELIMINARY STAGES

A desk study is the first stage in gathering data for a site investigation. Its purpose is to make an initial assessment of the ground conditions and to identify, if possible, any potential geotechnical problems (Herbert *et al.*, 1987). The desk study includes a search for, and review of, appropriate maps, documents, archival records, literature, imagery and photographs relevant to the area or site concerned (possibly including those gathered on the initial site visit), to ascertain a general picture of the existing ground conditions prior to field investigations. This begins the process of constructing an adequate geological model for the site, presented in one or more conceptual block diagrams (Fookes, 1997). The model should present all relevant aspects and terrain components within the karst, and may appear comparable to any one of the diagrams in Figure 2.11, but will normally have more details that are site-specific. Subsequently, and dependant on potential interaction between the proposed construction and the geological model, a ground investigation will be designed and implemented. Alternatively, a desk study can be undertaken to determine the factors that affect a proposed development, as an aid to feasibility assessment and project planning. In all cases, the terms of reference for a desk study need to be defined clearly in advance of its implementation. The amount of effort expended in a desk study should relate to the type of project, the geological and geotechnical complexity of the area or site, and the availability of relevant information.

A desk study for the planning stage of a project can encompass a range of appraisals from the preliminary rapid response to the comprehensive statement. There are some common factors within this spectrum that always need to be taken into account. Whether preliminary or exhaustive, an appraisal report should include a factual and interpretative description of the surface and geological conditions, information on previous site usage, a preliminary assessment of the suitability of the site for the planned development, an identification of potential constraints, and provisional recommendations with regard to ground engineering aspects. However, a desk study is a component of a site investigation, and should not be regarded as an alternative to adequate ground exploration prior to a construction project.

During or at around the same time as the desk study, preliminary work should include a site inspection that constitutes a reconnaissance or a walkover survey of the ground. This involves noting, where possible, distribution of soils and rocks, surface relief, surface drainage and associated features, locations and dimensions of any actual or likely sinkholes, ground cover and obstructions, and any signs of earlier uses of the site such as tipping or previous construction. The inspection should not be restricted to the site, but should examine adjacent areas to see how they affect or will be affected by construction on the site in question and also to recognise features significant to the concepts of karst development.

As water movement is the main process behind the development of subsidence sinkholes, it is essential that the groundwater conditions are properly understood at any potential development site on karst. Much of this understanding will normally develop during the preliminary stage from a thorough desk study and a perceptive walkover survey. An effective site investigation must determine the depth to the water table, its relationship to rockhead and how this changes with

Figure 9.1. New subsidence sinkholes in thick soils are the most widespread hazard in karst terrains, and the likelihood or potential for their development is one of the prime tasks of ground investigations in karst.
TW.

time in relation to rainfall, seasons and any abstraction. It may also need to estimate the direction and scale of groundwater flow, and perhaps the chemistry of the groundwater.

The ultimate importance of the preliminary investigation is that it should assess the suitability of a site for any proposed works. If the site appears suitable, the data from the desk study and the walkover survey will form the basis upon which the site exploration is planned. The walkover survey also allows a check to be made on some of the conclusions being developed within the desk study.

9.2 GROUND INVESTIGATION FIELDWORK

Investigation methods fall into two groups, those that are intrusive (probing, augering, boring, drilling, pitting, trenching, sampling and testing) and those that are non-intrusive (geophysics and aerial or satellite remote sensing). Some extent of drilling and sampling is a component of almost every ground investigation. It is employed most effectively when combined with, or following up, comprehensive desk study and appropriate non-intrusive investigations, especially in the complex, variable and unpredictable ground conditions that typify karst (Section 9.6).

The use of most remote sensing imagery and aerial photography is restricted where sinkhole subsidence features may be just a few metres across, but satellite imagery is becoming increasingly sophisticated, including radar measurement of millimetric ground movements in urban areas (Section 9.4). Over the past thirty years, the use of geophysical surveys has developed considerably for the location and delineation of voids and bedrock surfaces (Section 9.3). However, no one geophysical method has yet been developed that resolves all the problems of sinkholes and cavities in karst terrain. A variety of surface traversing techniques provide readings at close station intervals, mostly for the location of shallow voids with lateral dimensions that exceed the depth of burial. Borehole to borehole geophysical methods can be particularly useful in determining the shape and dimensions of open or infilled voids, and there is continuous evolution of useful new techniques, but cost is increased where they rely on the drilling of boreholes (Section 9.3.8).

Hydrogeological investigations may continue into the fieldwork stages of a site investigation in karst. Depth to the water table can be refined from observations in investigation boreholes, which subsequently may need to be screened if they are to be used for monitoring purposes. Multiple monitoring points are required to determine the direction of flow by constructing groundwater level contours, where flow is approximately in the direction of the steepest gradient. Groundwater movement can also be monitored by use of tracer dyes, including those that are collectible in sub-visible concentrations and fluoresce under ultraviolet light. Monitoring may be from boreholes or sinkholes to others of the same or to one or more springs. The design of a groundwater dye-tracing programme needs to be carried out by a specialist, as the results can be extremely complicated in karst terrain (Quinlan and Ewers, 1989). It is a characteristic of karst aquifers that flow is through discrete conduits, and flow destinations may change significantly where high-level conduits become active at high stage, perhaps generating flow to different suites of springs during summer and winter (Crawford and Ulmer, 1993). The selected dye, the locations, timing and methods of dye injection, the sampling strategy used and the analytical methods used are all critical to the success of a tracing programme. The results can be especially critical where there is the potential for underground transmission of pollutants, notably from stormwater run-off from new highways across karst (Bednar and Aley, 2001).

9.3 GROUND-BASED GEOPHYSICAL SURVEYS

One of the major advantages of geophysical investigations over intrusive explorations is that information is obtained for much larger volumes of ground at lower cost (McDowell *et al.*, 2002). This is an important consideration in sinkhole terrains, because the probability of finding a small target sinkhole or cave within a large volume of ground is very low using point-sampling methods. The probability of finding a target of $10\,m^2$ using 15 sampling points in a site of 0.5 ha is 3%, and this falls to 1.7% with 85 sampling points in a site of 5 ha (Hobson, 1992). That example is essentially 2-D, so uncertainty is further increased when the vertical dimension also has to be considered. However, geophysical surveys are not a replacement for drilling boreholes within ground investigation for engineering projects. They should be viewed as complimenting the boreholes, and perhaps guiding the borehole locations. Geophysical surveys are valuable because they provide an overview of ground conditions, of areas that may be small in specific applications, but are still large compared to the $0.005\,m^2$ of a site area that is examined in a typical borehole. Because the gathered geophysical data relates to variations in the physical properties of a volume of ground as a whole, it must be processed and then interpreted in the light of a previously created conceptual ground model. Data processing has been vastly improved by modern computer capacity and software, and has been responsible for the major recent advances in geophysical applications. However almost all geophysical surveys still require confirmation by drilling into their detected anomalies (ground control, or ground truthing).

In general, geophysical methods involve the identification of anomalies – where there are spatial changes in physical properties. These changes may relate to changes in the soil or rock (in lithological variations, structure or fracture densities), or to extreme anomalies (including voids wholly or partially infilled with air, water or soil), or to changes with time caused by groundwater movement (including the growth of pollution plumes). Whether or not a particular geophysical method is inherently capable of detecting a change in physical properties is dependent upon a number of factors:

- the required depth of penetration into the ground;
- the vertical and horizontal resolution required for the anticipated targets;
- the contrast in physical properties between the target and its surroundings; and
- signal-to-noise ratio for the physical property being measured at the site.

As an example, a spherical, air-filled cave 2 m in diameter would be detected by a gravity survey if buried at a depth of 2 m, but not if it were at a depth of 10 m (McCann *et al.*, 1997). The magnitude of the anomaly diminishes with depth, and in this case is close to measuring accuracy of the instrument where the cave lies 10 m deep. If the cave is infilled with soil having a density close to that of the surrounding rock, the cavity would probably not be observed at all because the gravity anomaly would be below the resolution of the instrument. Environmental noise can also reduce the effectiveness of geophysical methods; seismic measurements might be difficult to make near to a busy highway due to vibrations from the traffic, and electromagnetic surveys are affected by proximity to buried, or overhead, electrical transmission cables.

Selection of the most appropriate geophysical method, or methods, for the detection of a likely karst cavity relates to various factors (McCann *et al.*, 1987):

- The physical properties of the cavity and the surrounding rock should be known to within an order of magnitude, so that the contrast in physical properties can be assessed. The necessary data may be available from the literature or from initial site investigation boreholes.
- Other effects due to the presence of likely cavities such as changes in drainage patterns should be considered. In such cases, the altered properties of the rock mass are likely to be detected.
- When the depth of burial of the cavity is more than two to three times its diameter, surface methods may not work and cross-hole techniques are likely to prove more useful.

Two examples of the selection procedure are presented in Table 9.1.

While selection of the most appropriate geophysical method, or methods, is important, this aspect forms only a part of the planning and execution of a geophysical survey as part of the overall site investigation. Too often, geophysical investigations have failed to satisfy the expectations of the engineer, not because geophysical techniques are inherently poor, but because they have been wrongly applied or poorly managed. Fortunately, the complexities of geophysical science have reached a stage where nearly all engineering geophysics is carried out by specialist sub-contractors, but it is still important to have the appropriate team involved at all stages, from planning through execution to reporting, of a site investigation that involves a geophysical survey.

9.3.1 Geophysical methods

Geophysical techniques can be divided into two principal types:

- Passive geophysics, that make use of the earth's inherent physical properties – its gravitational, magnetic, electrical, electromagnetic and thermal fields.
- Induced geophysics, that utilise artificial sources whereby signals are transmitted into the ground from seismic, electrical or electromagnetic sources.

Table 9.1. Assessment of the most appropriate geophysical methods for cave detection. After McCann *et al.* (1987).

Geophysical method	Example A	Example B
	Air-filled cave, 5 m in diameter, at a depth of 5–10 m, in dry limestone above water table	*As in Example A, but in seasonal wet temperate climate, under clayey alluvium 1–2 m thick*
Electrical resistivity	Very little resistivity contrast	* Should be a large contrast in physical properties due to moisture in the limestone and drainage in the alluvium. Should detect cave by resistivity array; and delineate rockhead under alluvium by low-frequency electromagnetic survey
Seismic	*P*-wave surveys may be limited by attenuation. *S*-wave surveys possible but the wavelengths may be too long	* Closely spaced *P*-wave seismic refraction should show velocity and amplitude perturbations. Wave lengths for *S*-wave refraction may still be too long
Gravity	* May be a detectable anomaly if the host rock is homogeneous	Variation in overburden thickness may obscure any anomaly due to the cave
Ground penetrating radar	* Penetration of radar pulses would be >5 m and the cavity may be resolved	Radar pulse would be highly attenuated in the alluvium and saturated limestone
Magnetic	Only detectable if cave is part of old mine workings, with iron or brick debris	As for Example A

* Methods most likely to detect the cave under the specified conditions.

In both cases, the geophysical survey measures the vertical and/or lateral variation in a physical property in the ground. The data gathered must then be interpreted in terms of the ground conditions that are likely to give rise to the measured data set. A small void near the surface may create a gravitational anomaly of the same magnitude as that created by a larger void at greater depth. A conceptual model of the ground may help to resolve the interpretation. Alternatively, a more sophisticated data analysis, perhaps of an increased data set, may be able to distinguish the anomalies on the basis of their wavelength and profile revealed by Fourier analysis.

Table 9.2. Usefulness of geophysical methods for the detection of cavities.
After British Standards (BSI) (1999) and ASTM (1999a).

Geophysical method	Usefulness of method		Physical properties measured
	BSI (1999)	ASTM (1999a)	
Seismic refraction	1	B	Seismic velocity; largely related to variations in rock mass strength
Seismic reflection	2		
Cross-hole seismic	3		
Electrical resistivity sounding	2	B(A)	Electrical resistivity or conductivity; related to variations of porosity, fluid conductivity, degree of saturation
Electrical resistivity profiling	3		
Induced polarisation (IP)	0		
Electromagnetic profiling (EM)	3	A	
Ground probing radar (GPR)	3	A	*Same as electrical*
Gravity and microgravity	2	A	Density; related to lithology and fissuring
Magnetic	1		Magnetic field of ground materials
Downhole self potential	1		*Same as electrical*
Downhole resistivity	0		
Downhold neutron/gamma logs	0		Radioactivity; porosity, density, moisture
Downhole fluid conductivity	2		*Same as electrical*
Downhole sonic velocity	2		Seismic velocity (*see above*)

0 = not applicable; 1 = limited use; 2 = used, or could be used, but not the best approach, or has limitations; 3 = excellent potential but not fully developed.
A = primary method; B = secondary method.

It is essential that the geophysical interpretation be calibrated against information from previous investigations, boreholes and other sources, and efficiency is greatly improved if the survey is correctly targeted on the basis of an adequate geological model.

The usefulness of different, commonly available, geophysical methods can be summarised with regard to ground cavity detection, excluding lava tubes (Table 9.2). Within an overview of geophysical surveys in site investigation, none was generally considered as "excellent, with the technique well developed" for the specific task of cavity detection (British Standards, 1999). Overall, the most useful methods applicable in limestone karst are cross-hole seismic, microgravity, resistivity or

Table 9.3. Recommended methods for the geophysical location of specific dissolution features in karst.
After McDowell *et al.* (2002).

Karst feature	Dimensions	Recommended methods	Factors to consider
Pipes and hollows, with clay fill	Depth : diameter < 2 : 1 Depth < 30 m	Conductivity traversing Magnetic	Coil separation, *cf* depth Local magnetic gradient
Pipes and hollows, with sand fill	Depth < 5 m	Ground penetrating radar	Conductivity of cover and fill, and cover thickness
Small open caves	Depth : diameter < 2 : 1 Depth < 30 m Depth > 30 m	Conductivity traversing Microgravity Cross-hole seismic	Coil separation, *cf* depth Density and nature of fill Borehole spacing
Large open caves	Depth < 10 m Depth >10 m	Ground penetrating radar Conductivity traversing Gravity and microgravity Cross-hole seismic	Ground conductivity Cavity infill Cavity infill, terrain relief Borehole spacing

conductivity profiling and ground penetrating radar. Some other methods could be used but may have serious limitations. Most of the same methods were recommended for cavity detection twenty years ago (Owen, 1983), except ground probing radar that was not then well developed. Other methods generally are considered to be inappropriate for cavity detection. More detailed guidance has been provided on the suitability of geophysical methods to locate dissolution features that include both caves and soil-filled pipes (Table 9.3). The principles that lie behind each of these methods, including theory, instrumentation and data processing, are considered in detail in available publications on geophysics (Telford *et al.*, 1990; Hoover, 2003; Reynolds, 2005).

9.3.2 Surface seismic surveys

Surface seismic methods involve measuring the velocity of transmission of vibrational energy from a hammer, falling weight, air gun, explosive or other similar source to an array of geophones, usually placed in a line across the area of interest. The calculated seismic velocities are functions of the density and elastic properties of the transmitting soils, rocks or rock masses, and are therefore broadly indicative of strength. Intact rock, fractured rock masses and weak soils are readily distinguished. Repeated measurements at the same site create strong signals that stand out from random noise, but seismic surveys may not work well in urban areas or on sites where heavy equipment is being used. The transmitted signal may arrive at the geophones via a number of routes depending upon the elastic properties of the rocks and soils and the position of the water table, having travelled along the ground surface or by a range of refracted and/or reflected paths through multi-layered ground structures.

Surface seismic refraction methods have a depth of penetration around one-third of the geophone spread (Hoover, 2003). Generally on their own they are unlikely to detect limestone pinnacles, steep-sided buried sinkholes and voids in bedrock, unless features are near the surface and greater than about 6 m across (McCann *et al.*, 1987). However, they may successfully identify the profiles of sinkholes that have flatter sides and a large velocity contrast between the rock and the infilling soil, as where soft alluvium overlies strong limestone; they provided excellent results in the investigation of features in the chalk at the Mundford site, U.K. (Grainger *et al.*, 1973). Also, rockhead pinnacles have been profiled where a pilot conductivity survey enabled the seismic lines to be located directly across the suspected pinnacles (Jansen *et al.*, 1993).

Seismic reflection methods that use a high frequency source may detect cavities at greater depths. This use of surface seismic surveys for cavity detection and rockhead mapping is a relatively new field and only a few experiments have been carried out (Luke and Chase, 1997; Harrison and Hiltunen, 2003).

9.3.3 Electrical resistivity surveys

Electrical geophysics measures the resistance of the ground to the passage of an electric current. Resistivity is increased, or conductivity is decreased, by the presence of air-filled voids, but opposite characteristics are created where bedrock voids are filled by wet clay soils. The objective, therefore, is to identify and interpret areas of anomalous apparent resistivity, but surveys may not work well in developed sites where buried metal or electrical cables are present.

A resistivity survey is carried out by placing electrodes in or on the ground surface. Usually, a current is passed between two input electrodes while the induced voltage is measured between two others. The ratio of voltage to current gives the resistance and the apparent resistivity is derived by multiplying this by a factor that accounts for the electrode spacing. Modern equipment allows multiple electrodes to be placed on a grid, where sequences of input and measurement electrodes can be selected. Depth profiles are produced by increasing separation of the measuring electrodes, lateral variations are mapped by traversing with constant electrode separation, and a combination of measured patterns produces an apparent resistivity image along a section through the ground. There are many variants on the electrode configurations used.

Electrical surveys may have poor resolution that is no better than around 10% of the depth (McDowell *et al.*, 2002). On karst, they cannot readily distinguish between individual large dissolution features and zones of ground broken by multiple narrow fissures, as is demonstrated by the variable situations revealed by drilling into identified anomalies. Perhaps more significantly, a zone of hazardous dissolution cavities, where some are open and dry while others are filled with clay, may not create an anomaly because the electrical survey lacks the resolution to identify the individual features with opposing resistivity characteristics.

Resistivity surveys have been used to locate buried and incipient sinkholes in soils overlying the chalk in southern England, and there has been variable success with different electrode arrays in different situations (Case study #9). A site with 1–10 m of silty clay overlying limestone in Pennsylvania was surveyed with a traversing dipole–dipole electrode array, where success was influenced by the orientation of the electrode array, by the electrode spacing and by the line spacing (Roth *et al.*, 2002). Voids in the limestone were not detected, and anomalies associated with sinkhole formation were not clearly defined, but the rockhead surface could be mapped with moderate accuracy. It is clear that the selection of electrode array and its spacing requires detailed understanding of the various methods and how these will affect the results from any particular site. Some knowledge of the site conditions, competent interpretation and appropriate boreholes for ground truthing are all essential to electrical surveys.

9.3.4 Electromagnetic conductivity surveys

Ground conductivity surveying involves the energising of a transmitter coil with an alternating current, so that its generated electromagnetic field induces small currents in the ground, which are then sensed by a receiver coil located a fixed distance away. It is described as non-contacting because it avoids the use of ground electrodes. Coil spacing and operating frequency are selected so that a direct reading of the apparent ground conductivity is obtained. Depth penetration of 6 m is achieved with a coil spacing of 4 m, but depth can be increased to about 30 m by increasing coil separation. Electromagnetic conductivity traverses can be carried out very rapidly, as a single instrument with a 4 m coil spacing can be operated by one person carrying it in use. Equipment with larger coil separations is more efficiently operated by two people. The method is most appropriate on undeveloped ground, as electrical cables, wire fences and most buildings can provide interference, reducing or distorting the signal. The output of a survey is a conductivity map. Positive or negative anomalies may be correlated with the location of buried or incipient sinkholes, depending upon the nature of any infill material; clay has a higher conductivity than sand, and most limestone has very low conductivity. Soil moisture increases its conductivity, and sinkholes may be wetter where they collect drainage or drier where they efficiently drain the soil. Data interpretation compares to that of resistivity surveys, but the method cannot be extended to greater depth penetration.

A pilot conductivity survey used vertical coils with a separation of 10 m to attempt mapping anomalously shallow rockhead and buried pinnacles at a site in Wisconsin where dolomite is overlain by 6–12 m of clay-rich, residual soil (Jansen *et al.*, 1993). Profile lines were at 15 m separations with every 10 m along each line, on a grid that was designed as a compromise between cost and the likelihood of detecting the anticipated anomalies. Some areas of low conductivity were found and were interpreted as shallow or pinnacled rockhead (Figure 9.2), and some of these were subsequently proved by drilling. However, it was decided that the grid spacing was too coarse for the final survey, so the profile and station separations should be halved and different coil separations should be used to try to locate pinnacles more accurately.

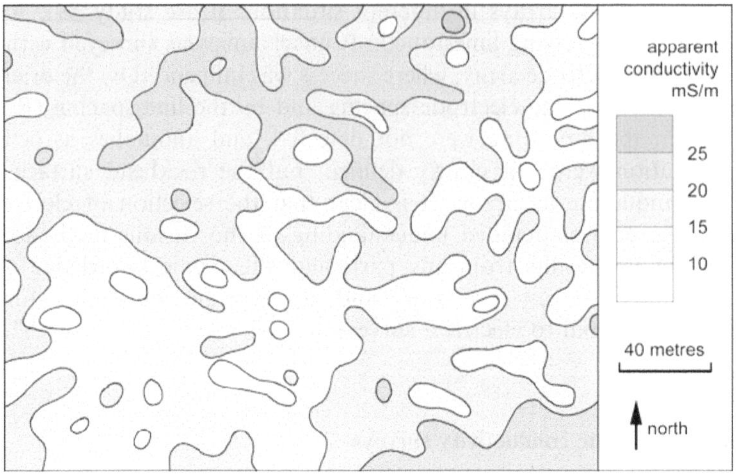

Figure 9.2. Apparent conductivity mapped over a site 250 m by 200 m in the covered karst of Wisconsin; areas of low conductivity, with light shading, are over shallow rockhead and dolomite pinnacles, while high values, shaded dark, relate to deep soil cover and buried sinkholes.
After Jansen *et al.* (1993).

9.3.5 Ground penetrating radar

The application of ground penetrating radar (GPR) involves the transmission of short pulses of high-frequency electromagnetic energy (25–1,000 MHz) into the ground through an antenna. Variations in the ground's electrical impedance produce reflections that are detected at the surface by the same or another antenna. A survey may trace a single line, as along a highway where the equipment can be conveniently towed behind a slowly moving car, or may cover a grid pattern of traverses. Variations in electrical impedance are mainly due to variations in the dielectric constant of the ground. Reflection of the input electromagnetic energy takes place where there are impedance contrasts. The radar signal is attenuated more in wetter materials that have higher conductivity, where depth penetration is therefore reduced. Similarly, clay soils have lower electrical impedances, and generally limit depth penetration to 6 m where dry or to only 2 m where saturated. The limited depth of penetration is one of the main drawbacks of GPR, though it is not always necessary to penetrate to bedrock; soil disturbance by movement or arching at shallow depths, that may precede development of a subsidence sinkhole, can create anomalous radar reflections that are identifiable. Soil cavities were detected at depths of 1 m in gravel overlying chalk in southern England, but the GPR could not detect voids at greater depths, probably due to the wet conditions (Case study #9).

In contrast, depth penetration reached 7 m in dolomitic limestone beneath a road in north-east England (Cuss and Beamish, 2002), and radar surveys have reached depths of 30 m in dry sandy soils in Florida. In profiling a site in central

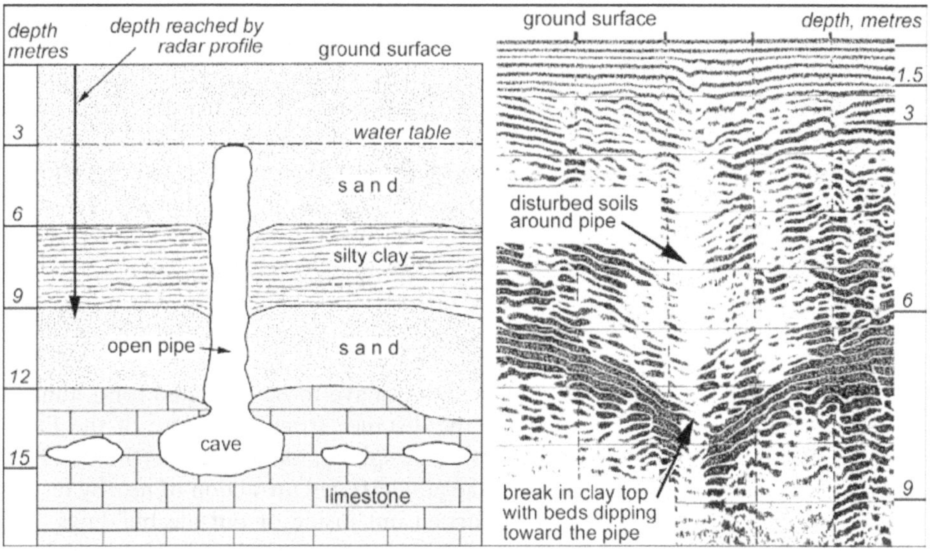

Figure 9.3. Profiles of pipes developing through soil over limestone in North Carolina; on the left, a conceptual geological section; on the right, an image from ground probing radar that could not reach below the clay; note that the ground section reaches deeper than the radar profile, on which the vertical scale is not linear, as it is time-dependant.
After Benson and Yuhr (1987).

Florida, where silty to clayey sands overlie 3 m of clay on top of rockhead 12 m deep on thick limestones, a GPR survey was able to identify both buried sinkholes and potential cavities in the limestone (Stangland and Kuo, 1987). At a site in North Carolina, limestone rockhead lies at a depth of about 12 m, but is overlain by a shelly sand and then by a silty clay with its top surface at a depth of about 6 m, beneath more sand (Benson and Yuhr, 1987). Strong reflections were only obtained from the top of the silty clay, but this allowed identification of small vertical piping features by depressions of this boundary and by disturbance of the overlying sands (Figure 9.3).

9.3.6 Microgravity surveys

Gravity and microgravity involves the measurement of small changes in the Earth's gravitational field that are caused by localised changes in soil and rock mass and density. They are particularly valuable investigations of karst, because negative anomalies represent "missing mass" which can then be interpreted either as open or water-filled ground cavities or as caves or sinkholes filled with soils of lower density than the surrounding rock. Measurements are made using extremely sensitive gravity meters, normally at a sequence of locations on a predetermined grid.

Early gravity surveys had very low resolution and were only applicable to large structures. The classic example in karst was the mapping of the very large buried sinkholes in the South African dolomite, whereby negative anomalies hundreds of metres across were, and still are, regarded as zones of hazardous ground (Kleywegt and Enslin, 1973). Subsequently, improved instrumentation and hugely refined computer analysis of the data has allowed the evolution of much more sensitive microgravity surveying. Stations spaced as closely as 1.5 or 2.0 m have been used in microgravity surveys, and yield increased benefits in that they provide a data bank from which cavity depths can be interpreted from the anomaly profiles. A gravity survey of a residential area in Kuwait used readings on a grid spacing of 7 m, as housing units were 14 m wide and readings could then be taken both inside houses and in their gardens (Bishop *et al.*, 1997). The search was for incipient sinkhole structures in 35–40 m of gravels and sands overlying the Dammam Limestone, but measurement stations on a 3-m grid were required in the areas of recorded sinkhole collapse.

Gravity measurements made at each station have to be corrected for a number of factors, including elevation (because the distance from the centre of the Earth varies), location (because the Earth is not a true sphere), ocean and Earth tides, drift in the calibration of the instrument and the gravitational attraction of nearby terrain features. Microgravity surveys can be carried out inside or outside buildings, and also in areas where electric cables and metal conductors limit the use of electrical and electromagnetic surveys. Along with GPR, they offer the only practical method for investigations in most urban environments. However, gravity surveys can become impracticably complicated by the excessive relief corrections that may be needed in mountainous regions.

A gravity survey was the best method of assessing flooded cavities beneath a limestone platform on Grand Bahama prior to grouting to stabilise the ground for construction of a container terminal (Case study #10). On a smaller scale, microgravity traverses around and beneath a building in Bowling Green, Kentucky, revealed the causes of structural distress arising from suffosion of the soil mantle into the karstic limestone bedrock 10–15 m down (Figure 9.4). The ground profile interpreted from the gravity data was confirmed with boreholes, and remedial grouting to fill the voids and compact the soils was directed to the negative gravity anomalies (Crawford *et al.*, 1999). A buried sinkhole in the gypsum at Ripon, England, was detected by a gravity low of −70 μgals. It was found in an area where bedrock is generally 11–14 m deep, and drilling encountered a sinkhole fill of loose sands, silts and clays that reached a depth of more than 40 m (Patterson *et al.*, 1995). However, there is no guarantee that a gravity anomaly will necessarily relate to a sinkhole. At a site with numerous fresh sinkholes in soils over strong limestone in North Wales, U.K., gravity anomalies were found, and were coincident with seismic refraction anomalies. However, drilling intersected only massive limestone, and the anomalies were thought to relate to either rockhead undulations or to variable lithologies in the drift cover (Nichol, 1998).

9.3.7 Magnetic surveys

Magnetic measurements record local variations and distortions in the Earth's magnetic field caused by the presence of underlying rocks with different magnetic

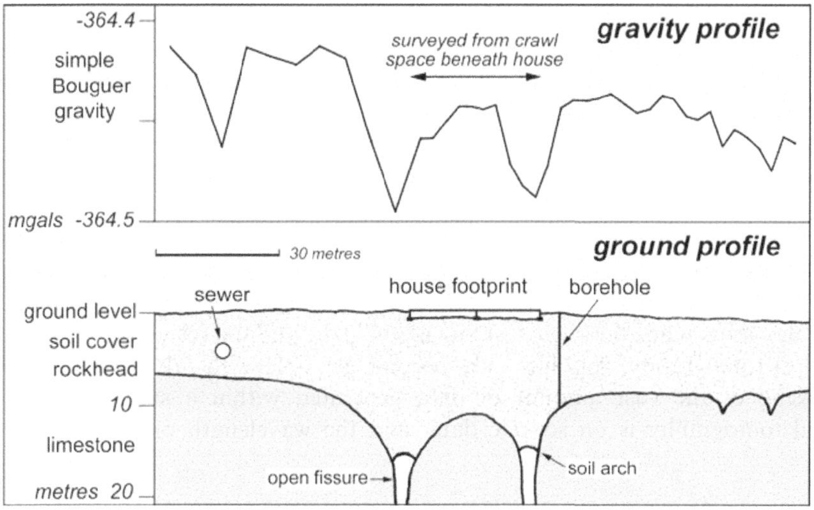

Figure 9.4. One profile from a microgravity survey, carried out in order to assess the subsidence sinkhole developing beneath a house in Kentucky; survey stations were at intervals of 1.5 or 4.5 m, and the data was calibrated and confirmed by boreholes to rockhead. After Crawford *et al.* (1999).

properties. They are quick, simple and economical, and the field data only requires corrections for diurnal variations in the Earth's field, normally monitored on site during the survey, though their integrity is reduced by nearby electrical and ferrous structures. Magnetic surveys are widely used for the detection of old and capped mineshafts, which usually have magnetic contrasts in their fill or lining. However, they are generally unsuitable for the detection of natural cavities and sinkholes, where magnetic contrasts are low or absent in limestones and soils. The exception is where small clay-filled buried sinkholes can be identified in pure limestone or chalk (McDowell, 1975).

Magnetic surveys have been used to detect lava tubes in magnetically conductive basalt. They have proved very effective at mapping systems of open tubes beneath rough terrain on the lava fields of Iceland (Wood *et al.*, 2002). GPR surveys on the same site were far less efficient, except that they could measure roof thickness over the tubes. Magnetic surveys have also been used to follow the evolution of tubes within active lava flows on the Etna volcano in Italy, but the detection of tubes containing hot flowing lava is barely applicable to most engineering sites (Budetta and Del Negro, 1995).

9.3.8 Cross-hole tomography

Most surface geophysical surveys can only be completed where the ground surface is not obstructed or disturbed by buildings, foundations, services or construction activity. Development of cross-hole geophysical methods, especially the technique of 3-D tomographic imaging, overcomes most of these problems, and can also provide far superior ground data. They do however require boreholes that are either available or purpose-drilled, though some costs can be saved by carrying out surface to borehole imaging. Pairs of boreholes are normally used to scan, electrically to seismically, from a transmitter in one borehole to receivers in another. A series of measurements are made by moving source and receiver up or down each borehole by a predetermined amount (usually 0.25 or 1.0 m) so that every possible ray path is scanned. Data manipulation then derives a physical property value for each of a grid of ground cells between the boreholes, and from these creates a 2-D tomography image in a vertical plane (Jackson and McCann, 1997). Multiple boreholes allow scans between every available pair, and the results can be combined into 3-D tomography; this has only become possible with advances in computer processing of the vast amount of data generated within a single survey. Most ground tomography is on seismic data, and the wavelength of the seismic signal

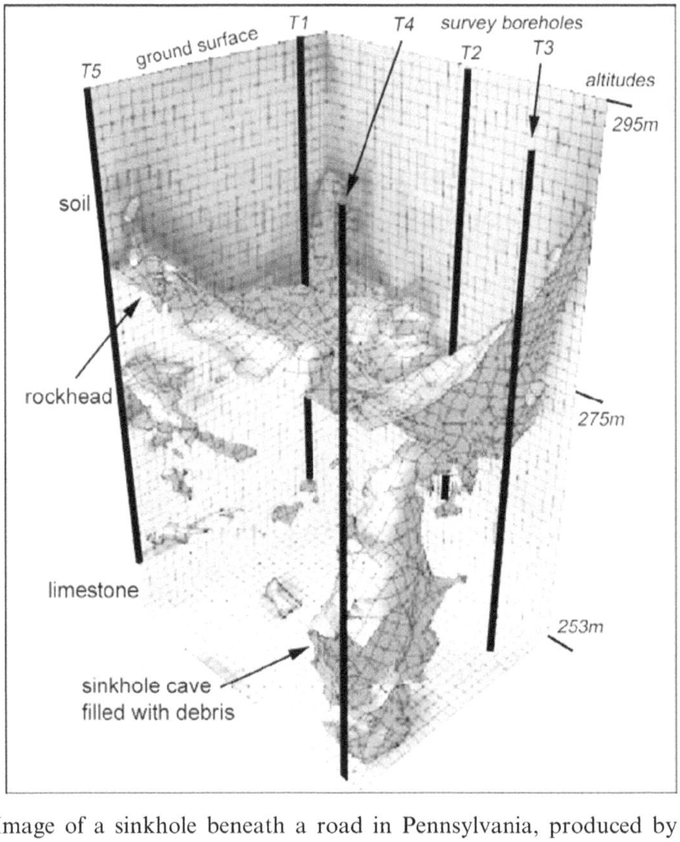

Figure 9.5. Image of a sinkhole beneath a road in Pennsylvania, produced by 3-D seismic tomography between five boreholes; the soil-filled cave that drains the floor of the sinkhole was verified by subsequent drilling.
Courtesy of 3dT/NSA Engineering.

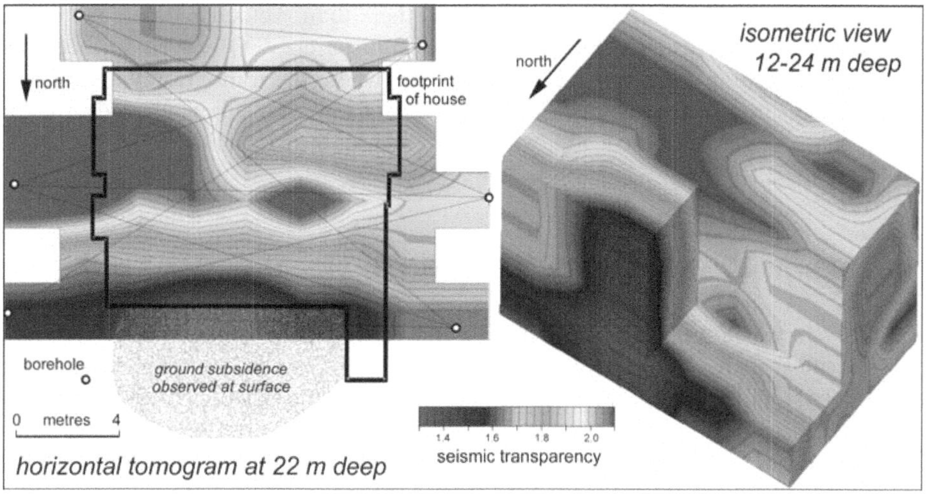

Figure 9.6. Seismic transparency tomography for ground beneath a house on soil-mantled chalk in southern England; both in the horizontal tomogram at 22 m below ground level and in the isometric projection of the 3-D tomography image covering depths between 12 and 24 m, the dark tones keyed to low seismic transparency show the zones of disturbed soil that are related to sinkhole subsidence.
After Jackson *et al.* (2001).

needs to be less than the average dimensions of the sinkhole target (McDowell and Hope, 1993). Comparable tomography can be based on electric resistivity measurements, and has been developed successfully beneath buildings threatened by sinkhole subsidence in karst terrains (Case study #16).

The quality of the tomography is a function of the nature of the ground, in particular contrasts in the physical properties, the number of borehole pairs, the distances between the boreholes and their location in relation to the target. In sinkhole investigations, tomography is usually only feasible where boreholes already form part of the site investigation or where a building or structure of sufficient value is so located that investigation from the surface is too difficult. Some 3-D seismic tomography has proved excellent, and the technique is perhaps the most useful, and most promising, that is currently applicable to sinkhole investigations where borehole access is available (Simpson, 2001). A ground image calibrated and presented in seismic velocities can provide a realistic model where intact bedrock limestone, open fissures, soil-filled caves, buried sinkholes, rockhead topography and disturbance zones in the soil cover are all identifiable (Figure 9.5). In the Chiltern Hills karst of southern England, a house 160 years old had suffered damage over a five-year period. It stands on sands and clays that overlie chalk, and the damage was caused by subsidence into a buried sinkhole. Ground conditions beneath the house were investigated by a 3-D seismic tomography survey (Figure 9.6), in which over 5,000 rays were scanned between 17 pairings among 7 boreholes that were sunk around the building (Jackson *et al.*, 2001). Because the ground was so disturbed, seismic amplitude was used, rather than velocity data, to create tomograms of empirical seismic (or acoustic) transparency. Observed surface subsidence had been at the north side of the house, where the tomography identified a deep zone of ground that is seismically opaque (of low transparency), and this was interpreted as the disturbed soil within or over a buried sinkhole in the chalk (Figure 9.6).

9.4 AIRBORNE AND SATELLITE REMOTE SENSING

Remote sensing data from both aeroplane and satellite platforms has been used as a part of site investigation for many years, but its use in the detection of sinkhole subsidence is mainly restricted to rural areas, and scale is then a critical factor. The resolution necessary for the detection of relatively small subsidence features (1.5–3.0 m across) is provided by aerial photographs with scales between 1 : 25,000 and 1 : 10,000. Colour photographs may be more useful than black and white ones since they can reveal subtle changes in vegetation related to subsidence and changes in moisture conditions. However, tones on monochrome photography are generally darker on healthy vegetation and wet ground, and these tonal contrasts can sometimes prove to be valuable indicators of soil water movement. False-colour infrared photography maps thermal emission, and has been used for both the identification of stressed vegetation, which might indicate problematical ground conditions, to locate wetter or drier areas, and also to detect hot or cold spots that might relate to cave entrances. Detail obtained from all aerial photographs should normally be represented on a site plan at a scale of 1 : 2,500 or larger.

False-colour infrared and black-and-white aerial photographs were used for hazard mapping of a freeway corridor across karst in Florida that is prone to solution and subsidence sinkholes (Padgett, 1993). On the infrared images at a scale of 1:40,000, vegetation around sinking streams appeared bright red and around active sinkholes it appeared dark red. Tonal variations could be used to determine the extent of enclosed drainage features associated with relic sinkholes and recharge zones. However, black-and-white photographs at a similar scale were not useful for determining the extent of closed drainage basins. At scales nearer to 1:10,000, aerial photographs can be used as stereoscopic pairs to identify subtle variations in the morphology of the ground surface, particularly if photography was in a season when vegetation is reduced and a low sun angle creates clear shadows to accentuate even the smallest of surface depressions (Figure 4.17). On false-colour infrared film, bright or deep red colours represent growing, healthy vegetation, probably related to wetter areas of poorer drainage that might be associated with sinkhole depressions that have a soil cover. However, if these soils are freely drained or absent, and vegetation is stunted or absent, the image shows pinkish to yellow and grey colours.

Radar and laser sensors on airborne platforms are being used to produce high-resolution (centimetre to metre) digital terrain models that are already finding application in floodplain studies, but may also be applicable for locating topographic lows associated with sinkholes where the depth of the depression is within the resolution of the technique. The LIDAR (LIght Detecting And Ranging) system sends a laser pulse from an airborne platform to the ground and measures the speed and intensity of the returning signal, in order to map ground elevation. Radar systems can produce results similar to those from laser. Satellite imagery has gradually improved in its resolution over time so that its use has extended into detailed geomorphological mapping and geohazard identification. The original LANDSAT images were limited by their low resolution, but new satellite imagery is more applicable to sinkhole studies in the style of airborne photography.

Satellite radar measurements are becoming increasingly sophisticated, and a technique known as PSInSAR or PSI (Permanent Scatterer Interferometry) uses radar data collected by satellites 800 km out in space. The PSInSAR method exploits a dense network of natural reflectors that can be any hard surface such as a rock outcrop, a building wall or roof or a road kerb. These reflectors are visible to the radar sensor over many years, typically in urban regions but also in partially developed rural areas, and are known as permanent scatterers. They are derived from the analysis of a stack of 30 or more radar scenes derived from repeated satellite passes, and the density of recognised permanent scatterers is about 1 per hectare in urban areas. Using this dense network of points common to all 30 images, corrected for contemporary atmospheric conditions, PSInSAR produces maps showing rates of displacement, accurate to a few millimetres per year and over extensive time periods. Data since 1992 is available from three satellites launched by the European Space Agency.

The PSInSAR process provides the millimetric displacement histories for each reflector point across the entire time period analysed, as calculated at every individual radar scene acquisition. Small incremental ground movements, that might be caused by gradual sinkhole subsidence, can therefore be detected. There are some disadvantages with the technique. If movement of a permanent scatterer is too great, coherence between one image and the next is lost, as the reflection point effectively vanishes because it has moved too much. Also, the full time series of movement since 1992 can only produce data along the line of sight from the satellite, which is at an angle of 20–30° to the vertical. It is possible to resolve movement only into vertical and north–south components, but this requires utilising both forward and backward images of the point on different passes of the satellite and requires a greater degree of computer processing. PSInSAR is currently too expensive for use in most site investigations, but it is likely that cost will come down as processing software improves and larger computers become available.

9.5 DIRECT INVESTIGATIONS

No single method of investigation is appropriate for locating and quantifying sinkholes in all circumstances. The most effective approach to a site investigation on karst is a combination of methods, usually involving those that are both indirect and direct. Some extent of drilling and probing is always likely to be required, and is also critical to confirming almost all geophysical surveys. Pinnacled rockheads and highly cavernous ground, in karst of classes kIV or kV, can require very large numbers of boreholes to adequately define ground able to bear construction loadings. In the notoriously difficult ground of Kuala Lumpur, Malaysia, geophysical surveys may not be successful in defining a founding surface for the piles for high-rise building foundations (Tan, 1987; Bennett, 1997). It is not uncommon to drill as many as 100 boreholes for each high-rise building to map out the variation in the limestone rockhead profile (Figures 5.3 and 5.7); this borehole density is ten times what would be expected to locate rockhead on schist or sandstone.

With particular regard to buried, suffosion and dropout sinkholes, the aim of an intrusive investigation is normally to provide evidence of ground truth in relation to the bedrock profile, particularly the shape and dimensions of the sinkhole and any ravelling zone, the geotechnical and hydrogeological properties of the soil and bedrock, and the groundwater conditions that may alter the character of the sinkhole in the future. Selection from the available techniques should be appropriate to the scale and nature of the immediate situation.

Among the various methods of direct investigation, there is an extra option that is specific to karst, because its ground voids are commonly large enough to be physically explored by a person. Though cavities in soil may be so unstable that direct entry is unsafe, caves in bedrock limestone may be perfectly safe for exploration by competent cavers, preferably by those with experience in exploration, mapping, geology and engineering (Figure 9.7). Physical examination and mapping are undoubtedly the most cost-effective means of investigating any mature cave passage or cave system that happens to lie beneath a construction site.

Pitting and trenching are commonly used in shallow soil investigations, to allow block sampling and visual inspection. Reachable depths are limited by safety considerations, and rarely can be adequate for useful investigation of sinkholes. However, a backhoe can often dig a hole that does not have to be descended to locate bedrock at depths of up to 4 m for less cost than deploying a drill rig.

9.5.1 Soil probing

Because the most widespread sinkhole hazard is the development of new subsidence sinkholes entirely within the soil profile, a large proportion of ground investigations on karst focus on the stability or potential failure of the soil cover. One concern is to locate soil cavities (referred to by the regolith arches over them in most of the American literature), that may migrate upwards to form a dropout sinkhole. The second concern is to find ravelling zones, where soil is disturbed and unstable due to losses into limestone fissures beneath, and may evolve into either a suffosion or a dropout sinkhole.

Soil voids can be located by the simplest form of probing, involving the manual pushing of steel rods, usually 12 mm in diameter, into the ground. Penetrations of as much as 6 m have been achieved in Florida, and these could be increased by use of a drop hammer, but results of such probing may be regarded as subjective (Handfelt and Attwooll, 1988). Conventional soil probing uses a light percussion rig with

Figure 9.7. Direct exploration: an engineering geologist, who is also an experienced caver, abseils from an excavator bucket into a sinkhole that collapsed into an open cave during road construction in Slovenia.
Photo: Martin Knez.

capability of either driving a shell or turning an auger. Soil voids may also easily be recognised during a probing operation, either by the loss of end resistance, or by complete or partial loss of circulating fluid. However, the loss of flush return can be disastrous, as increased water flow through the soil profile is the most effective means of inducing sinkhole activity (Chapter 8). There have been multiple cases in Florida alone, where drill rigs deployed on sinkhole investigations have created their own subsidence sinkholes and thereby self-destructed. In the worst cases, drilling investigations at sites of modest ground subsidence under or adjacent to houses have created large new sinkholes, and thereby have caused major damage to the buildings under investigation. Where a potential hazard is recognised, by appropriate desk study, dry augering or air drilling becomes appropriate when direct investigation is essential.

The main type of probing in the less cohesive soils is the standard penetration test (SPT). A split sample tube is driven into the ground by means of a fixed weight dropping a fixed distance onto a drive head connected directly to the drilling rods (British Standards, 1999; A.S.T.M., 1999b). The number of blows to drive 300 mm is recorded and quoted as the N-value, usually measured at depth intervals of 1.5 m. The method is crude but effective. It is widely used, so test results are well understood, and the split sampler also produces a disturbed sample. Ravelling zones are widely identified by their lower N-values that reflect the disturbed and unstable nature of the soil. In the soil-mantled karst of Florida, ravelling is described as an isolated, continuous vertical zone of cohesive soil having N-values of 2 or less, or non-cohesive soil having N-values of 4 or less, and this zone forms a pipe surrounded by firmer, stiffer, or denser soil, to distinguish it from a laterally continuous layer of very soft or very loose soil (Zisman, 2003). This move towards a more specific definition of a sinkhole in Florida has been driven by the inclusion of sinkhole coverage in homeowners' insurance policies (Chapter 9) and by an increase in the number of disputes over whether damage has been caused by a sinkhole or by another process. Significantly this represents a narrowing of the definition of a sinkhole, by greatly reducing the threshold N-values from those cited previously by the same author (Zisman, 2001). However, some practitioners still regard the use of SPT in the recognition of sinkhole hazards as potentially misleading (Kannan, 1999).

More appropriate to investigations of cohesive soils, the Dutch cone or cone penetrometer test (CPT) involves continuously pushing a so-called friction cone into the ground by means of hydraulic rams (A.S.T.M., 1998). The cone resistance (Q_c) and the friction (F_s) on a sleeve immediately behind the cone are both measured to produce a continuous graphic log with depth. The cone can also be fitted with a porous sensor to measure fluid pore pressure. The ratio F_s/Q_c is known as the friction ratio (R_f), which can be used to recognise changes in soil lithology and density. Ravelling zones are indicated by low cone resistance, high friction ratio and negative values of a corrected pore pressure measurement (Wilson and Beck, 1988). The CPT is relatively cheap and easy to carry out because full-time supervision is not required, and results are simple to interpret with respect to identifying the depths to voids and associated weak zones. At a site of 200 ha in Pennsylvania, over 300 CPT soundings were completed as they were considered to be the most effective intrusive technique for investigating small sites for proposed building foundations (Pazuniak, 1989).

SPT and CPT results were compared at four sinkhole sites in Florida (Bloomberg *et al.*, 1988). The conclusion was that CPT is a superior technique because it produces more information, is sensitive to minor lithological variations and is particularly useful for detecting potential conduits and piping failures. For these reasons it may be regarded as a more cost effective method for sinkhole investigation. However, it does have a significant drawback in that progression of the cone can be stopped by relatively small stones or pieces of rock. With the SPT, run on a conventional light percussion rig, boring could remove the obstruction so that further tests could continue at greater depths.

9.5.2 Rock probing and boring

Rock is only penetrated by rotary drilling. This can produce an intact core inside a bit armoured with diamond or tungsten carbide. Alternatively, probing (or destructive drilling) simply bores a hole without retaining any rock core, and relies on flushed cuttings and penetration rates to interpret the ground conditions. Probing is quicker and cheaper, and is generally adequate for simple cavity searches in karst bedrock, once strata control has been established by a smaller number of cored holes. All rock drilling requires the use of a flushing medium to cool the drill bit and to bring cuttings to the surface. Loss of drilling fluid can be a valuable indicator of sinkholes or caves, especially where the fluid escapes through a narrow fissure that drains into an adjacent cave missed by the borehole. Uncased boreholes can be inspected by means of cameras or echo-sounders, especially where they penetrate a void. Rotary holes that breach an open or water-filled cave may have to be terminated where a steeply inclined cave floor prevents the drill biting in to continue the hole; if deeper exploration is required, it is often cost-effective to drill a second hole. Flush loss does not create a hazard in limestone, as it may only wash loose sediments out of any caves and is unlikely to induce any sinkhole failure. However, care is needed when drilling in salt due to the possibilities of very rapid dissolution, either by the normal water flushes or by chemically aggressive groundwater that is able to flow from another aquifer via the new borehole. Drilling in salt can use a brine flush, and all boreholes in salt and gypsum should be sealed after use; failure to complete the latter may lead to new sinkhole development shortly afterwards (Figure 9.8).

Figure 9.8. A man standing on the collar of a borehole that had dropped into a sinkhole over salt in the Israeli desert near the Dead Sea; though dissolution, cave development and sinkhole formation were already active in the area, the location of this sinkhole was determined by the borehole that was drilled to investigate the subsidence problem.
Photo: Mark Talesnik.

The optimum spacing and depths of investigation boreholes is particularly difficult to prescribe for the extremely variable ground conditions of karst. With respect to rock boring, both parameters must relate to potential cavity size and hazard. Minimum borehole depths are defined in terms of rock roof stability over caves (Table 7.1). Borehole spacing must be appropriate to specific site conditions relating to the potential cave size and the unsupported span that can be safely bridged by any proposed construction, and economies can usually be made where boreholes can target recognised geophysical anomalies. The frustrations of cavity searches were demonstrated by the unfortunate case of the Remouchamps Viaduct in Belgium (Waltham *et al.*, 1986). The five pier sites on limestone were investigated by 31 boreholes, all of which missed two caves at critical locations only found during excavation for the footings; the project was then halted to allow a second phase of investigation, but 308 new probes found no more caves. Minimising risk is one of the hardest tasks for the geotechnical engineer working on karst.

10

Hazard and risk assessment of sinkholes

10.1 HAZARD AND RISK

Where a hazard impinges upon human activity, it involves a degree of risk, the elements at risk being life, property and possessions; the natural environment may also be affected, though this may be considered as part of natural change. Risk involves quantification of the probability that a hazard will be harmful, and the tolerable degree of risk depends upon what is being risked, life being more important than property. The frequency of a particular hazard event can be regarded as the number of events of a given magnitude in a particular period of time at a certain location. As such, a recurrence interval for such an event sometimes can be determined in terms of the average length of time between events of a certain size. The risk to society can be regarded as the magnitude of a hazard multiplied by the probability of its occurrence and by the cost of its impact. If there are no hazard mitigation measures for an area that is subjected to a recurring hazard, then such an area has the highest vulnerability.

Karst areas present a variety of risks to people and their environment (Wilson and Beck, 1988). These include the subsidence that may be rapid or slow, resulting in damage to buildings and infrastructure. It may also cause the loss of surface water and groundwater contamination, as at Altura, in Minnesota, when effluent from a newly constructed sewage disposal pond disappeared down nine sinkholes that opened in a line across its floor (Alexander and Book, 1984). Although the degree of risk may be minimised by thorough investigation, it is never likely to be eliminated.

Risk should take account of the magnitude of the hazard and the probability of its occurrence. It arises from uncertainty that is due to insufficient information being available about a hazard, and to incomplete understanding of the mechanisms involved. The uncertainties prevent accurate predictions of hazard occurrence. The total risk in a particular period of time can be expressed as:

$$\text{Total risk} = \text{Hazard} \times \text{Elements at risk} \times \text{Vulnerability}$$

where the hazard is the probability of a potentially damaging phenomenon occurring within a given area and within a specified period of time, the elements at risk are the population, properties, assets and activities at risk in a given area; and the vulnerability is the degree of loss (on a scale of zero to total) to a given set of elements at risk from the natural event (Varnes, 1984). The hazard may be considered to comprise two elements – the physical scale of the hazard (the size of a sinkhole), and the probability of an event occurring within a given time; however sinkhole size may not be appropriate where even quite small surface expressions of subsidence might represent subsequently larger movements. The risk analysis should mention all the assumptions and conditions on which it is based and, if these are not constant, the conclusions should vary according to the degree of uncertainty, a range of values representing the probability distribution. The confidence limits of the predictions should be provided in the analysis.

The literature on karst hazard and risk can be confusing, as the terms hazard and risk are commonly used in ways other than those defined above. A published strategy for assessing the risk of karst subsidence (Benson *et al.*, 2003) assesses the hazard according to the definitions above. A sinkhole risk map of Florida (Wilson, 2000) shows sinkhole density, and is therefore a form of hazard map. The scheme for hazard and risk assessment in South Africa (Buttrick *et al.*, 2001) describes the hazard, the inherent risk (which is the probability of the hazard) and the development risk (which is the true risk).

10.2 RISK MANAGEMENT

The aim of reducing risk involves the three components of risk analysis, risk assessment and risk management (Fell and Hartford, 1997). Risk analysis involves identifying the degree of risk, then estimating and evaluating it (Figure 10.1); analysis can vary from the use of quantitative statistics to a qualitative evaluation. A risk analysis in terms of karst terrain requires data on the types of hazard (including sinkholes and subsidences), their frequency of occurrence, the consequences of the hazard, the geological and hydrogeological conditions, and the possible mechanisms involved.

Data should be gathered in a uniform manner over a sufficient length of time to allow any conclusions to be applicable to the specific situation. Records for Chiefland, a town in Florida, were examined for a 35-year period, during which there were only a few sinkhole collapses; then, in one weekend of torrential rainfall, more than 100 sinkholes appeared (Beck, 1999). Hundreds of sinkholes have been mapped over the gypsum in the Ripon area in northern England (Figure 10.2). These sinkholes have been developing probably since the end of the last glaciation, around 10,000 years ago, and some of them have been infilled by Holocene sediments (Cooper, 1998). If it is assumed that the mapped sinkholes have occurred over this period of time, a return period of around 30 years might be expected. However, dating from newspaper and local records shows that at least

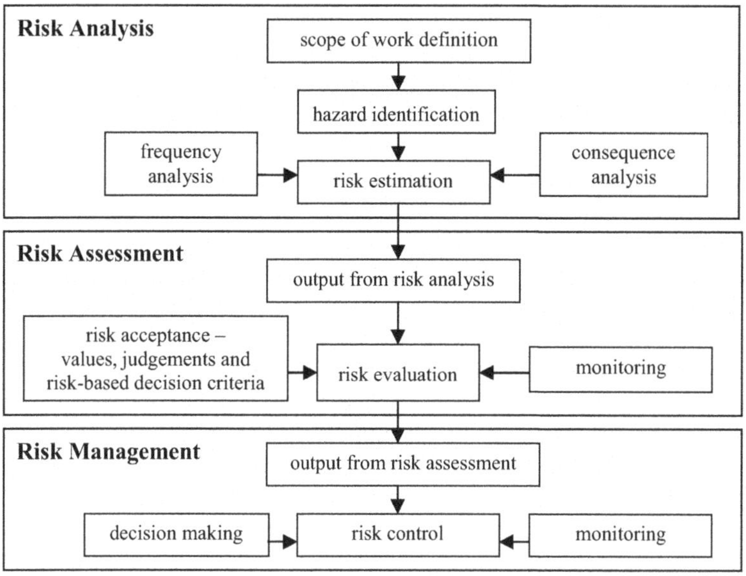

Figure 10.1. Staged processes towards risk management.
After Fell and Hartford (1997).

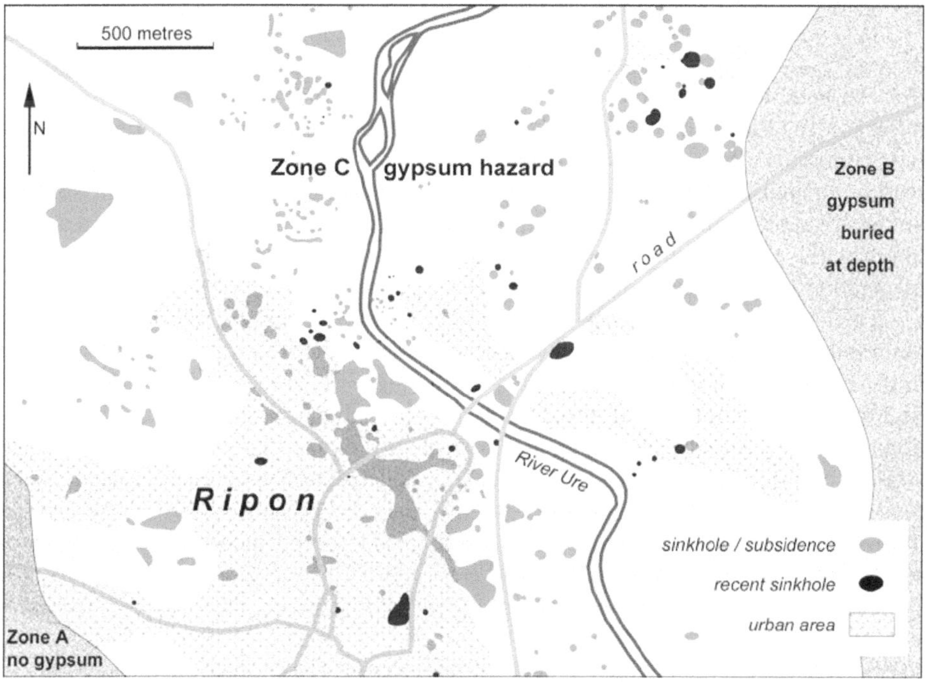

Figure 10.2. Distribution of gypsum sinkholes around Ripon, northern England; sinkholes marked as recent are those recorded since 1834 (after Cooper, 1998). The hazard zones refer to the planning guidelines (Thompson *et al.*, 1996).

30 major sinkholes (10–30 m wide and up to 20 m deep) have occurred in the last 150 years, and future collapses might therefore occur every 2–3 years. The time period for determining the hazard must be fully representative for the processes that trigger the sinkholes. If the trigger includes torrential rainfall, as at Chiefland, the sample time must relate to the return period of storms. If the trigger is flowing groundwater, as at Ripon, the sample time should relate to any changes in groundwater conditions through Holocene times and into the modern environment.

The risk from a specific sinkhole hazard can be quantified in terms of the resultant damage to property; deaths and personal injuries are generally absent or subordinate in sinkhole events. Damage to property in Ripon, due to the gypsum sinkholes in the 10 years up to 1998, was estimated at about £1,000,000, and there were no injuries or deaths (Cooper, 1998). Building damage in the city of Calatayud, northern Spain, caused by sinkholes over gypsum-rich silts within an alluvial fan, was assessed using a damage classification modified from that developed to describe mining subsidence (Table 10.1), and its distribution was mapped across the city (Figure 10.3). Because there is not always a direct relationship between subsidence and building damage, such a map should be regarded with some caution. This was recognised at Calatayud, but the characteristics of most of the damage were diagnostic of sinkhole subsidence (Gutiérrez *et al.*, 2000). The mapping identified the main unstable area where the gypsum is thickest down the axis of the alluvial fan which is followed by the Calle de la Rua (Figure 10.3).

It is relatively straightforward to assess financial losses resulting from sinkhole damage using data on property values and rebuilding costs. The losses involved can then be compared with the costs of hazard mitigation. However, losses are also influenced by the proposed type of foundation, likely building use and the construction contract to be used (Rigby-Jones *et al.*, 1994). Not all risks and benefits are readily amenable to measurement or estimation in financial terms. Even so, risk analysis is an important objective in decision making by urban planners because it involves the vulnerability of people and the urban infrastructure on the basis of probability of occurrence of an event.

Having determined by risk analysis how safe a location is, the next stage is to determine by risk assessment how safe it needs to be (Figure 10.1). This should improve the planning process. The final stage of risk management involves deciding how to control and mitigate the risk by reducing the hazard or the vulnerability, or at least not allowing them to get any worse. This can use a designed strategy for decision making with regard to construction where sinkholes may be present (Figure 10.4). The outcome of this may be preventive action by ground improvement or sinkhole remediation (Chapter 11) or by incorporation of appropriate construction methods (Chapter 12).

A number of problems are inherent in risk management. The degree of risk does not increase linearly with the length of exposure to a hazard. Moreover, the response to risk can change over time, so that mitigation and risk reduction can change. In addition, the dichotomy between actual and perceived risk does not help attempts to reduce risk, and it may promote a conflict of objectives. Risk management has to attempt to determine the level of risk that is acceptable. This has been referred to as

Table 10.1. A classification of subsidence damage to both buildings and infrastructure. After N.C.B. (1975) and extended by Tony Cooper, pers.comm.

Category	Typical building damage	Damage to roads and pavements
1 Very slight	Fine internal cracks, <1 mm *Generally not visible from outside*	Not visible
2 Slight	Typical crack widths up to 5 mm Doors and windows may stick slightly *Difficult to record from outside*	Generally not noticeable
3 Moderate	Typical crack widths 5–15 mm, or multiple Doors and windows sticking, slight tilts to walls, service pipes may fracture *Visible from outside, remedial works needed*	Slight depressions in roads Hairline cracks to pavements Repairs generally superficial
4 Severe	Typical crack widths are 15 to 25 mm Requires replacing sections of walls Windows/door frames distorted, floor slope Walls leaning or bulging, services disrupted *Noticeable from outside*	Significant depressions in roads Surface cracking, some open holes Highway repairs generally require excavation and reconstruction
5 Very severe	Typical crack widths are greater than 25 mm Structural damage, requires major repairs involving partial or complete rebuilding Beams loose bearing, walls lean badly Danger of instability *Very obvious from outside*	Rotation or slewing of the ground Significant depressions and cracks Open craters form over voids Disruption of services under roads Significant repair required
6 Partial collapse	Partial collapse	Collapse of ground or highways Significant open voids Services severed or disrupted
7 Total collapse	Total collapse	Large open void or landslip scar Rotation or slewing of the ground

risk balancing. Public and private resources then may be allocated to meet a level of safety that is acceptable to everyone concerned. Cost-benefit analyses can be used to develop a rational economic means for risk reduction expenditure, though obtaining the necessary data can be difficult and time-consuming.

10.3 KARST HAZARD ANALYSIS

Karst terrains can present to engineering structures a number of hazards, including highly variable rockhead surfaces with gaping fissures between pinnacles, open cavities in both bedrock and soil cover, and also sinkholes. All can cause ground

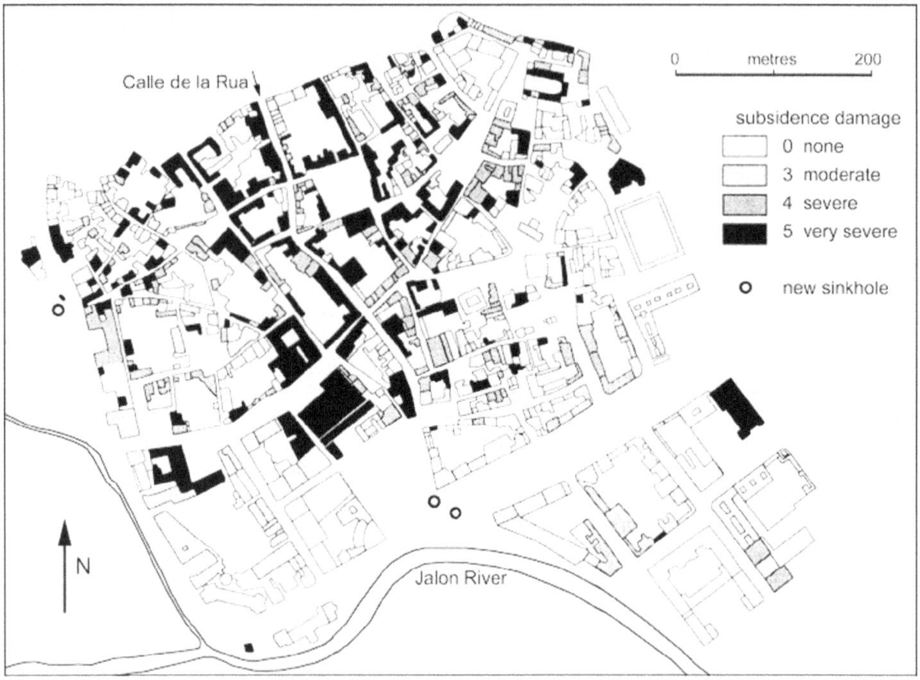

Figure 10.3. Distribution of subsidence damage to individual buildings in Calatayud, Spain, due to sinkholes and dissolution in underlying gypsum; classes of damage are as in Table 10.1. After Gutiérrez *et al.* (2000).

subsidence, which varies in amount and in the time taken for development, from slow differential settlement to sudden collapse. The consequences can range from cracking in a building to the collapse of part or all of a built structure. Occasionally, individuals may be injured or lives may be lost. A total of 38 people have died in sinkhole collapse events in the South African Rand, and the cost of property damage amounted to over £100 million, though 29 deaths occurred with the single collapse of a mine crusher plant in 1962 (Box 8.1).

The possibility of sinkholes and other karst features affecting any new development must be evaluated in order to assess the associated degree of risk. Such an assessment might take place at the planning stage to determine the viability of development or after the site investigation. Alternatively, once the ground investigation has been concluded, and the earthwork and design parameters have been determined, a risk assessment can be made as an integral part of the foundation design (Destephen and Wargo, 1992). This may consist of a qualitative assessment of the probabilities of sinkhole development, based on a weighted evaluation of both the geological and development hazard factors. In addition, the possible impacts that the development may have upon the ground conditions must be assessed. These impacts include changes in ground stress due to loading by construction activities

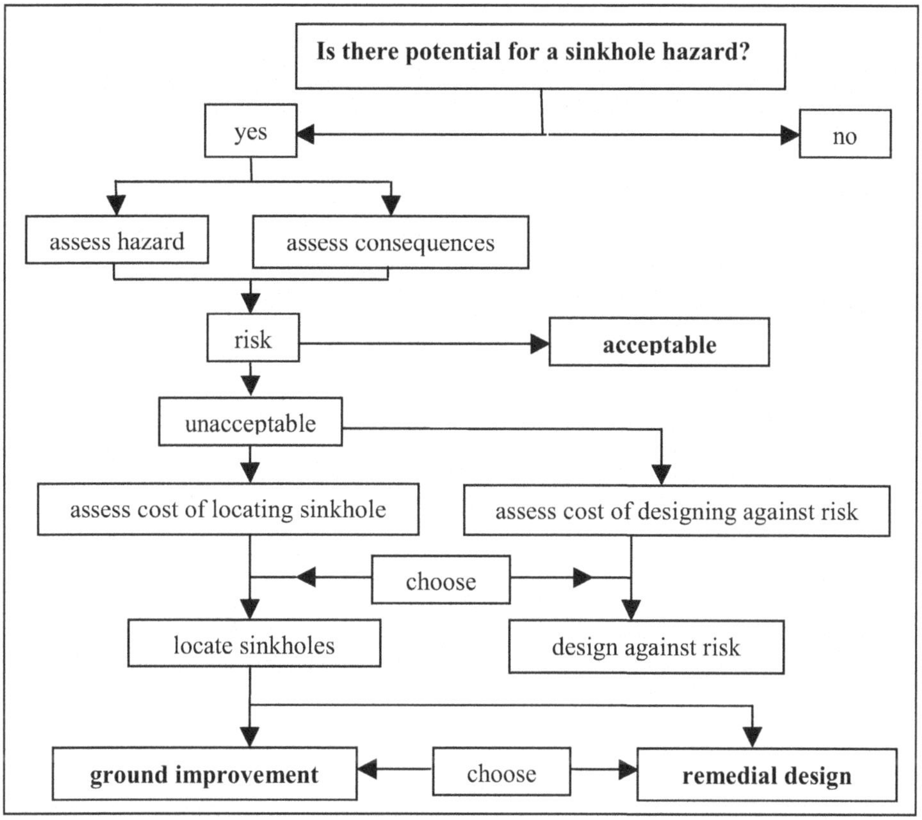

Figure 10.4. A strategy for dealing with sinkhole risk.
After Rigby-Jones *et al.* (1994).

and by built structures, and, most importantly, any changes in surface water and groundwater conditions (Chapter 8).

The likelihood of subsidence in areas of karst could be assessed by a structured programme that first obtains appropriate site-specific data, then understands the site-specific mechanisms of collapse and third interprets the data to make a hazard assessment (Figure 10.5). Though this should be inherent in an adequate site investigation, the geological and hydrogeological data are a prerequisite for any hazard assessment of sinkhole activity, and should be gathered in sufficient density to develop a reliable model of the site conditions. This may require the use of ground investigation methods that are specific and appropriate for karst sites (Chapter 9). It may be argued that hazard predictions must be inherently general because of the numerous variables involved, and that attempts to quantify the hazard or predict the time of collapse are usually inappropriate. In these cases, a simple, qualitative categorisation may define hazards that are low (not likely to occur in the project lifetime), medium (may occur if aggravated by construction activity) or high (active or very likely to occur) (Benson *et al.*, 2003).

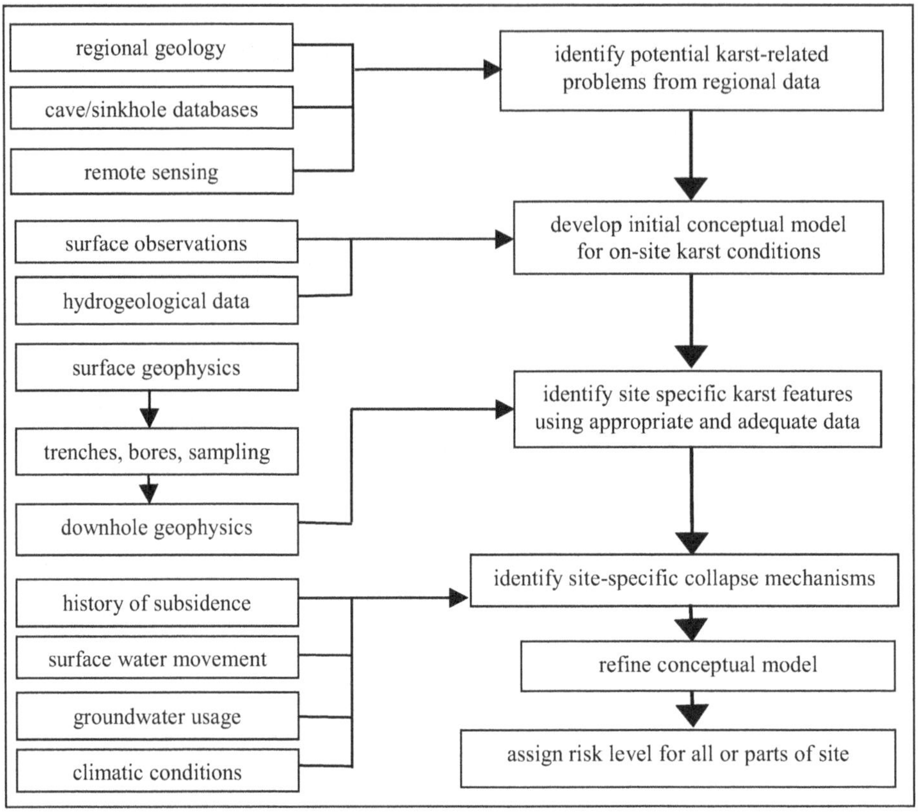

Figure 10.5. Stages within a site investigation for karst hazard assessment.
Modified from Benson *et al.* (2003)

The likelihood of sinkhole development can be evaluated statistically for particular areas by analysing the historical frequency with which new sinkholes appear per unit area in a given unit time, with historical data providing a basis for the risk analysis modelling. A sinkhole frequency map may then be produced from the analysis. The sinkhole hazard should take account of the expected maximum width of new sinkholes, as this determines the maximum distance that has to be bridged to prevent collapse of any building or structure (Wilson and Beck, 1988). Hazard mapping of the chalk karst in southern England includes sinkholes and incorporates this factor (Figure 10.6).

10.3.1 Parameters in sinkhole hazard assessment

The greatest sinkhole hazard lies in areas of soil-mantled, cavernous karst where subsidence sinkholes can develop widely and with little or no warning. Any hazard assessment must therefore recognise the key processes that cause these sinkholes

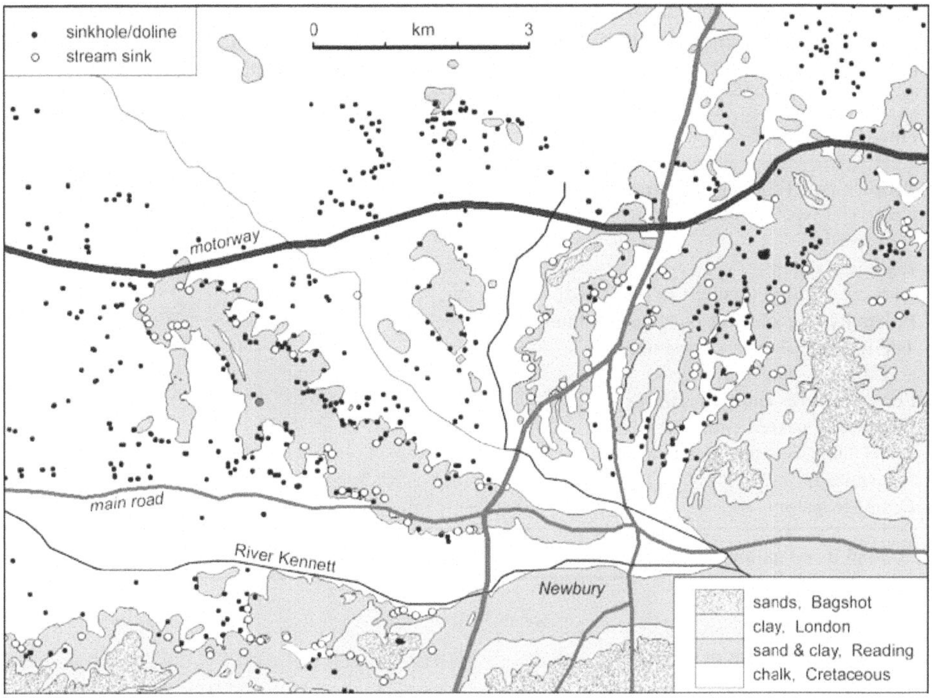

Figure 10.6. An inventory map of sinkholes and stream sinks in the chalk, overlain on a geology map of the fluviokarst near Newbury, in southern England.
After British Geological Survey.

(Chapter 4), and should then assemble data on the critical parameters and possible indicators related to those processes. Table 10.2 provides a generalised list of the key factors in terrains prone to sinkhole development. The full list should be applicable to any karst site, though many parameters have greater or lesser significance in specific karst terrains; it is significant that published hazard assessments refer to only elements from the tabulated list. Van Rooy (1989) and Venter and Gregory (1987) both refer only to the dolomite karst of South Africa where mine de-watering was a major factor in inducing sinkhole development (Section 10.3.2).

Zhou *et al.* (2003) assessed the sinkhole hazard for a highway over limestone near Frederick, in Maryland, where they found that sinkhole events were increasing in response to intensive land development in the area. They developed a semi-quantitative assessment that, for each 30 m length of road, identified a greater hazard where an existing sinkhole was nearby, where the soil cover was thinner and where other ground factors had lesser influence.

Zisman (2001) referred only to subsidence sinkholes in the thick cover soils over the karst of Florida. He presented a somewhat complicated rating system based on eleven observed ground characteristics, with ratings for each that could be added to produce overall hazard ratings (Table 10.3). An alternative was to use a flow chart approach: if the standard penetration test (SPT) average N value of the top 5 m of

Table 10.2. Ground conditions and parameters that should be considered in assessment of hazard and risk in karst terrains prone to sinkhole development; columns on the right indicate those parameters that have been incorporated in published assessment schemes.

Observable parameters	Zhou *et al.* (2003)	Zisman (2001)	Edmonds (2001)	Van Rooy (1989)	Venter *et al.* (1987)
Bedrock type (lithology and fracturing)	✓		✓	✓	
Cover soil type (structure and strength properties)	✓	✓	✓	✓	✓
Rockhead depth	✓				
Rockhead nature (fissured or pinnacled)		✓			✓
Proximity to geological boundary or fault	✓		✓		
Topography (slope and position relative to thalweg)	✓		✓		
Local sinkholes (number, size, type)					
Distance to nearest sinkhole	✓				
Water table elevation (relative to rockhead)			✓		✓
Drainage history			✓	✓	✓
Ground investigation data if available					
SPT N-value of soils		✓			
Known geophysical anomalies	✓			✓	✓
Loss of drilling fluid, or caving of bored holes		✓			
Variable depths to water table in boreholes		✓			
Recorded ground settlement		✓			

SPT = standard penetration test.

ground is < 7, and if the average N value for the depth interval 5–12 m is <15 or less, then the other nine ground conditions from Table 10.3 are assessed. Higher N values in either of the first two steps indicate that there were no soil voids that could cause immediate distress and sinkhole failure.

Edmonds (2001) referred specifically to the subsidence sinkholes and buried sinkholes in the chalk of southern England. He examined the spatial distribution, metastable character and subsidence triggers associated with sinkhole occurrence, and assessed eight geological, hydrogeological and geomorphological factors that might affect sinkhole distribution (Table 10.4). A subsidence hazard rating was then obtained by multiplying the topographic factor by the sum of the other seven factors, and this rating was related to the subsidence hazard, ranging from none anticipated to extremely high. The output of the methodology was checked in areas where sinkholes had been mapped in detail and a good relationship was found between sinkhole occurrence and the higher hazard zones that had been predicted. However, the numbers derived for each factor, and the final rating, were all of local significance, and the system may not translate to sinkhole hazards on stronger cavernous limestones.

Table 10.3. Ground conditions observed in sinkhole investigation in west Florida, with their maximum ratings applicable to a hazard assessment; N refers to SPT values (after Zisman, 2001).

Ground condition	Rating
Average N < 7, at depths 0–5 m in granular soil	5
Average N < 15, at depths 5–12 m in granular soil	3
Loss of drilling fluid	7
Variations in depth to water table in short distance	11
Caving of drill hole or drop of drilling tools	7
Substantial decrease in N values with increasing depth	10
Loose of soft material overlying highly fractured rock	11
Absence of clay confining layer over limestone bedrock	12
Highly variable limestone rockhead in short distance	11
Settlement near or adjoining structure	11
Soils not in normal sequence or uniform thickness	12

Table 10.4. Ranges of values for each of the factors influencing sinkhole formation in the chalk of southern England.
After Edmonds (2001).

Influencing factor	Range of values
Topographic factor – relief, infiltration and surface drainage	1–10
Chalk lithostratigraphy	0–20
Cover soil or sediments – presence and position	1–14
Cover soil and sediment lithology	0–20
Relationship to feather margins of cover soil	0–4
Water table level and fluctuations, related to chalk rockhead	0–10
Pleistocene drainage paths	0–10
Effects of glaciation	0–5

Matching the emphasis on topographic aspect on the chalk, a probabilistic approach was adopted to assess the sinkhole hazards under a bauxite tailings pond in the cockpit karst of Jamaica (Day, 2003). It was developed from previous studies of collapses in the region, and showed that sinkholes were prevalent along the lower slopes and cockpit depression floors, diminishing in number with increasing elevation. The distribution of 205 mapped sinkholes matched that of the 35 collapse events recorded in 25 years, and suggested that the 100-year probability of collapse declined from about 9% on the lower slopes to about 0.5% on saddles between the cone hills. However, such hazard assessments are highly site specific and are unlikely to be readily applied elsewhere.

10.3.2 Sinkhole hazard assessment in South Africa

Catastrophic sinkholes developed in response to mine de-watering in the dolomite karst of the South African Rand (Box 8.1). Because the de-watering was essential for operation of the hugely important gold mines that lie beneath the dolomite, the sinkhole hazard was ongoing, and considerable attention was paid to its assessment.

A multivariable method of hazard classification considered the relative importance of factors contributing to and against ground instability, with reference to the subsidence of roads (Venter and Gregory, 1987). The strength, resistance to erosion, competence and thickness of the overburden materials were considered the most important factors likely to prevent instability while the overall slope of the upper surface of the bedrock, the incidence of pinnacles and extent of cavities in the bedrock or residuum were regarded as the primary disturbing forces. Steeply sloping rockhead aids surface instability due to the higher groundwater flow velocities that increase erosion of the weak residual soils in contact with the bedrock, and pinnacles provide temporary arching support to the same soil, thereby increasing the likelihood of sinkhole formation. Over and above these factors, the original position of the water table and its new position relative to rockhead were rated very important in relation to the formation of both subsidence and compaction sinkholes in the Rand karst. Preventative and disturbing factors were all rated and summed, to see which total was the greater. A final hazard assessment was obtained by combining the outcome of the multivariable analysis with data from residual gravity surveys (Kleywegt and Enslin, 1973) and with data on variations in depth to water table.

A rather intricate multiple factor hazard classification system was created for planning township development on dolomite areas (Van Rooy, 1989). This evaluated the geology, the drainage history of the site (partly by airborne thermal infrared imagery), the gravity data, multiple parameters of the soil structure and properties and records of past subsidence damage, all with regard to their influence on sinkhole formation. A high hazard exists over negative gravity anomalies and steep gravity gradients that identify the sites and edges of large buried sinkholes (Kleywegt and Enslin, 1973). There is also a high hazard for small subsidence sinkholes over positive gravity anomalies, where thin soils overlie rockhead fissures too small to be detected by the regional geophysical survey. The hazards are low where the water table is deep and the overburden is more than 15 m thick. The methodology involved an initial zonation based on geology, drainage and gravity profiles, then confirmation with boreholes, before combining the data to produce a sinkhole hazard map that identified potentially unstable areas, but there was no quantification of hazard for this final stability rating. This type of data could be used in a modern GIS system to produce a hazard map.

A more recent classification of ground conditions in the South African karst is the method of scenario supposition (Buttrick and Van Schalkwyk, 1995; Buttrick *et al.*, 2001). This has now evolved into zoning land in any one of eight classes, based on the potential hazard from various sizes of subsidence and compaction sinkhole. The system is summarised in Table 10.5, which has been modified from the original to eliminate inconsistencies in terminology and also a few parochial factors. Each class has its recommendations for planning and construction that can maintain acceptable levels of development risk. Low-rise housing is restricted to the lowest risk category as units are normally placed on the soil profile that is prone to the

Table 10.5. Classes of sinkhole hazard defined by the method of scenario supposition, as used in the South African karst (where they are known as classes of inherent risk), with recommendations for appropriate development in respect to perceived risks from various sizes and types of sinkholes; the sinkhole hazard is quantified as the rate of new sinkhole occurrences (NSH) in events per km^2 per year; heavy industry development is excluded as it is preceded by specific ground investigations and design.

After Buttrick *et al.* (2001).

	NSH for subsidence sinkholes of sizes					
Class	0–2 m	2–5 m	5–15 m	>15 m	Compaction sinkholes	Recommended appropriate development
1	< 0.5	< 0.5	< 0.5	< 0.5	< 0.5	Any development
2	0.5–5.0	< 0.5	< 0.5	< 0.5	0.5–5.0	Residential with precautions
3	0.5–5.0	0.5–5.0	< 0.5	< 0.5	0.5–5.0	Any development with precautions
4	0.5–5.0	0.5–5.0	0.5–5.0	< 0.5	0.5–5.0	Any development with strict precautions
5	>5	< 0.5	< 0.5	< 0.5	>5	Non-residential with drainage precautions
6	>5	>5	< 0.5	< 0.5	>5	Non-residential on special foundations
7	>5	>5	>5	< 0.5	>5	Car parks and yards only
8	>5	>5	>5	>5	>5	No development

development of subsidence sinkholes. Larger developments can be placed in areas of higher risk when they are designed on the basis of detailed foundation investigations, and are founded on bedrock where necessary. An increased sinkhole risk may demand special controls on industrial water use, while heavy industrial developments, including any with major usage of water, are restricted to land of the lowest risk category, alongside housing units placed without prior investigation. The factors used to evaluate the possible occurrence of sinkholes included the nature and thickness of the soil mantle, the position of the water table, the presence of voids in either the soil or bedrock into which material from above can move, soil mobilisation by water seeping into the ground from a surface source or due to water table decline and the maximum possible size of sinkhole as a function of slope profile and soil thickness (Buttrick *et al.*, 2001). Rates of new sinkhole occurrences are preferably based on at least 20 years of local records, but have to be interpolated from local assessments of the hazards at many greenfield sites. Most importantly, all factors are assessed in terms of the appropriate de-watering or non-de-watering scenario (Case study #15).

These methods provide means of evaluating the stability of the ground surface in limestone terrains. Characterisation of the potential stability of a site in such an area requires evaluation of the effects of human impact during the lifetime of a development. In the case of an undeveloped area the potential stability initially is reviewed in terms of whether or not the area is being de-watered. A number of factors are chosen to help determine the likelihood of whether or not sinkholes or subsidence depressions would then occur. This allows an assessment to be made of the inherent likelihood of the hazard – the event of a new sinkhole making its appearance.

Over and above this, there are the financial risks related to the damage to property and to the risk of loss of life. These risks are either acceptable or unacceptable for a particular type of development. Because the sinkhole development is commonly related to human activity, then that activity may be controlled or regulated to reduce the likelihood of a sinkhole occurring (though not in the South African case, where the value of the mining exceeds the costs of the sinkholes). If a site is to be developed, then it should be done in such a way that the risk to human safety and property is reduced to acceptable levels wherever possible throughout the lifetime of the development.

10.4 HAZARD MAPS AND LAND-USE PLANNING

The presentation of geological data in the form of a hazard map represents a useful tool for planners and developers in that such a map indicates areas of potentially suitable and unsuitable land in relation to development. However, it is significant that urban, commercial or industrial development can affect the nature and frequency of a hazard. This particularly applies to karst since construction works may lead to changes in hydrogeological conditions that can induce development of subsidence sinkholes in any soil cover.

Any spatial aspect of a particular geohazard, such as the incidence of sinkholes, can be mapped, providing there is sufficient information on its distribution. The various levels of mapping that can be carried out for landslides (Einstein, 1997) have been modified for sinkhole mapping (Table 10.6); Figure 10.6 is an example of a level 2 danger map. However, these maps, like other maps, do have disadvantages. Depending on their scale, they may be highly generalised. They also represent a static view of reality, and therefore need to be updated periodically as new data becomes available. It is also important that development itself can induce sinkhole formation, thereby increasing both the hazard and the risk.

The use of geographical information systems (GIS) provides a means by which the power, potential and flexibility of mapping may be increased. Land-use planning represents an attempt to reduce the number of conflicts and adverse environmental impacts both in relation to society and nature (Bell *et al.*, 1987). In the context of hazards in karst terrains, the role of geologists should be to provide planners and engineers with sufficient information so that they can develop an area that is relatively safe for those who live or work there. Most local governments exert some control over land use and so have the opportunity to monitor and mitigate sinkhole impacts. A local government in a karst area can develop a sinkhole database and enter its data into a GIS that can produce up-to-date maps of the area, where sinkholes that may be of special concern can be flagged for added surveillance. A suitable assessment scheme (Section 10.3.1) can be developed into a sinkhole hazard map where adequate data is collected (Figure 10.7).

A sinkhole database could include fields covering distribution (location and site details), sinkhole topography (morphology, number and trends), overburden (description, type, thickness), bedrock (description, type, nature of karst), hydro-

Table 10.6. Levels of hazard mapping in sinkhole terrains.

Modified from Einstein (1997).

Level 1: Environment maps	Analytical maps that provide data as a base for later maps: topography, solid and drift geology, geotechnical properties, hydrogeology, surface drainage, climate and vegetation
Level 2: Danger maps	Inventory maps that show the locations of mapped sinkholes and provide information on size, age and other relevant parameters
Level 3: Hazard maps	Information on the spatial and temporal probabilities of mapped sinkholes; combinations of the inventory and the probabilities of occurrence, either quantitatively or qualitatively; these are also known as hazard susceptibility maps; where zoning will be in the form of none, low, medium or high, these terms must be clearly defined
Level 4: Risk maps	Combinations of hazards and consequences; several risk maps may be needed if there are various consequences to property, life and environment; the simplest combines a sinkhole hazard map with a land use map to evaluate severity of the hazard for various land uses
Level 5: Management maps	Mitigation maps showing engineering works and procedures required; regulation maps, perhaps as part of the planning process, to prevent or control development in some areas

geology and hydrology, site investigation (details of what has been carried out) and remedial works (methods of treatment), as proposed for works in Ireland (Beese and Creed, 1995). A karst database and GIS are being developed for Britain, and incorporate information on sinkholes and subsidence features, stream sinks, springs and resurgences, caves and hydrology (Cooper *et al.*, 2001). The main database fields for sinkholes include the type, shape, dimensions and fill, the date of occurrence and any damage caused. The format allows data to be continuously revised. The sinkhole database of the Pennsylvania Geological Survey includes fields that record the location, geology and description, and also provides links to digitised aerial photographs showing the sinkholes (Kochanov and Kochanov, 1997). The local government of Carroll County, Maryland, has a GIS to map vulnerable areas by combining sinkhole locations with the geology and fracture trace analysis (Devilbiss, 1995).

Sinkhole density and distribution have often proved to be useful indicators of potential subsidence. As solution voids are developed as secondary openings on inherent geological structures, data on fracture orientation and density, fracture intersection density and the total length of fractures have been used to model their distribution within a cavernous rock. Locations of areas of high probability of sinkhole collapse have therefore been estimated by plotting the intersections of

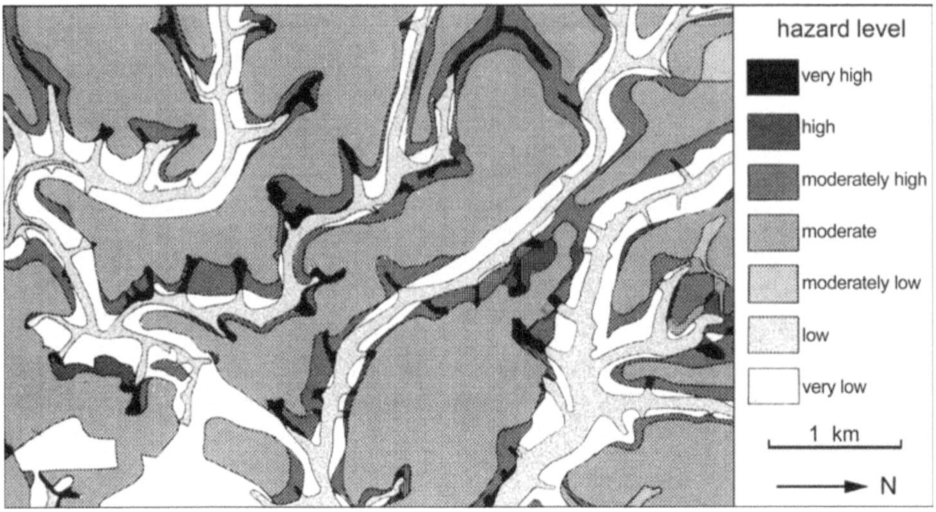

Figure 10.7. Sinkhole subsidence hazard map of part of the chalk fluviokarst forming the Chiltern Hills, just north of Reading, England; the area covers a number of dry valleys grading down to the south and south-east. In addition to the indicated zones, the feather edges of alluvium soils along the valley floors constitute linear zones of high hazard along the boundaries between the very low hazard areas on the valley floors and the moderately high hazard areas on the valley sides.
After Edmonds *et al.* (1987).

lineaments formed by fracture traces and lineated depressions. However, these map interpretations are largely based on aerial photographs showing subsidence sinkholes within the soil cover, and their validities are open to serious doubts (Section 4.4). More reliable predictions of new subsidence sinkhole events are based on distributions of soil cover thickness, though these do tend to be more generalised than specific (Section 4.4.1). Even more successful are hazard assessments based on the immediate relationships of the levels of rockhead and water table, especially when water table decline is also considered. In the covered karst around Tournai, Belgium, hazard zoning related to new dropout sinkholes that developed by reactivation of buried palaeokarst features when the water table declined in response to overabstraction from the limestone aquifer (Kaufmann and Quinif, 2002). A series of single-element maps were produced, with the first recording former collapses in order to map sinkhole distribution, which was clearly clustered. Rockhead levels were determined from hand-drawn maps, as these were preferred to a map produced by strict interpolation, and a third map showed the falling piezometric level. From the latter two maps, a fourth map was drawn to show the elevation of rockhead above the water table. The thematic maps were then combined in a GIS to delineate zones of low, moderate and high collapse hazard that were defined by sinkhole density and the relationship between water table and rockhead (Table 10.7).

Table 10.7. Defining parameters of the sinkhole hazard zones identified in the Tournai area of Belgium.
After Kaufman and Quinif (2002).

Sinkhole hazard zone	Zone definition
Low hazard	Sinkhole density <1 per km^2; limestone rockhead <10 m above water table
Medium hazard	Sinkhole density 1–15 per km^2; and/or limestone rockhead >10 m above water table
High hazard	Sinkhole density >15 per km^2

10.5 LEGISLATION IN RESPONSE TO SINKHOLE HAZARDS

Various ordinances have been promulgated by municipalities in New Jersey and Pennsylvania for land use development on karst terrains (Fischer, 1999). These range from the "do not build" type to those that required a multi-phased geotechnical study prior to development. The latter can involve a literature survey, aerial photograph interpretation, site investigation, municipal review and construction inspection. However, the occurrence or likelihood of sinkholes can significantly impede local economic development. Consequently, once the degrees of risk have been assessed, methods whereby it could be reduced need to be investigated and evaluated in terms of public costs and benefits. The risks associated with karst hazards might be reduced by restrictions on development of land, by specific mitigation measures (Chapter 11) and by the use of appropriate construction methods (Chapter 12).

Subsidence in areas underlain by gypsum that is undergoing dissolution can present severe problems. A hazard analysis of the gypsum karst around Ripon, U.K., was based on recorded sinkhole occurrences and an understanding of the processes causing sinkhole formation (Cooper, 1998). However, the area could only be divided into three zones (Figure 10.2), one in which no known gypsum occurs, one where some gypsum is present at depth, and one where gypsum is present and susceptible to dissolution (Thompson *et al.*, 1996). The zoning was devised for planning purposes. No special planning constraints are placed on developments within the first zone. A ground stability report is required for development in the second zone as there still may be a small risk of subsidence. If development is planned in the third zone, then a site investigation is required so that suitable foundations may be designed and any mitigation measures implemented. A specialist geotechnical engineer with experience of gypsum subsidence problems must sign off the investigation and foundations design before the granting of planning permission to proceed can be considered. In reality, there is no specific advice on appropriate investigation methods, and the constraint does little more than pass responsibility from the local government to the specific engineer. While the research did not produce sophisticated hazard zoning, it did result in much greater awareness of gypsum sinkhole problems among planners, engineers and the local people.

The approach taken in Ripon is similar to that taken in Clinton, New Jersey, where a local planning ordinance requires developers to consider problems associated with the limestone karst by means of multi-discipline and multi-phase investigations that are reviewed by an experienced geotechnical consultant (Fischer and Lechner, 1989). Clinton's Township Planning Board must also be satisfied with design, construction and operational information for the proposed development. In South Africa, the Development Facilitation Act 67 of 1995 places an obligation on land development applicants to carry out investigations to assess the suitability of the proposed site for development in relation to a number of specified geohazards, including sinkholes (van Schalkwyk, 1998). In 1998, this legislation was extended by the Home Consumers Protection Measures Act 95, to require hazard assessment of all land over dolomite followed by utilisation of appropriate foundation structures for all new buildings (Buttrick *et al.*, 2001). A major rolling programme to review and manage all sites on hazardous dolomite karst is currently underway at the behest of the South African government.

With respect to construction and any threats of structural collapse, regulations and guidelines related to sinkhole hazards are perhaps best addressed at the local level. There are major difficulties in presenting sensible legislation to cover such a hazard as complex and variable as sinkholes. Any sets of regulations or guidelines should only be applicable at local level, so that they can be specific to a local environment. Clinton provides one such example, and the South African laws apply to the unusual situation in the dolomite karst of the Rand. As demands for building land increase in densely populated karst terrains, increasing numbers of local governments, including many at town or county level in the U.S.A., are developing sinkhole databases into construction guidelines and then into planning regulations. In most cases the progress is not inappropriate, as it generates awareness and experience in good construction practice on karst.

Although special building and development techniques can be expensive in terrains prone to sinkhole development, planning guidelines or legislation to avoid the worst of the hazard zones and to limit aggravation of subsidence problems can certainly be cost effective, and can lead to less planning blight. In the gypsum karst in Lithuania, it is simply recommended that building should not take place near, or between, existing sinkholes and subsidence hollows, as these still may be unstable or undergoing dissolution that could bring about collapse (Paukštys *et al.*, 1997). This advice comes on the back of zoning that has been defined to control agriculture and pollution (Case study #13). Lithuania is just one of many countries, and many states within the U.S.A., that do already have well-established zoning and legislation with respect to groundwater pollution. In karst areas these controls include many restrictions specifically related to sinkholes, but only relate to construction by their impacts on drainage and pollution potential.

10.6 SINKHOLES AND INSURANCE

For individuals and small businesses, the most economical way to mitigate financial losses due to property damage caused by sinkholes is by means of insurance. In Britain, "subsidence" (including that caused by sinkholes) is usually included in standard insurance policies covering domestic buildings. This provision was introduced in 1971. This addition to policies is possibly regretted by many insurers, as losses since the late 1980s have been substantial, though mainly because of damage caused by shrinkage of clay soils in southern Britain. In high hazard areas, policyholders have to pay a significant part of any claim for subsidence damage. This is usually around the first £1,000 but can be as much as £5,000. Also, insurers may decline to renew policies where a property has a claims history or, surprisingly, may refuse insurance to new purchasers moving to a property that has been remediated following subsidence damage. A significant proportion of companies insuring against subsidence damage in the U.K. have used a national geohazard assessment digital information system based on information at scales of 1 : 50,000 and 1 : 250,000, both to assess risks and also to set premiums (Culshaw and Kelk, 1994). The British Geological Survey replaced this system in 2004 with a new one based entirely on information at a scale of 1 : 50,000.

Insurance on buildings in France does not cover the damage caused by geological hazards such as sinkholes. Instead, a small levy is raised on the sale of each house or building, and goes into a government-controlled national fund. In the event of a natural hazard causing damage, the relevant local government can declare a "natural disaster", which triggers the release of monies from the national disaster fund to cover part or all of the cost of remediation. However, it is unlikely that a single sinkhole event would be regarded as severe enough to be declared a "natural disaster", and the building's owner would therefore have to bear the entire cost of repair.

Under Florida State Statute 627.706 (2003), authorised or licensed insurance companies are required to make insurance against damage caused by sinkholes available within domestic buildings insurance cover. This will cover the cost of a site investigation to confirm the cause of damage, and, if this is the result of sinkhole activity, the cost of repair to the property. An insurer cannot refuse to renew a policy because a sinkhole damage claim has been paid out. However, if the claim exceeded the limits of coverage, the insurer can refuse to sell a new policy. An insurer can refuse cover to a new owner of a property in an area prone to sinkholes. In some counties, insurance companies may refuse to sell homeowner's insurance policies for any property within one mile (1.6 km) of a sinkhole and some use a five mile (8 km) restriction. Sellers must inform potential buyers if damage caused by sinkholes has occurred in the past.

Elsewhere in the U.S.A., insurance cover is less easily obtained. In Kentucky, insurance companies are not required to offer such policies, and sinkhole insurance may be hard to find. In Pennsylvania, sinkholes are not covered on standard

Figure 10.8. A house and road destroyed by sinkhole development in Centurion, South Africa, during a period of water table decline within the underlying dolomite aquifer; major costs will have to be borne, wholly or partly, by the home owner, his insurance company, the highways authority and perhaps by whoever was responsible for the groundwater abstraction. Photo: Fred Bell.

household policies, but cover may be obtained by payment of an additional premium. Elsewhere in the world, sinkhole insurance is either not available or is additional to standard buildings policies. Property owners are then expected to pay all the costs for sinkhole losses, except where local or central government pays part of the costs in circumstances that they deem to be exceptional.

The purpose of insurance is to spread the costs of isolated hazard events from the few to the many. But this aim is lost to some extent when private insurance companies selectively decline cover in order to minimise their exposure to risk. Even in the light of extensive geological research, most sinkhole events remain unpredictable, particularly with respect to the exact location of the next one. In terrains of soil-covered karst new subsidence sinkholes will occur, and spreading the risk by insurance may be the best means of avoiding large-scale land blight. The same applies to the rapid development of solution sinkholes in soil-covered salt karst, where some scale of subsidence is almost inevitable over large areas of salt outcrop. Appropriate construction practices can reduce the impact of most small sinkhole collapses, and are essential to ensure that new sinkholes are not induced by the land development. Beyond these provisos, insurance that is paid for by everyone and protects everyone is the most logical approach. Its classic application is in a karst terrain where a scatter of sinkhole failures can have massive immediate impacts, but imposes costs that are miniscule in comparison to the economic values of the entire area. Huge swathes of land in Kentucky, Pennsylvania, Florida, China, Slovenia and many other countries fall into this concept, and merit levels of insurance protection that are commensurate with their levels of development. Legislation may be appropriate to create insurance cover that is available to everyone, either with or without regional weighting that recognises a geohazard. In far too many cases, monies spent on legal arguments over who is responsible would be far better spent on concrete to improve or remediate structures that are threatened or damaged by sinkhole activity.

11

Prevention and remediation of sinkholes

Design and construction of buildings and infrastructure in karst terrains are influenced not only by the character of the bedrock but also by the nature and thickness of the soil cover. Voids and cavities in both the bedrock and the soil are significant as potential sites for collapse. The possibility of subsidence due to them, whether slow or rapid, therefore has to be assessed prior to the commencement of construction operations. Hence construction must involve the total environment that influences those processes that aggravate sinkhole development in karst and its overlying soils, both on-site and in the immediate environs (Sowers, 1996). Once an assessment has been made, it may be possible to change the layout of a site to avoid the potential hazards. Ground preparation on sites within karst terrain can include remedial work on existing sinkholes and dissolution features, and also works to prevent or minimise the impact of their future development. Landfills on karst provide a special case where the nature and behaviour of the cover soils is critical; any potential hazard of new sinkhole development beneath landfill threatens its integrity with the added risk of serious pollution of local groundwater resources.

Most ground failures in karst are as subsidence sinkholes developed entirely within the soil (though related to fissures within the underlying bedrock). Stabilisation of the soil is therefore a most effective means of preventing future ground failures, besides being the objective of remediation of sinkholes that have already developed in a site.

11.1 SOIL TREATMENT AS SINKHOLE PREVENTION

Many of the conventional techniques of soil improvement, including various methods of grouting and densification, are applicable to overburden on karst (Terashi and Juran, 2000). However, it is essential to note that such soil treatment is only effective when accompanied by appropriate drainage control (Section 12.1.1) so that new sinkhole development is not induced by increased water flows, regardless of whether the soil has or has not been improved (Section 8.1).

11.1.1 Grouting in karst

Innovation within the civil engineering industry has produced a great variety of grouting techniques for the purpose of improving various parameters within ground conditions (Bruce, 1994; Grouting Committee, 1998). Permeation grouting with very mobile grouts is the oldest method, and is still applied widely for soil improvement, along with its variants of hydrofracture grouting (*claquage*) in soils of low permeability and, more recently, jet grouting. Any soil improvement by permeation grouting has to offer the benefits of a more stable soil that is less prone to development of subsidence sinkholes. Techniques are generally not specific to karst situations, but may be limited in application by the scope for major grout losses offered by open fissures in a karst rockhead beneath the soils to be treated. Compaction grouting, which also compacts the surrounding soil, is generally more useful in karst (Section 11.2).

Bulk grouts are relatively cheap slurries consisting of mixtures of cement, sand, gravel and pulverised fuel ash (PFA), that can be pumped via bored holes into ground cavities. They have been widely used to treat areas of potential subsidence, but only have limited application in karst situations. Typically, the amount of cement in a bulk grout is small, as low as 4% of the proportion of sand and gravel, while rock paste is cement-free colliery spoil mixed with water; many PFAs are used in grouts as they are inherently cementitious. Foam grouts are useful for filling large voids, and may be either gaseous emulsions in an ordinary grout or expanding polyurethane foams. All these materials have only very low strength, and their prime purpose is to provide just enough support within a cavity to inhibit small-scale failure of roof slabs, and thereby prevent cavity migration towards the ground surface. Their main use has therefore been in stabilising old mines with thinly bedded and unstable roof spans. Most natural caves in strong karstic limestone have stable arched roofs in massive rock, where low-strength bulk grouting offers little benefit.

Boreholes for the injection of bulk grouts are commonly drilled on a grid pattern with a spacing of 3–10 m. This is good practice for the treatment of abandoned, partially collapsed, pillar-and-stall mines, but is not generally applicable to cavernous bedrock in karst terrain. Natural caves tend to have fewer, more widely spaced, individual conduits than in a typical mine network, and are therefore better treated by site-specific grouting schemes developed to target individual caves. Most natural caves extend far beyond the zone of influence under any engineering site, and uncontrolled filling can lead to enormous grout losses. Shuttering inside a cave can prevent grout waste, and efficient filling may only be possible after works to gain access where there is no natural open entrance. Installation of shuttering allowed the filling of only the critical half of a cave that lay beneath an abutment of Belgium's Remouchamps viaduct (Waltham *et al.*, 1986). However, pattern boreholes have been used to fill soil cavities over karst in Florida (Kannan, 1999).

An added problem with bulk grouting in karst is the inadvertent blocking of natural underground drainage conduits. This can cause back-flooding in adjacent ground or even diversions of drainage flows that then induce new sinkhole development elsewhere; installation of drainage pipes through any kind of grout fill or backfill may be essential. Any grouting of a cave system in gypsum is generally inadvisable (Cooper, 1998). Unless the caves are small, abandoned and very dry, injecting grout into them could divert drainage flows elsewhere. Very high dissolution rates of gypsum in groundwater mean that this could lead to rapid cavity development in surrounding ground. Remediation of a sinkhole in gypsum is therefore especially difficult if it is to avoid threatening adjacent land (Case study #1).

11.1.2 Compaction grouting over cavernous karst

Compaction grouting is defined as the staged injection of low slump grout to improve soil properties. It uses highly viscous grout mixtures of cement, sand, clay and PFA to displace and compress loose soil around an expanding grout bulb. The grouting increases overall density of the soil, and so improves its strength and bearing capacity, and also reduces the soil permeability. Although compaction grouting can be applicable in any type of soil, it has been used most frequently in soils finer than medium-grained sand (Bell, 1993; Grouting Committee, 1998). The benefits of compaction grouting are revealed by increased resistance in an electronic cone penetration test (CPT) carried out before and after grouting (Figure 11.1). The technique has been used to improve soft soils over bedrock of karst limestone, either prior to construction or as a remedial measure to underpin a subsided building. Once soil densification is achieved during remediation works, the growing grout bulb can be used to lift subsided buildings back to their original positions in a karst terrain (Henry, 1987). At many sites, compaction grouting can seal cavities in limestone and heal cover soils more economically than can conventional grouting, and is less expensive than using deep foundation methods (Welsh, 1988).

A site in Pennsylvania had 5–20 m of clay soils over a pinnacled limestone rockhead. Compaction grouting was applied prior to construction, and mean grout take was 6.5 m^3 in more than 800 holes (Stapleton *et al.*, 1995). There was a good correlation between grout take and rockhead depth within a number of buried sinkholes (Figure 11.2), and also with zones of soft clays previously identified by a CPT. Also in the Pennsylvania karst, two low-rise buildings for a business centre in King of Prussia were sited on soft residual soil over cavernous limestone (Welsh, 1988). It was thought that large-diameter bored piles would have to be used for their foundations, but installing the piles at varying depths, socketing them into limestone and proof sub-drilling each one would have proved extremely expensive. Compaction grouting was chosen as an alternative, to fill any voids, compact the cover soil from bedrock up to 1.5 m beneath ground level and to form a column of grout (in essence, a mini-pile) around each grout-hole from rockhead up to formation level. The maximum depth of grouting was 18.3 m, and grout pressures and quantities

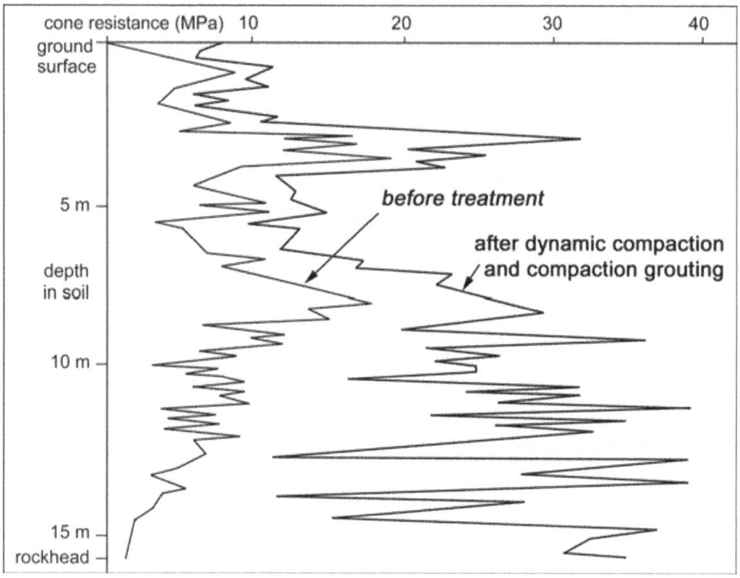

Figure 11.1. End resistance in an electronic CPT in fine sands, before and after treatment by compaction grouting, and also dynamic compaction, at a site over limestone near Jacksonville, Florida.
After Henry (1987).

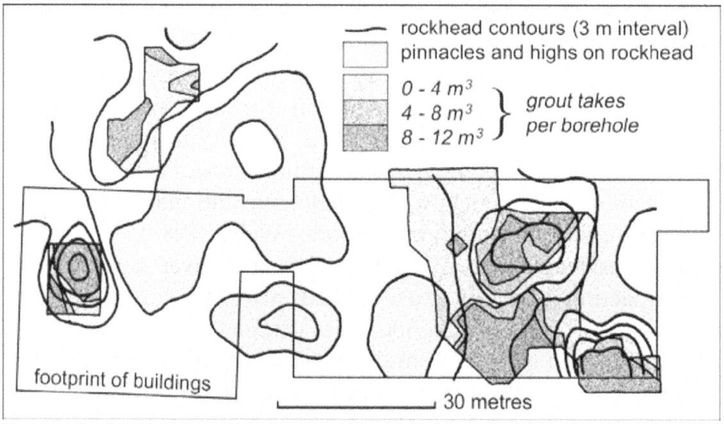

Figure 11.2. Grout takes per borehole during compaction grouting of clay soils within buried sinkholes in a pinnacled rockhead in Pennsylvania.
After Stapleton *et al.* (1995).

were controlled to minimise or eliminate heave; achievement of suitable bearing capacity was checked by a standard penetration test (SPT).

Early in construction of a highway interchange, also near King of Prussia in the Pennsylvania karst, a large number of subsidence sinkholes appeared in the soil mantle after a drought that was followed by a period of intense rainfall (Petersen *et al.*, 2003). This suggested that unpredictable sinkhole activity would continue, even with drainage measures in place. As the depth to bedrock precluded exposure of the throats of sinkholes for direct repair, it was decided to grout near-surface voids in the sinkhole zones, which also would consolidate the soft soils over the rockhead epikarst. A grid pattern was used for grouting, with primary holes at 3-m centres, followed by grouting from secondary and tertiary holes where deemed necessary. The grout-holes extended 3 m into competent limestone bedrock and were grouted upwards in 600-mm stages, taking a total of 48,000 m^3 of low-mobility grout of cement and fly ash. The works were successful but added 10% to the project costs.

After cracks appeared in the driveway and walls of a house in Tampa, Florida, investigation boreholes indicated that the lower layers of a sandy soil had been loosened by ravelling into voids within the underlying limestone (Henry, 1987). Some 218 m^3 of low-slump grout was injected via five holes extending to depths of 9–18 m. Formation of large grout bulbs at depths of 10.5 and 13.5 m then created a controlled lift of a zone 9 m across at the ground surface. The compaction grouting therefore plugged the cavities, densified the soil and restored the distressed building to its original position. Compaction grouting also was used to remediate a sinkhole in chalk that began to develop in overlying soils in the central reservation of England's M2 motorway. Eighty grout-holes were sunk over an area that extended into both carriageways, the work taking five weeks to complete. Compaction grouting sometimes has been used to form plugs in the throats of small sinkholes that are only partially choked with clay soils. As it is rarely possible to determine dimensions for the plug, a generous quantity of grout should be emplaced (Sowers, 1996).

Cap grouting is a means of reducing or sealing rockhead fissures immediately below a soil that may develop subsidence sinkholes by ravelling or collapsing into them (Sowers, 1996). It is similar to compaction grouting as it involves injecting a viscous grout (of cement, sand and PFA mixtures) to form a more or less continuous layer over the karst rockhead, and is also known as closure grouting (Fischer and Fischer, 1995). Primary grout-holes are bored on a grid at 3–6-m centres, and a grout is injected with a pressure of 1.0–1.5× the overburden pressure at the relevant depth (Figure 11.3). Grouting from secondary or even tertiary holes may be necessary to complete the grout cap. Grouting is terminated when refusal occurs at the required pressure in a grout-hole. The grout should set rapidly so that it does not continue to flow and thereby be lost into bedrock cavities; accelerators, such as calcium chloride, alkali carbonates and hydroxides, may be added to hasten setting, especially during cold weather (Bell, 1993). If grout continues to be pumped without any build-up of pressure, a temporary stop can allow some setting and blocking in the outflow zones, so that injection can be resumed the next day to achieve full pressure.

11.1.3 Soil stabilisation for sinkhole limitation

Dynamic compaction, by dropping a 30-tonne weight from a crane, may be used to compact soils that are prone to significant settlement under structural loads. It can

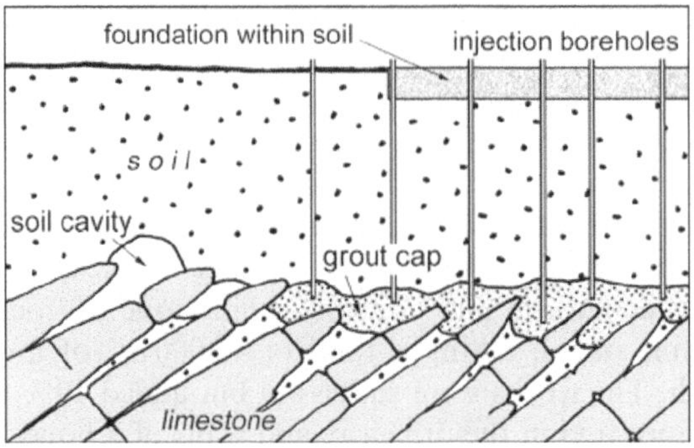

Figure 11.3. Creation of a cap of viscous grout across rockhead fissures by injecting coalescing bulbs of compaction grout through a grid of boreholes.

be used on soil cover over karst, where it can also collapse any voids migrating through the soil (Guyot, 1984). However, though this may retard void formation in soils, it does not necessarily eliminate the cause, so a new void, and then a subsidence sinkhole, may develop in the soil in the future if the conditions are right (Sowers, 1996). Dynamic compaction has also been used to collapse small caves at shallow depths in weak limestones, including the Miami Limestone in southern Florida. However, dynamic consolidation is inappropriate where vibrations could reach adjacent sites to induce new sinkholes (Section 8.3).

Precompression, by preloading a site with a temporary surcharge, is an economical means to consolidate weak and compressible soils. The method has been used on soft soils in shallow solution dolines and filled sinkholes (Sowers, 1996), but its application is limited notably by the long times required for effective pre-loading. Vibrocompaction (or vibroreplacement) can be applied to densify most types of soil within an annulus around a steel shaft that is vibrated into soft ground. As the shaft is withdrawn, coarse granular backfill is mixed into the soil to create a consolidated column about a metre in diameter. Stone columns were used to enhance the performance of soft cohesive soils in wide buried sinkholes within the chalk beneath road embankments in Berkshire, England (Rhodes and Marychurch, 1998). Vibrocompaction can also be used to collapse voids in the soil, but its ground vibrations are smaller than those created by dynamic compaction.

Lime treatment or cement stabilisation can be applied to improve the performance of many soft clays (Rogers *et al.*, 1996). Unconfined compressive strength of clay soils can be roughly doubled by the addition of small proportions of either cement or lime to a clay soil; a rule of thumb for the amount to be added is 1.0% by weight for every 10% of clay minerals present (Bell and Coulthard, 1990). A lime-stabilised layer 150 mm thick usually gives satisfactory performance, and is very effective when created beneath a raft. The site chosen for a motel in Allentown, Pennsylvania, had eight depressions, each about a metre deep, in its clay soil overlying karst limestone (Qubain *et al.*, 1995). Backhoe excavation into some of the depressions revealed that these were old subsidence sinkholes within the overburden soil. There was no surface evidence of sinkhole activity, though soil voids up to 1.2 m across were found. All the depressions were therefore exposed, to a maximum depth of 5.5 m, and all the soft wet clay within them was removed until firm soil was reached. The excavated clay was treated by addition of 5–10% lime, and used to backfill the exposed depressions in compacted lifts of 300 mm, to establish an allowable bearing pressure of 150 kN/m for conventional strip footings.

Some residual soils, derived by weathering from the insoluble components of a limestone, contain appreciable amounts of unstable clays that are subject to swelling on wetting and shrinking on drying. A thin layer of dissolution residue is commonly found immediately above pinnacled rockheads on the limestones of Kuala Lumpur, Malaysia (Tan and Komoo. 1990). Even though this may underlie 70 m of stiffer or harder residual soils, it commonly has negligible SPT N-values; some of these extremely soft clays are probably protected from consolidation by their locations between tall pinnacles. Fortunately, this weak residual soil is irrelevant where bored piles pass through it into bedrock sockets. Soft clay soils, capped by a chert residuum, were also found over a deeply pinnacled dolomite rockhead at a site for a uranium plant at Stillfontein, South Africa (Partridge *et al.*, 1981). The soil and some pinnacled crests were removed to a depth of 3 m, and replaced by a compacted soil raft formed of the stable residual chert. Excavation and fill involved 60,000 m^3 of soil, with compaction by an 8-tonne impact roller, heavy grid rollers and smooth-tyred vibrating rollers, to develop considerable densification of the soil satisfactory for foundations.

11.2 SINKHOLE REMEDIATION

Location of buildings and structures away from open and active sinkholes is always the preferred course of action, but may not be possible on a constrained site or where a sinkhole develops during or after construction. In these situations, sinkholes require complete remediation. The cheap and easy option of simply filling with any available material is almost never stable in the long term. There are far too many cases where small subsidence sinkholes have opened up, to be promptly backfilled, covered and forgotten, until they fail again some years later (Figure 11.4). Unless soil drainage is eliminated or controlled, reactivation of backfilled sinkholes is commonly inevitable.

A programme of land rehabilitation in South Africa's Wonderfontein Valley involved filling some of the large sinkholes that had developed catastrophically in response to de-watering of the karst (Box 8.1). The deep sinkholes were simply filled with available mine tailings, but this was during a period of rising water tables, after mine pumping had ceased (Swart *et al.*, 2003). The fills exhibited small amounts of subsequent compaction, but most performed well until some years later when subsidences were reactivated after a period of heavy rainfall. One sinkhole failed during filling, and was only stabilised when coarse rock debris was mixed with the mine tailings on a second phase of filling.

Figure 11.4. A small sinkhole that has been repeatedly filled, without proper treatment, under a much-patched parking lot in Bowling Green, Kentucky.
TW.

The two basic alternatives in most cases of sinkhole remediation are either to completely seal the outlet conduit at its base or to fill with graded material that will remain stable when storm-water drains through it. Sinkhole throats can be sealed with concrete plugs or capped with reinforced concrete where they enter bedrock; the use and type of reinforcement depends on the size of the sinkhole and the type of structure, if any, that will be constructed above or near it. With the drainage outlet sealed, the material used to fill the sinkhole should not be critical, and the only ongoing hazard is from any new sinkhole developed by the diverted infiltration flows. This should be prevented by adequate control of surface drainage, preferably removing all storm-water flows away from the site. Installation of geogrids in sinkhole repairs can be beneficial, especially in the manner in which they effectively delay any subsequent subsidence movements (Section 12.2.1).

A novel method of plugging stopes in abandoned gold mines in South Africa, is called the balloon technique (Parry-Davies, 1992). This consists of drilling 150 to 200-mm holes to intercept the void, and inserting strong inflatable polyethylene balloons via the holes. The balloons are then filled with polyurethane foam to form a lightweight barrier across the void, and this is capped with low lifts of fibrecrete to form a plug thick enough to support overlying backfill or bulk grout. This could be adapted to plug the throats of sinkholes but may encounter problems with cavity shapes that are less uniform than a mined stope.

Sinkhole remediation that allows continuation of its natural drainage is commonly preferable, and simple inverted filter fills can be appropriate to rehabili-

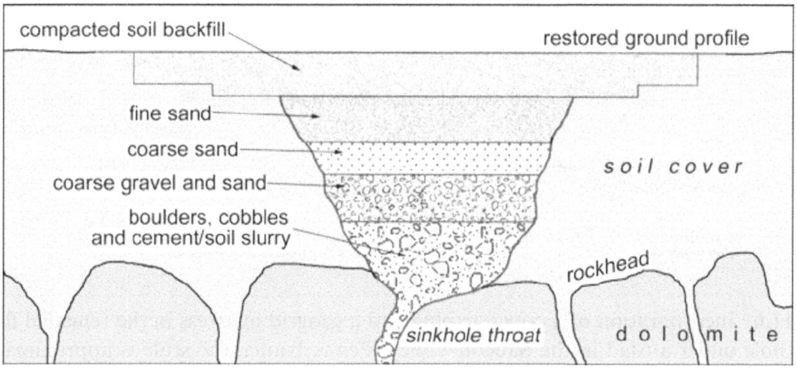

Figure 11.5. Inverted filter fill to rehabilitate a small subsidence sinkhole over dolomite.

tate subsidence sinkholes that are only a few metres deep (De Bruyn and Bell, 2001). The sinkhole is initially filled with boulders and soil-cement slurry to choke its throat (Figure 11.5). Once the slurry has set, unstable material on the sidewalls is removed by backhoe, and further backfilling uses a coarse gravel–sand mixture over the top of the boulders, followed by progressively finer sands. All backfill is placed in layers 150 mm deep that are each densified by small mechanical compactors. Soil around the sinkhole is excavated to a radius of 3–5 m, before being replaced and compacted, with or without incorporation of anchored geogrid.

Small subsidence sinkholes are remediated on appropriate scales. On a site of 57 ha to be covered by a large single-story commercial mall and surrounding pavement, near Clarksville, Tennessee, treatment of numerous sinkholes was based on the depth to bedrock and recognition of a definable sinkhole throat (Vandevelde and Schmitt, 1988). Where a throat was located, it was exposed by excavating the overburden to bedrock, and was then plugged with concrete or rock, before the excavation was backfilled. Where a throat was identifiable but the depth to bedrock was excessive, the suspect area was excavated to a depth based on the immediate ground conditions, and a concrete cap was placed above the throat area. Where the throat was near buildings, a larger excavation was made and the complete floor of the excavation was covered by geogrid, before placement of the concrete cap. At all sites, the excavation was brought up to grade with compacted clay soil. Surface drainage was then redirected to selected sinkholes where disposal wells were constructed into bedrock.

11.2.1 Repairs to large sinkhole failures

A large subsidence sinkhole that develops in the soil profile beneath a road may require a complete remediation to allow the road to be returned to use. The classic sinkhole repair is based on variations on the three-fold concept, that blocks its throat into the fissure in cavernous bedrock, then fills the bulk of the hole then caps the fill with a reinforced structure to carry the road (Waltham, 1989). That model is based

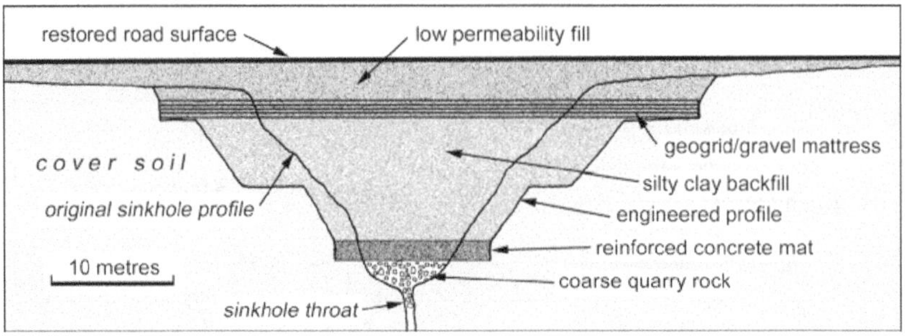

Figure 11.6. Incorporation of a concrete plug and a geogrid mattress in the remedial filling of the sinkhole under a road in the Saucon Valley, Pennsylvania; the scale is approximate. After Bonaparte and Berg (1987).

on a repair to a road in the Saucon Valley in Pennsylvania (Bonaparte and Berg, 1987), where a sinkhole nearly 30 m across had opened in a soil cover at least 15 m deep over karstic limestone where the water table had declined due to nearby mine drainage. The throat of the sinkhole was blocked with quarry rock of 150–300 mm size. This was capped by a small concrete slab that supported the main fill of locally available silty clay (Figure 11.6). A gravel mattress 35 m long and 1.2 m deep contained 13 layers of polyethylene geogrid (with a unit strength of 70 kN/m), and was designed to support the road by sagging while retaining integrity over a void up to 12 m across that may develop during any reactivation and renewed suffosion within the sinkhole. Unfortunately for this "textbook" repair, the sinkhole did open up again in the following year, destroying the road. A second repair placed a reinforced concrete slab 30 m long, which still supports the road today. It appears that bedrock was not exposed in the sinkhole throat, so the plug of undersize rocks sat in the soil profile, where ongoing drainage caused renewed suffosion of both the soil and the plugging material.

More successful, and only marginally smaller, was the repair of a caprock sinkhole under a road in the town of Macungie, also in Pennsylvania (Dougherty and Perlow, 1988). This sinkhole opened up to 26 m across and 13 m deep, though the karstic dolomite is 30 m down beneath its cover of shale and soils (Figure 11.7). It occupied the site of a large sinkhole containing a pond, which had been filled with mixed soil and rubbish, and then forgotten, when it was in farmland about 20 years before the road and houses were built over it. The repair was based on experience at the nearby Saucon Valley sinkhole (which had occurred 30 months previously, and at the time had its first repair intact). The throat was choked with 800 m^3 of large dolomite boulders (up to 1 m in size) tied with a lean-mix concrete (Figure 11.8), though the bedrock fissures could not be exposed because adjacent apartment blocks prevented any deeper or wider excavation of the unstable hole (Figure 11.9). This conical plug was capped with a reinforced concrete slab 900 mm thick (Figure 11.10), before soil was placed to restore grade. The replaced road remains stable today, with only the slightest of settlement – which can be ascribed to compaction of the soil fill.

Sinkhole remediation that does not accommodate drainage is likely to require

Figure 11.7. The unstable sides of the Macungie sinkhole, in Pennsylvania, below an investigation drill-rig that found rockhead at a depth of 30 m.
Photo: Percy Dougherty.

Figure 11.8. Chunk rock being placed at the lowest exposed point in the Macungie sinkhole.
Photo: Percy Dougherty.

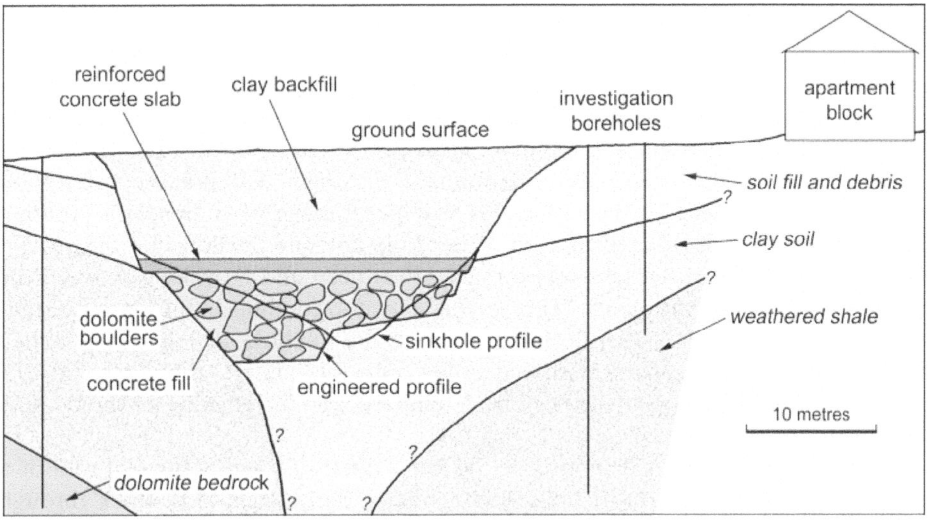

Figure 11.9. Cross section through the Macungie sinkhole with its engineered remediation and backfill.
After Dougherty and Perlow (1988).

Figure 11.10. Pouring the last of the concrete into the plug of chunk rock in the Macungie sinkhole, before rebar was laid as reinforcement to a cast concrete cap.
Photo: Percy Dougherty.

subsequent repair. When a sinkhole 30 m long and 12 m wide appeared in 1997 beneath a railway in Missouri, it was filled with rock as an emergency response; this was followed shortly by 40 m^3 of grout injected through 12 bored holes that reached just to rockhead 15–20 m down (Abkemeier and Stephenson, 2003). The track subsided again in 1999, and 84 m^3 of grout (of cement, fly ash and sand) was injected through boreholes that reached down to cavities in the bedrock limestone. Further reactivation occurred in 2002, when a new sinkhole grew to about 35 m across. At the same time, a large flow of sand-laden water entered an adjacent quarry, the deepening and drainage of which had induced infiltration through the soils beneath the railway. A new programme of compaction grouting in the sandy soils immediately above rockhead did not stop the flow into the quarry. Consequently, a grout curtain was formed 110 m deep and 60 m long between the railway sinkhole and the quarry. This reached down to conduits in the limestone that would appear to be the outlet for suffosion from the sinkhole, and were only sealed by injecting molten asphalt. With water flow into the quarry visibly reduced, the cause of the sinkhole appears to have been removed, and its remedial fill should now be stable.

These three examples demonstrate the importance of treating the real root of a subsidence sinkhole failure by adequately choking its outlet into bedrock through either one or more fissures. Engineered fill with a 150 mm grain size will rarely prove adequate in any reasonably mature karst; chunk rock of 1 m diameter is more appropriate. An unbonded plug of large rocks allows the natural drainage through to the bedrock fissure, while a massive cemented plug diverts drainage elsewhere, with the potential for opening a new fissure and adjacent sinkhole. This is not an issue where a road structure carries drainage away from the plugged sinkhole. While geogrid installed beneath a road may be invaluable for eliminating the instant appearances of open holes, it does not constitute a repair in itself. In a long-term repair of a large sinkhole in thick soil, geogrid can only serve as a warning mechanism where the perceived risk of subsequent reactivation does not warrant the costs of a large concrete slab.

11.3 LANDFILLS IN SINKHOLE KARST

The ideal landfill site should be hydrogeologically acceptable, should pose no potential threat to surface water or groundwater quality when used for waste disposal, and should have a sufficient store of material suitable for covering each individual layer of waste. With their dissolutionally opened ground fissures, karst terrains afford the least protection to groundwater from pollution by leachate, and any proposal for locating a landfill on them requires that the site investigation determine the full character of the karst before the project goes ahead (Hall *et al.*, 1995).

A site near Madison, Florida, demonstrated the crucial need for thorough investigations prior to landfill emplacement within a karst terrain (Hoenstine *et al.*, 1987). Operation of the site began in 1971 after a very cursory investigation that consisted of a number of shallow auger holes and examination of the appropriate local soil map and descriptions. Variable thicknesses of the Hawthorn Group, with an upper sand and a lower clay, overlie cavernous Suwannee Limestone beneath the site. Agricultural and industrial hazardous wastes were deposited in the landfill, even though no plastic liner had been installed under the site, and leachate soon contaminated some of the local wells. Investigation revealed that the clay unit had been breached by the ravelling soil voids that were migrating upwards from the limestone and into the sand (Figure 11.11). These were incipient subsidence sinkholes, but prior to any surface failure they created flow paths from the overlying sand into the underlying limestone, thereby allowing contaminants from the landfill access to the limestone aquifer. Pollution had therefore reached the limestone, and had migrated for 2 km through the aquifer, with slugs of contaminants moving from the landfill after periods of heavy rainfall. Construction of that landfill site had been inadequate by modern standards.

A landfill was placed over karst near Chattanooga, Tennessee, after an extensive ground investigation with boreholes and a resistivity survey found no voids within the soil cover 3–15 m thick (Tinjun *et al.*, 2003). Twelve sinkholes were cleaned out to expose their bedrock throats, which were then choked with rock and soil, and capped by graded backfill. Limestone pinnacles were found across the site in greater numbers than had been anticipated from the geophysical survey, and all were removed to 1,500 mm below formation level. The entire site was then sealed with

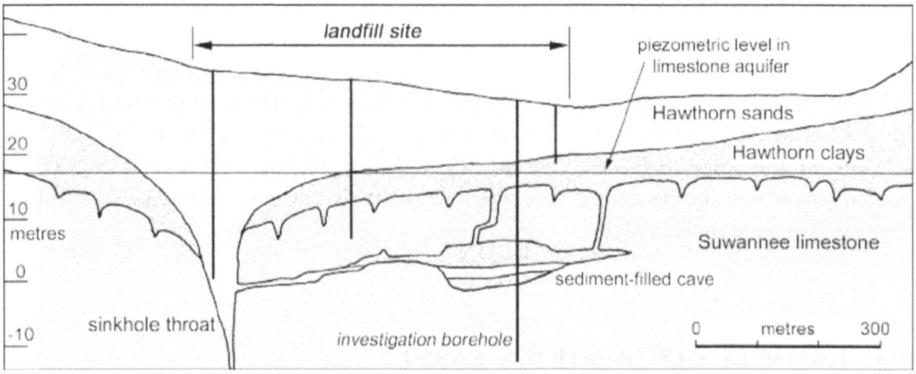

Figure 11.11. Profile through a buried sinkhole beneath a landfill site in Florida, with subsurface details conjectured from minimal borehole data.
After Hoenstine (1987).

a clay liner 600 mm thick. Liners constructed with one or two drainage layers, to convey contaminants that have percolated into the liner to a leachate collection system, have been recommended for use in karst terrains (Mitchell, 1986; Davis, 1997).

Any landfill installed in karst terrain is exposed to risk from sinkholes. Development or enlargement of a sinkhole beneath a landfill could mean that the bottom liner becomes unsupported and eventually collapses into the sinkhole. Geosynthetic reinforcement can be incorporated in liners to mitigate this risk, particularly at sites that have a thin soil cover (Siegel *et al.*, 2001). A landfill in Missouri was closed with an impermeable cap that was required to extend over a sinkhole almost encircled by the landfill (Stelmack *et al.*, 1995). Bedrock was sandstone overlying limestone, and the sinkhole was a caprock failure with a partial fill of natural soils and inert wood waste. A double layer of geogrid was installed across an area 30 m square beneath the composite layered cap that covered the entire site. This was designed to resist cap failure by spanning a new soil void 2 m across or by surviving 300 mm of settlement by the entire sinkhole fill 15 m in diameter.

An alternative to geogrid emplacement may be grout improvement of the soil, especially where needed for remediation of an existing site. A landfill in the Appalachian Great Valley of Pennsylvania stood on alluvium overlying residual clay totalling 4–18 m thick over cavernous limestone, and there was concern that sinkholes could develop at the toe of the cap after the landfill was closed (Mirabito *et al.*, 2001). Consequently, compaction grouting was used as a preventative measure against development of dropout sinkholes in the soils between the landfill and the limestone. The purpose of the grouting was to densify and strengthen the sediments so that they would be less likely to collapse. Low-slump grout, consisting of cement, fly ash and sand, was injected into holes at various depths, staged upward from 1.5 m below rockhead and into the soil cover.

11.3.1 Sinkholes as pollution sources

The conduit flow that characterises karst aquifers provides potential for rapid and unattenuated pollutant transmission, thereby reducing the effectiveness of dilution and dispersal. Karst aquifers have strongly anisotropic permeability, and exhibit major variations in flow patterns and aquifer properties that reflect the maturity of the karst, the lithology and fracture spacing, and the position within the groundwater flow system. Sinkholes constitute pathways for contaminants to enter an aquifer, thereby adversely affecting groundwater quality. Flow velocities through a conduit system can be measured in kilometres per day. Contaminants entering a sinkhole will therefore pollute the associated aquifer very rapidly. Counts of faecal coliform units increase in the groundwater of cavernous limestone in Kentucky after storm events where surface water entered the aquifer rapidly through sinkholes (Ryan and Meiman, 1996). Over a longer timescale, a pollution plume from a landfill has been recorded to extend within 3 months over a distance of 30 km through karstified limestone (Hagerty and Pavoni, 1973).

Groundwater recharge via sinkholes bypasses the purifying processes that normally takes place within the soil, and attenuation of pollutants is reduced significantly. The sinkhole plain of south-western Illinois, in which there are some 10,000 sinkholes, has chronic and widespread bacteria-related groundwater quality problems due to the rapid recharge to its shallow karst aquifers (Panno *et al.*, 1997). The major source of pollution has been private septic systems, of which 10% illegally discharge raw sewage directly into sinkholes or sinking streams.

An additional impact of pollution is created by acids, derived from landfill leachate or any other pollutant, that can attack carbonate rocks and enlarge solution features. A milk factory at Allansford, in Australia, disposed of dairy waste with little or no pre-treatment into a series of injection holes directly into karstic limestone (Shugg, 1998). About 300,000 m^3 of this waste had been disposed of annually for over 50 years, creating a pollution plume that extends more than 2,500 m through the upper 30 m of the aquifer. The waste is a mixture of soluble and suspended organic material that ferments and produces an anaerobic, odorous black sludge, along with methane and carbon dioxide. Initially, the pH of the effluent is 3 or 4, which may be further reduced by the CO_2 produced, so that it has potential to dissolve the limestone. Consequent structural collapse of the aquifer was considered, but total dissolution by the waste is unlikely to have exceeded 500 m^3, which even if concentrated around the injection wells should have minimal impact on ground stability.

11.3.2 Landfill within sinkholes

Though landfill over sinkholes may be a necessity in an extensive sinkhole karst, there can be very few cases where landfill or waste disposal that fills up inside an existing sinkhole can be considered appropriate. Unfortunately, sinkholes fall to natural temptations to improve land usability and value by filling any inconvenient depressions. They offer the added benefit of very convenient disposal sites, and

Figure 11.12. Uncontrolled landfill in a small subsidence sinkhole in the covered karst of Newfoundland, Canada.
Photo: Derek Ford.

numerous sinkholes around the world are filled with uncontrolled waste of all kinds (Figure 11.12). Under any attempt at control, engineered liners would prove excessively uneconomical in most sinkholes, if they were intended to achieve long-term stability. Even if a liner could be installed in a sinkhole, leachate drainage is impossible without expensive maintained pumping.

An even greater pollution hazard is created by illegal dumping of waste into open sinkholes. Sinkholes are frequently filled by farmers with relatively inert rubbish, though too many have a rich mixture of animal carcases that contribute foul and undiluted liquids to the underlying aquifer. Even more hazardous is the illegal practice of fly-tipping, where all manner of waste is dumped out of sight, usually during nocturnal visits. Sinkholes adjacent to country roads are especially prone to fly-tipping (Figure 11.13), and a tanker-load of cyanide waste dumped down a roadside sinkhole in the Mendip Hills karst of southern England is only the most horrific of many examples. In such situations, the best hope is that the target site is an old solution sinkhole or a mature subsidence sinkhole that has a substantial thickness of soil on its floor and choking its drainage outlets. After 14,000 litres of diesel spilled into a sinkhole from a traffic crash on the freeway across Kentucky's sinkhole plain, treatment of the sinkhole soil recovered only part of the lost fuel (Stephenson *et al.*, 2003). Flow paths and conduit routes are well known in this karst (Quinlan and Ewers, 1989), but none of the diesel reached

Figure 11.13. Fly-tipping of waste in a deep caprock sinkhole that is conveniently adjacent to a moorland road across the Llangattwg interstratal karst, U.K.
TW.

down-flow monitoring sites. It would appear that the unrecovered fuel was attenuated within the thick soil of the sinkhole, but the same could not be anticipated for the great majority of karst sinkholes.

A sample study in the karst of Virginia found more than fifty dumps in sinkholes, even though state law specifically prohibits dumping refuse, garbage and dead animals in caves or sinkholes (Slifer and Erchful, 1989). The statistics suggest that there are more than 1,300 sinkhole waste dumps in Virginia, of which over 600 are likely to represent a serious threat of aquifer pollution. While such statistics are not available for karst regions worldwide, the Virginia figures would appear to indicate the scale of this sinkhole hazard in many or all karst terrains, especially those that are semi-urbanised or densely populated.

12

Construction in sinkhole terrains

Lowland terrains of soil-mantled cavernous karst provide some of the worst conditions possible for civil engineering and construction. The soil cover may not offer a sound foundation medium because it can fail slowly or rapidly by suffosion into the underlying karst, with consequent development of destructive subsidence sinkholes. Foundations carried to bedrock avoid this sinkhole problem, but encounter the twin difficulties of extremely irregular rockhead profiles and the possibilities of open caves that may collapse within the bedrock. An engineering classification of karst ground conditions (Waltham and Fookes, 2003) offers a general view of what sort of difficulties may be anticipated or encountered within a particular karst terrain (Chapter 10). Rockhead in juvenile karst is generally sound except for isolated fissures or shallow caves: in youthful karst, rockhead only gives rise to minor problems due to its irregularities that usually can be dealt with by the use of piles or rafts; in both karst classes, small sinkholes can be spanned by rafts or piles with reinforced ground-beams. Pinnacled rockhead in complex and extreme karst may mean that piles are required to found buildings or structures in competent rock; caves commonly extend to 10 m or so in width, and therefore necessitate adequate probing below pile tips and may need filling with mass concrete.

For any construction site, the engineer has to choose between founding structures within the soil profile or on bedrock. The options are usually dictated by the thickness of the soil cover, the complexity of the karst with respect to cavity and rockhead profiles, the drainage conditions that may influence the site's susceptibility to development of subsidence sinkholes, and the scale, value and sensitivity of the built structure.

12.1 CONSTRUCTION ON SOIL OVER KARST

In the densely populated lowland terrains, where the largest share of construction activity proceeds, most areas of karst are mantled by soil covers more than a few metres deep. In these areas most small buildings and nearly all roads and railways are founded within the soil, by reason of economics. Large or sensitive structures are founded on bedrock, regardless of its depth, and most karst terrains with thinner soils are in upland regions where construction activity is less. There are various types of ground treatment that can be used to enhance the behaviour of the soil cover (Chapter 11), but engineering works have to continue on the soil profile while ensuring that new subsidence sinkholes do not and can not undermine or reduce the integrity of the new structures.

12.1.1 Control of drainage

The vast majority of new sinkholes, formed as subsidence features within the soil profile, are induced by man's activities, largely by disturbing the natural ground drainage (Chapter 8). The prime concern of every engineer working on karst is therefore to control the site drainage. The key factor is proper disposal of all storm-water, especially that off built structures, so that it is carried off-site or is channelled directly into bedrock. Where drainage is allowed to infiltrate the soil profile, it will eventually and inevitably drain down into bedrock fissures, thereby causing suffosion of the soil, and ultimately the development of subsidence sinkholes. On any site in a karst terrain, except in some juvenile types of karst, design codicils should ban soakaway drains, require the use of flexible lines and junctions on all water and drainage elements and require diversion of all inbound drainage flows. However, soakaways or dry wells can be used if they are sealed into open fissures and cased below rockhead. Sites should be landscaped in such a way as to prevent concentrated ingress or ponding of water. The backfill in service trenches should be compacted so that it has a permeability close to that of the surrounding ground. Storm-water ditches and canals should be lined over critical areas and should discharge well away from any development, and water-retaining structures should be underlain by impermeable membranes to prevent accidental infiltration into the ground.

The large areas of impermeable black-top on any major highway can shed massive quantities of run-off that must be captured by effective drainage to avoid infiltration to the soil all along the highway shoulders. Engineered options include paving of drainage ditches, lining the interface between subgrade and sub-base, and also the sides of the pavement box, with an impermeable membrane, and installing sub-base drainage into managed catch basins (Moore, 1984, 1988). Natural sinkholes can be utilised to dispose of storm-water, as long as they are regularly cleaned of any debris that might impede drainage into them. Many are lined or filled with chunk rock to maintain their stability when large flows drain into them, and many others are engineered with debris traps and conduits through the soil cover (Figure 12.1); it is essential to ensure that their subsurface structure carries drainage directly to bedrock without scope for diversion into the soil (Figure 8.4). It is also better that disposal wells constructed in existing sinkholes should be remote from buildings.

On highways and building projects alike, pipelines, storm sewers and culverts should not be bedded in permeable material such as crushed rock since this provides

Figure 12.1. A shallow sinkhole in Bowling Green, Kentucky, engineered to take storm-water drainage from an urban area directly into an underlying limestone cave without eroding the soil cover.
TW.

a conduit for adjacent surface drainage. Inadequate or inappropriate site drainage is usually disastrous over time (Section 8.1). Flexible plastic pipes are appropriate for many modern service installations over karst, as the breakage of old clay-pipes by the smallest movements induced by small-scale suffosion creates a water input, increased suffosion and an inevitable sinkhole within the soil profile. Soil voids and incipient subsidence sinkholes are commonly intercepted during trench excavation for wastewater, storm-sewer and fuel pipelines. As soon as a void is exposed, it should be inspected so that remediation can be designed to maintain the integrity of any buildings or structures on site, while at the same time preserving the hydrogeological character of the void to minimise impact on the quality of groundwater resources (Pope, 2001). Appropriate remedial measures include sealing the face of a trench with concrete, installing a durable PVC pipe across the base of the trench to permit continued drainage flow across, or encasing wastewater or storm-sewer pipes in concrete along the length of a void plus 1.5 m at either end.

12.1.2 Foundations within the soil over pinnacled bedrock

Spread footings, as either pads or strips, distribute the load of a structure to the subsoil over an area sufficient to suit the ground properties; their sizes are therefore increased for higher loading and on weaker soils. They usually provide the most economical type of foundation structure, but the allowable bearing pressure must be chosen to provide an adequate factor of safety against shear failure in the ground and also to ensure that settlements are not excessive. Spread footings are commonly appropriate on firm to stiff residual soils and also on alluvial soils that floor many solution sinkholes, but they can cause excessive settlement on some soft, compressible, residual soils on karst. The major hazard on karst is where they can lose integrity if a soil cavity migrates upward to develop below them.

A problem that may arise due to loading by a building or structure where the rockhead is pinnacled is that of differential settlement, especially where large pinnacles extend upwards into a soft soil. A stiff soil between tapering pinnacles forms a down-facing wedge that may offer additional resistance and support. However, where pinnacles or floaters (broken pinnacles forming blocks within the soil) approach the surface, differential subsidence may be significant. If the soil's potential compaction suggests that differential movement may be 10–25 mm, a building of flexible construction may be appropriate. Many houses on the soil-covered karst west of Tampa, Florida, stand safely on conventional shallow spread footings, notably where the soil cover has been shown by ground radar survey to be free of any ravelling failures.

A single-story commercial mall covering 70,000 m^2 on karst near Clarksville, Tennessee, provides a fine example of construction on shallow foundations (Vandevelde and Schmitt, 1988). Within the 57 ha site, 29 sinkholes included a few wide solution dolines, many suffosion sinkholes and some small dropouts; surface run-off was towards these sinkholes, until it was re-directed to a number of them away from the buildings, where disposal wells were installed. The buildings were designed with steel frames and masonry walls on spread footings bearing on firm to stiff residual soils or on compacted structural fill. After the site had been stripped, the subgrade in the building and pavement areas was proof-rolled with a pneumatic-tyre roller, which was also intended to reveal any unsuitable ground conditions and incipient dropouts. The subsidence sinkholes were cleaned out, plugged and back-filled (Section 11.2), and the excavation was brought up to grade with compacted residual soil. Potential settlement of most of the buildings was estimated to be < 20 mm, but would be greater where compacted fills had been placed; where fill was more than 15 m deep, settlement of 75–130 mm could take place. At sites of maximum settlement, it was recommended that walls and floors had flexible joints, service pipes were assembled with extra care and detailing was appropriate for such subsidence (I.C.E., 1977).

12.1.3 Extended foundations on rafts and mattresses

A raft permits construction of a satisfactory foundation in materials whose strength is too low for the use of spread footings. Its chief function is to spread the building load over a maximum area of ground and thus reduce the bearing pressure to a minimum. In addition, a raft provides a degree of rigidity that reduces differential movements in the superstructure, and is capable of spanning small new subsidence sinkholes that may develop in the soil beneath it. Critical to its integrity in spanning new sinkholes is the extent of reinforcement within the concrete (Figure 12.2); this is designed in conventional manner with respect to the anticipated load and to the maximum unsupported span that can be anticipated. A raft can provide adequate support for a building founded on variable soil conditions across natural fills in old

Figure 12.2. Double weldmesh reinforcement for a concrete slab, cast as foundations for an ornamental lake with a bridge and pagoda, on deep soils over pinnacle karst at Shilin, southern China.
TW.

sinkholes or across placed fills in younger sinkholes, though it is still desirable that fills should be properly compacted when placed.

To be most effective, rafts should be laid on beds of granular material to reduce friction between the ground and the structure. Post-tensioned rafts have been found to tolerate more than 25 mm of differential movement under individual houses built on unstable soils over karst limestone around Tampa, Florida (Kannan, 1999). These were placed on soil stabilised to a nominal depth of 3–5 m. This scale of movement would not have been tolerated by conventional strip footings, but excessive movements can also break a light raft. Some of the worst sites, where ravelling and suffosion of the soil indicated significant potential for new sinkhole failure, were given additional treatment by compaction grouting above rockhead.

A reinforced raft will be able to span or cantilever over unsuspected small cavities and sinkholes that develop beneath them. For low buildings, up to four stories in height, it may be possible to use an external, reinforced, ring beam around a central raft that is only a lightly reinforced raft, as a practical and economical foundation structure. Stiff reinforced concrete rafts were used as foundations for houses at a site in southern England, where a number of pipes of different diameters occur in the chalk (Rhodes and Marychurch, 1998). The rafts were designed to span 3 m if a pipe collapse occurred within the footprint of a house, and to cantilever 1.5 m at the edges. Even so, a house plot generally was rejected if a pipe was found within the plot that was greater than 2 m in diameter.

Extensions to foundations can improve their ability to span small new sinkholes that may develop beneath them. Beams of reinforced concrete connected spread footings, to form a rigid frame supporting a large house on karst in Nashville, Tennessee, after three small subsidence sinkholes developed during the ground works phase of construction (Mishu *et al.*, 1997). The same concept can extend to bridge foundations. A new bridge carries the Ripon bypass across the River Ure, on the gypsum karst of northern England, and was therefore designed and built with protection against any future ground subsidence (Cooper and Saunders, 2002). The bridge has a strengthened heavy-duty steel girder construction on supporting piers with oversize foundation pads that can span a future small subsidence feature. Furthermore the design incorporates a sacrificial concept, so that it will withstand the loss of any one pier without collapse. Some highway bridges in Pennsylvania stand on shallow foundation pads within the soil, after cap grouting of the rockhead to mitigate against subsidence due to soil suffosion into open fissures in the limestone (Knott *et al.*, 1993). These bridges are stable, but others in the same state have failed where sinkholes have undermined shallow footings (Box 8.2).

A cellular raft can be built to a sufficient depth that the weight of excavated ground equals the weight of the building, which therefore imposes no structural load. These foundations are described as buoyant, compensated or floating, and should give rise to no meaningful settlement. They can be used in karst terrains where there are deep residual soils that are soft and highly compressible, but their ability to span a new sinkhole opening beneath a building is largely a function of the extent of reinforcement within the structural concrete.

Engineered mattresses of compacted soil have been used where the natural soil mantle is of variable nature and/or thickness above pinnacled karst bedrock. A mattress limits both total and differential subsidence by spreading the imposed load to an acceptable level on the underlying soil. It constitutes a relatively impermeable layer, thereby limiting the ingress of water, and so reduces the risk of subsidence sinkholes developing. It also forms a relatively strong layer that may bridge any small cavity that develops beneath it, especially if the compacted soil is reinforced by geogrids. Mattress design relates to the thickness and geotechnical properties of the soil, and also to the sensitivity to subsidence of the planned structure. The method of construction depends on the site conditions and the fill material available.

Experience in South Africa has shown that sites over shallow pinnacled rockhead, where gravel or waste rock is available as fill, are best treated by excavat-

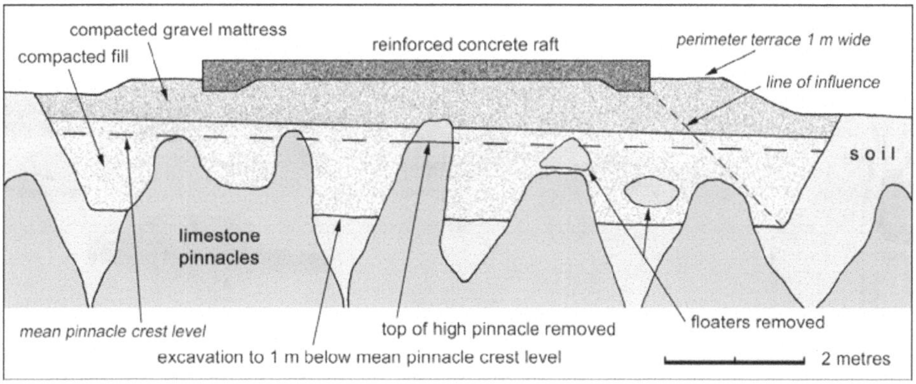

Figure 12.3. An engineered gravel mattress that extends over bedrock pinnacles and intervening soil-filled fissures.
After Wagener (1985).

ing the soil cover to a depth of 1 m below the tops of pinnacles and removing all floaters (Wagener, 1985). At some sites it may be necessary to remove by blasting the tops of pinnacles that protrude above their general level. Additional excavation may be required where pinnacles are spaced far apart, or where soils are very loose. The base of the excavation is first levelled, and then lifts of fill are placed and compacted with a small vibratory roller. A layer of waste rock is placed about 200 mm above the tops of pinnacles and similarly compacted. Uncontrolled ingress of water must be avoided during compaction, and ten passes of the roller is usually sufficient. Soilcrete or lean mass concrete can be used as an alternative to waste rock. Then, the mattress is built up to the required level (above ground level to provide good drainage) preferably using selected gravel compacted to 95% modified dry density (Figure 12.3). Alternatively, houses can be founded on light rafts with thickened edge beams on mattresses of lesser thickness. On sites with a thick soil cover over the pinnacles, mattresses consist of properly compacted gravel or waste rock, provided the latter is capped by a layer (greater than 1 m thick) of less permeable material to limit water ingress into the underlying karst.

12.1.4 Sinkhole flooding

Storm events can cause the temporary flooding of sinkholes where their capacity to drain to bedrock is exceeded. This can threaten infrastructure and development that occupies the wide solution dolines, depressions and sinkholes in lowland polygonal karst. It is a significant hazard on Kentucky's Sinkhole Plain. Most sinkhole flooding occurs where the internal sinks are choked with debris and/or sediment, where catchments have been increased by development works, or where back-flooding emerges from conduits impeded by sediment or breakdown. The latter may be uncontrollable without implementation of major engineering works; some large

Figure 12.4. A shallow sinkhole within a car park in Bowling Green, Kentucky; drains within the road surface carry storm water directly into the limestone, but their limited capacity makes flooding inevitable after heavy rainstorms.
Photos: TW and Alan Glennon.

karst depressions and poljes in both western China and Bosnia are now drained by tunnels where they can feed into adjacent depressions at lower altitudes. The first two causes of sinkhole flooding may be reduced or eliminated by appropriate drainage management that is initiated during site development.

In most terrains of polygonal karst, storm drainage cannot be completely removed, and it must be directed into available sink points within the natural sinkholes. Selected sinkholes can be engineered to behave as retention ponds (Figure 12.1), as topography generally limits the scope for built retention ponds. These and all other sinkholes may require engineering of their internal sinks so that they become efficient disposal wells that transfer drainage directly to the karst conduits; they must avoid imposing flows through the soil cover that could promote development of new subsidence sinkholes (Figure 8.4). Failure to maintain the drainage outlet can allow frequent flooding of the sinkhole, which may be acceptable in agricultural land or in car parks (Figure 12.4) but cannot be tolerated where buildings are placed.

Increasing urbanisation accelerates run-off from impermeable surfaces, thereby increasing flood frequency in sinkholes, and requires detailed planning to protect new and existing structures (Kemmerly, 1981). Measures that have been used to control sinkhole flooding include the construction of channels to drain into sinkholes, the installation of drain pipes into sink points, the transfer of storm-water from one sinkhole to an adjacent drainage basin or sinkhole, the installation of grids and gabion filters to prevent debris blockages and the enlargement of sinkholes by excavation to increase drainage or retention capacity (Crawford 1984, 2001; Moore and Amari, 1987; Barner, 1999). Urban development on karst around Springfield, Missouri, led to the city adopting an ordinance that would protect the drainage capacity of sinkholes and prevent their flooding. This included the requirement for a hydrogeological report, prior to any proposed construction works, that assesses the susceptibility of the cover soil to erosion, the relationship between sinkholes and the overall drainage area, the capacity of sinkholes to take storm-water and the impact of the construction works on sinkholes. It also prohibited buildings within the sinkhole floodplain areas, prohibited installation of basements, banned the use of heavy plant that may compact soils and reduce the permeability of sinkhole floors, demanded reinforcement of foundations and required wider easements between buildings. Where measures were not implemented correctly, periodic sinkhole flooding has persisted (Barner, 1999).

12.2 ROADS AND RAILWAYS ON KARST

Most highways, both road and rail, across lowland karst are founded on the soil cover, as economics generally preclude extensive foundations to bedrock. In mountainous karst terrains, highways are commonly cut into rock. They may therefore encounter almost any kind of foundation requirements and may require precautionary works with respect to both subsidence sinkholes in the soil profile rock cavities and open cavities within karst bedrock.

Construction of a series of motorways in the classical karst on the limestone of Slovenia encountered many difficulties with caves and sinkholes in bedrock (Case study #3). Many caves were found when they collapsed during blasting operations; they were either capped with concrete slabs (Figure 12.5), or were filled after any interior sediments had been removed and replaced by compacted rubble (Slabe,

Figure 12.5. Solution sinkholes exposed during construction along a highway footprint in Slovenia, one cleaned out ready for filling, the other with a concrete plug forming its floor. Photos: John Gunn.

Figure 12.6. A sediment-filled cave in the wall of a road cutting in Slovenia, seen when it was first exposed and later with a masonry wall across its sediment fill.
Photos: Tadej Slabe.

1997; Knez and Slabe, 2002b). Caves exposed in the sides of road cuttings were walled up to avoid rock fall (Figure 12.6).

Ground investigation for a major road improvement in England revealed pipes in weathered chalk, both at outcrop and beneath the cover of Tertiary and Quaternary sediments. The pipes, up to 20 m in diameter and 7 m deep were filled with clays, sands and gravels derived from cover sediments (Rhodes and Marychurch, 1998). Where pipes of less than 1 m diameter occurred beneath bridge sites, the infill was excavated and replaced by structural fill or mass concrete. Pipes up to 8 m across were spanned by reinforced concrete rafts, with diameters double that of the pipes, transferring the load to the adjacent chalk. Where a pipe was wider than 8 m, a bridge was founded on piles into chalk no closer than two pile diameters beyond the pipe to avoid pile–pipe reaction. A geogrid-reinforced granular blanket was used to support embankments over small pipes filled with medium dense sands and gravels. Smaller pipes with a soft cohesive fill beneath embankments were covered by a concrete cap, and soft fills in larger ones were improved with vibro-replacement stone columns capped by geogrid mattresses.

Deep soil-filled fissures (grikes) in strong limestone may require grouting or filling with mass concrete. Alternatively, a graded rock-blanket formed beneath subgrade level throughout a fissured zone can prevent the loss of sub-base material and minimise the effects of strains in the surfacing materials. While grouting, sealing or capping fissured karst may stabilise the ground immediately beneath a highway, it may also divert local drainage to initiate formation of new subsidence sinkholes in adjacent land. Placing of a new highway embankment may act in a similar manner, and may also cause its own undermining by new sinkholes if it is permeable and has no controlled internal drainage. Successful design was applied to a road through cone karst in Puerto Rico that had to cross 12 large solution dolines between the cones; each of these had active drainage through its soil floor into the limestone beneath (Vasquez Castillo and Rodriguez Molina, 1999). It was appreciated that the new road embankments should not impede the natural drainage, so the dolines were investigated individually, by drilling, geophysics, dye tracing and pump testing, to estimate their drainage capacities and storage requirements. Soft sediments were then cleared from the road footprint across the doline floors, and were replaced by geotextile-lined filters that transferred highway drainage efficiently into the limestone. Unstable cavities in shallow bedrock were collapsed by surcharge loading prior to construction of the embankments, which have subsequently retained integrity.

At some sites on mantled karst, adequate soil stabilisation may be unrealistic, and the expensive alternative is founding a road or railway on deep bedrock. For many years a railway trackbed stood on soils 10–20 m thick over karst limestone in the coastal plain of North Carolina. A new reservoir, on the soil cover and behind a 10 m high earth dam adjacent to the railway, then modified the local drainage regime and within two years of impoundment initiated numerous subsidence sinkholes around and underneath the trackbed (Erwin and Brown, 1988). As more sinkholes developed through the soil cover, temporary repairs and diversions over a period of 14 years, involved sinkhole filling, bedrock grouting, dynamic compaction of the cover and temporary draining of the reservoir. The railway is of military strategic importance, so a permanent solution was sought; it was rebuilt on a land bridge, 1,240 m long, across the entire sinkhole zone. This stands on bored piles (caissons) 1,200 mm in diameter that were founded on rockhead or were drilled up to 5 m into bedrock; probing established 1.5 m of sound rock under each end-bearing pile and no voids more than 150 mm high in the next 1.5 m of underlying limestone. Despite drilling dry for the piles, eight sinkholes, the largest 9 m in diameter, developed during construction works; these were simply filled with compacted soil, as the piles did not rely on soil pressure for lateral support.

Pipelines are increasingly being built as efficient modes of bulk transport. When buried in the soil they impose very modest additional loads on ground, and their main impact on a karst terrain is likely to be the disturbance during their construction. The new oil pipeline from the Caspian Sea to the Mediterranean crosses the gypsum karst near Sivas, Turkey, where a sinkhole hazard was recognised and evaluated during route selection. Small subsidence sinkholes pose no risk to the pipeline, but a few large natural collapse sinkholes are known in the area (Figure 3.18). Though the likelihood of a large collapse was considered to be very small, the consequence of such an event would be very serious, because a pipeline fracture over a new sinkhole would be economically disastrous and environmentally catastrophic. Maximum collapse width in a single event was considered to be 25–40 m, and the pipe, 1.067 m in diameter, was modified to a steel thickness of 22.6 mm as a precautionary measure for the entire route across the karst, so that it could safely span 44 m, even with a worst-case soil wedge balanced on the exposed pipe (Arthur *et al.*, 2004).

12.2.1 Geogrid as a sinkhole defence

Synthetic plastic reinforcement incorporated within a road structure offers an economical alternative to placing concrete slabs where soil and sub-base undermining by new subsidence sinkholes is a perceived threat (British Standards, 1995). Though geosynthetics are available in various forms, heavyweight geogrid is most applicable

Figure 12.7. Heavy paralink geogrid being rolled out onto a road sub-base as a protection measure against sinkhole development in the underlying soil.
TW.

to bridging sinkhole voids (Bonaparte and Berg, 1987); it is produced with tensile strengths up to 400 kN/m along its length (and typically 40 kN/m across its length). Geogrid can be rolled out during sub-base construction to cover any area prone to sinkhole failure (Figure 12.7).

A programme of full-scale tests and numerical modelling showed the value of geosynthetic sheets with a tensile strength of 200 kN/m in spanning dropout sinkholes up to 4 m across (Villard *et al.*, 2000). The material is now used under sections of trackbed for France's TGV high-speed railway. The same study also found that the surface subsides as a shallow dropout where the geosynthetic lies beneath a cover thickness less than 75% of the underlying void width, while soil arching supports an undisturbed surface over a thicker cover (Figure 12.8). In either process, the geogrid delays surface collapse, and thereby provides a window of opportunity for remedial action before any suffosion sinkhole can increase in size, though it cannot be relied on to span large dropout failures.

The new Ripon bypass was built across gypsum karst in northern England, where new sinkhole events average one per year within an area of a few square kilometres. The road line was constrained by topography and existing land use. Grouting of cavities proved not only impractical, because of their size, but also undesirable, as it would have led to accelerated dissolution of gypsum under adjacent ground (Cooper and Saunders, 2002). Geogrid was therefore incorporated in the roadbed structure, designed to provide support for at least 24 hours after being

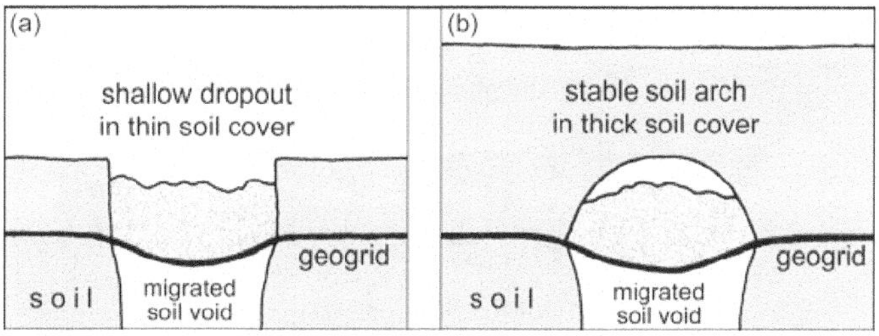

Figure 12.8. Sinkhole prevention by geogrid installed within soil profiles: (a) development of only a shallow dropout where cover is thin; (b) development of a soil arch in a thick soil over the geogrid.
After Villard *et al.* (2000).

undermined by any new sinkhole subsidence. A cluster of 51 small subsidence sinkholes in glacial drift were found during upgrading of Britain's North Wales Coast Road across a limestone outcrop (Nichol, 1998). Loose material was removed from sinkholes within the road footprint and was replaced by crushed rock or concrete, and drainage outfalls were directed into surface watercourses away from the zone of subsiding ground. Heavy geowebbing was then installed 1 m below the road surface along the complete section affected by sinkholes, and this has ensured that two subsequent subsidence events developed so slowly that remediation of the causative new sinkholes could be taken before they threatened road traffic.

While geogrid installed beneath a road may be invaluable for eliminating the sudden appearance of a sinkhole, it does not constitute a complete repair of a new or existing sinkhole (Section 11.2). Within a long-term repair of a large sinkhole in thick soil, geogrid can only serve as a warning mechanism where the perceived risk of subsequent reactivation does not warrant the cost of a large concrete slab.

12.3 FOUNDATIONS ON KARST BEDROCK

A characteristic of karst terrains is that the rock surface where exposed, or the rockhead beneath a soil cover, may be extremely irregular, with deep fissures, cutters or grikes between residual blocks, clints or pinnacles of intact rock. Typical cavernous limestones are strong enough that structural loads can be placed on most pinnacles, as long as they are not detached, undercut or cavernous. Where numerous pinnacles occur near the ground surface, a raft of reinforced concrete can be used to span between pinnacles. This extends the concept of the soil mattress (Figure 12.3), except that the entire structural load may be borne by the rock pinnacles. Such a raft is normally only feasible where the soil cover is less than 3 m thick. A reinforced raft also retains its integrity when soil is subsequently washed out of the fissures by unintended input of drainage (Figure 12.9). Most bridge footings within Pennsylvania are founded on spread

Figure 12.9. A concrete foundation, directly on dolomite bedrock in South Africa, that has retained integrity after an open fissure has been exposed beneath it when its soil fill was washed out.
Photo: Dave Haskins.

footings or mats placed directly onto limestone where it lies under less than about 3 m of soil (Knott *et al.*, 1993). Prior to construction, any irregularities in the bedrock surface had been filled with lean concrete, to provide more uniform support for the shallow foundations.

Where soil is 3–7 m thick, a structure may rest on beams that transfer loads to short piers down to the tops of individual pinnacles. If this type of foundation is used for larger and heavier structures, it is necessary to proof-drill the pinnacles (Partridge *et al.*, 1981). Pinnacles that are potentially undersized or unstable may require assessment by probe drilling splayed at 15° from the vertical from the point of foundation (Foose and Humphreville, 1979). Extension beams can be incorporated into a raft to transfer its load to available pinnacles where they are lacking at critical locations (Figure 12.10). A television station in South Africa was founded on a grid of beams that included extensions to reach stable pinnacles outside the building's footprint (Figure 12.11). Fortuitously, the pinnacle tops were at a uniform level on

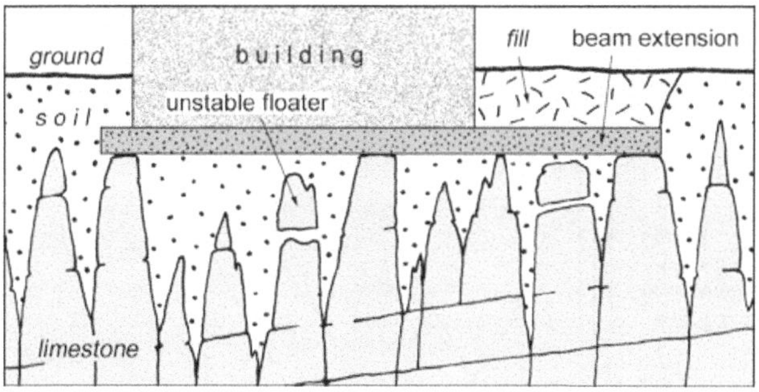

Figure 12.10. Foundation raft or ground beams with extensions to reach stable rock pinnacles in a deeply fissured limestone rockhead.

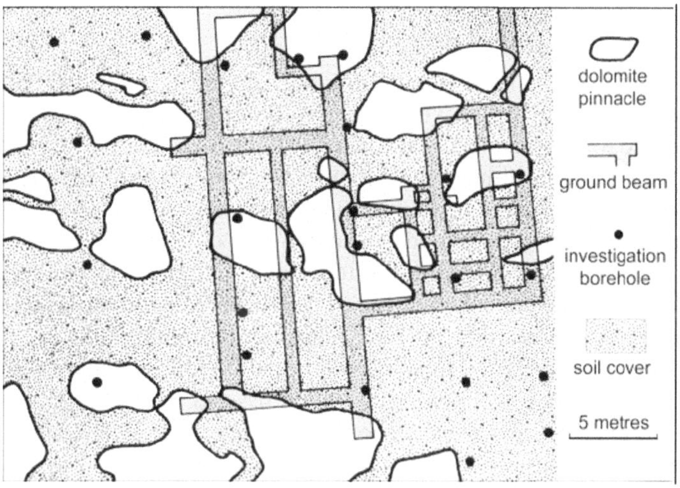

Figure 12.11. Plan view of ground beams designed and extended to reach stable dolomite pinnacles to support a building at Zeerust in South Africa; soils reached more than 20 m deep between the pinnacles.
After Brink (1979).

an old erosion surface, and were completely exposed by removing just 1 m of soil cover, leaving soils that reached depths of at least 20 m between the pinnacles (Brink, 1979). Massive concrete rafts, reinforced with steel girders, support a power station in Illinois by bridging over fissures and weak soil zones between pinnacles of sound limestone (Swiger and Estes, 1959).

The alternative to spreading loads onto available pinnacles is to improve the ground beneath individual footings or piles. The most common means of rock reinforcement is by dentition, whereby concrete or masonry infill is placed in open fissures and cavities in or just below rockhead (Figure 12.12(a)). Deeper fissures may be cleaned of soft soils before being packed with permeable backfill prior to capping

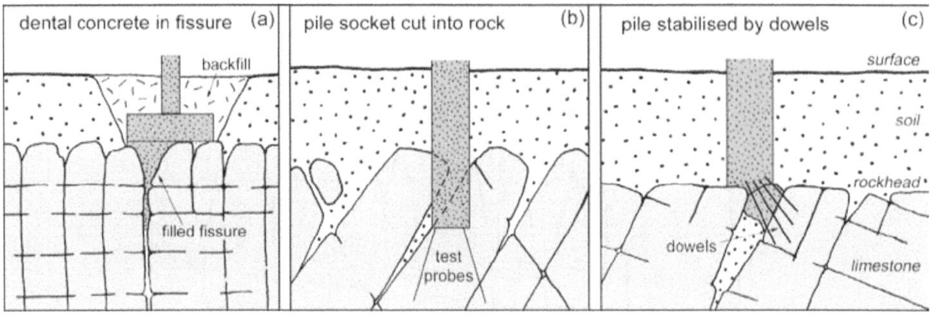

Figure 12.12. Pile integrity in karst achieved by treatment of rockhead fissures: (a) dental contrete filling in a wide fissure; (b) pile socket cut into rock and probed beneath; (c) pile stabilised by dowels through concrete fill.
After Sowers (1986).

with concrete or stonework; adequate drainage should be provided within the engineered fill. Steel dowels up to 3 m long may be drilled into adjacent bedrock and grouted in place to reinforced dental concrete (Figure 12.12(c)). They can be useful in thinly bedded limestone with complex fissure patterns where low loads are needed to increase stability and where the discontinuity surfaces are at least moderately rough. Deformation in the rock mass loads the untensioned dowels until sufficient stress is developed to prevent further strain. Pre-tensioned rock bolts, up to 8 m long, may be used as reinforcement to enhance the stability of discontinuous carbonate rock masses, however, their normal purpose is to induce compression in the rock mass and thereby improve shearing resistance on potential failure planes, so their application is limited in karst with open or soil-filled fissures. Every karst site is different, and many sound structures are site-specific adaptations (Sowers, 1986).

Integrity of individual footings that bear onto bedrock is only assured where sound rock has been proven beneath their loading points to a depth adequate to span any unseen cave within the karst rock. Depths to which drilling or probing must prove intact rock relate to the intact and mass properties of the karstic rock, and guidelines related to rock types can only be approximate due to the extremely variable nature of karstic ground (Table 7.1). Though rigorous proving of rock is essential under pad foundations that impose high point loads on the rock, the same does not apply to strip or raft foundations that are adequately reinforced. These distribute loads onto sound rock and are able to span small voids, and the same applies to structures on multiple pad footings that incorporate sacrificial design, whereby any one can fail while the structure retains integrity. Wide caves at shallow depth create the main hazard where stable arches cannot develop in their roof profiles, and these may require filling with concrete (Section 11.1.1). Where the cave has to continue to carry drainage, partial filling may be required, and this is normally only practicable where the cave is accessible, either naturally or by an engineered entrance (Figure 12.13).

The poorly cemented Miami Limestone, which is widespread in southern

Figure 12.13. A masonry wall built beside the underground stream inside the St. Augustine's Cave in Ireland in order to reduce the unsupported span beneath a main road that stands on a concrete slab.
TW.

Florida, may lose much of its strength in the weathered zone, and so not present the strong rockhead that is typical in most karst terrains. Low-rise concrete buildings have exhibited major settlements with foundations on this soft, porous, oolitic limestone where it has been subject to dissolution, leaving it leached and highly porous, with little induration remaining in the 2 m below rockhead (Sowers, 1975). One building was surrounded by a crack in the more intact surface crust and settlement exceeding 100 mm was recorded. Standard penetration tests (SPTs) carried out at the site revealed blow-counts of 1–2 for the weakened limestone. Reinforced rafts can be used at such sites, but end-bearing piles would require founding on a deeper horizon of higher strength. Though these conditions do not apply on limestones with high intact strength that form most karst terrains, they are replicated in the softer chalks of northern Europe and also in some younger and poorly indurated coralline limestones that form coastal karsts and raised terraces.

12.3.1 Driven piles and pin piles

Where the soils beneath a proposed structure cannot provide adequate support, the structural load can be transferred to the underlying bedrock by means of piles. Driven piles of steel H-section or small-diameter concrete can be hammered through the soil profile, until they meet refusal at or close to rockhead. Alternatively, non-displacement piles of steel or concrete can be grouted or cast into pre-drilled

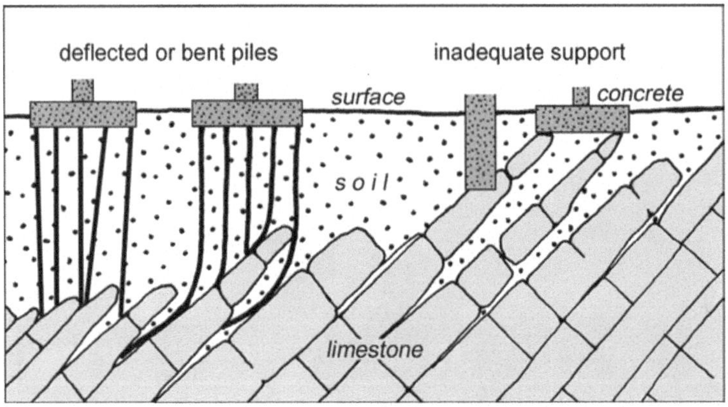

Figure 12.14. Unsafe foundations on pinnacled rockhead in karst due to deflected or bent driven piles and to footings on inadequate pinnacles.
After Sowers (1986).

holes. A third alternative is the use of large-diameter bored piles (Section 12.3.2). The choice of pile normally depends on the local ground conditions, but any type can present special difficulties in karst terrains (Wyllie, 1999).

Steel piles driven to rockhead successfully support numerous bridges on variable karst terrain in Pennsylvania (Knott *et al.*, 1993), and H-piles 350 mm in section can carry loads of over 1 MN. However, rockhead relief, with or without pinnacles, that is characteristic of karst, commonly creates difficulties in placing end-bearing piles on rockhead of strong limestone. Piles frequently have to be founded at different elevations, while floaters and detached pinnacles can both make pile diving difficult and also threaten pile integrity where driving refusal on them leads to their mistaken interpretation as sound rockhead (Figure 12.14.). Driven piles may fail to find a sound footing on sloping rockheads, where a tendency to move along open fissures or pinnacle faces can cause bending of steel piles and deflection of concrete piles, especially in steeply dipping limestones (Sowers, 1975). There may be occasions on driving high capacity piles, when the pile continues to penetrate slowly under repeated hammer blows. It is difficult then to determine whether the pile is progressively fracturing, the pile tip is crushing or the pile is bedding into the rock, though dynamic pile driving analysis, which measures wave propagation generated by each hammer blow, may help interpret what is happening (Sowers, 1996).

Non-displacement piles are formed by drilling holes, then inserting steel H-beams or narrow rebar structures, and filling with concrete. Also known as pin piles, minipiles or micropiles, they can be used in karst ground where it is uneconomic to install driven piles. They normally have diameters of 125–300 mm, consist of steel reinforcement with cement grout, and gain their supporting capacity largely from skin friction (Tarquinio and Pearlman, 2001: Dotson and Tarquinio, 2003). Steeply sloping rockhead surfaces can easily deflect a drilling bit, so drilling should progress slowly and carefully to allow the bit to cut into the bedrock, after which a normal drilling rate can be resumed. Holes may be lined with casings that are or are not left in place. Casings can prevent loss of concrete into karst voids when casting or grouting a pile into a drilled hole. Alternatively, uncased holes can be useful where voids require to be filled simultaneously with the pile grouting. Holes through soil and floaters may collapse during drilling, thereby preventing any casing being installed down to rockhead, so an eccentric bit can be used to produce a larger hole with the casing being installed simultaneously.

A benefit of micropiles is that they distribute structural loads and may therefore perform well where karst offers non-uniform and potentially cavernous ground conditions. Heavy silos were to be placed on bare limestone karst in Greece, but a programme of compaction grouting failed due to heavy losses of grout with minimal benefits (Sotiropoulos and Cavounidis, 1979). They were therefore successfully founded on 1,256 micropiles of 800 kN capacity, 142 mm in diameter, averaging 11 m long and each taking 1–4 tonnes of grout. They were also used in covered karst in Tennessee, where soft soils, 6–9 m thick, overlie strong limestone that contains open caves (Heath, 1995).

It may be argued that piles should only be used in karst areas when other foundation structures are not feasible. Though piling may be difficult in karst, end-bearing piles have the benefit of transferring structural loads to bedrock and thereby avoiding the major hazard from subsidence sinkholes in unstable soil cover. Piling through ground in which there are notable voids may precipitate collapse of the voids and in turn lead to piles being buckled or even sheared, but any piling in bedrock creates safer foundations than those found in the soil over the same cavernous ground. However, it may be inadvisable to use piles in ground that contains beds of cavernous gypsum, because its relatively rapid dissolution could create new voids and so destabilise a piled structure (Cooper, 1998).

The lateral stability of piles passing through collapsed zones or through large voids is also open to question. Cast pin piles have been placed successfully through caves in Greece by using expanding sleeves to contain concrete columns within the voids, while not filling the entire cave systems at unnecessary expense (Sotiropoulos and Cavounidis, 1979). The pin piles were used to transfer loads through caves found within 4 m depth beneath the base of large-diameter caissons supporting bridge piers (Figure 12.15). A sleeve was lowered down the bored hole and expanded to twice its diameter when filled with concrete to create a wider unconfined column with a bearing capacity matching that of the narrower, confined pin pile. Comparable textile sleeves, without the widening capacity, have been used on micropiles (Heath, 1995). There must always be an element of doubt over the integrity of such sleeved piles through inaccessible caves, and the larger capability of modern drilling rigs means that they can generally be replaced by cased holes of larger diameter.

Piles that are placed through soil cover to a pinnacled rockhead on limestone will require to be of greatly variable lengths. Steel H-piles that are easily cut or welded can be advantageous, as pre-cast concrete piles are difficult to shorten or splice. It has been found that the mean length of end-bearing piles can be about 30%

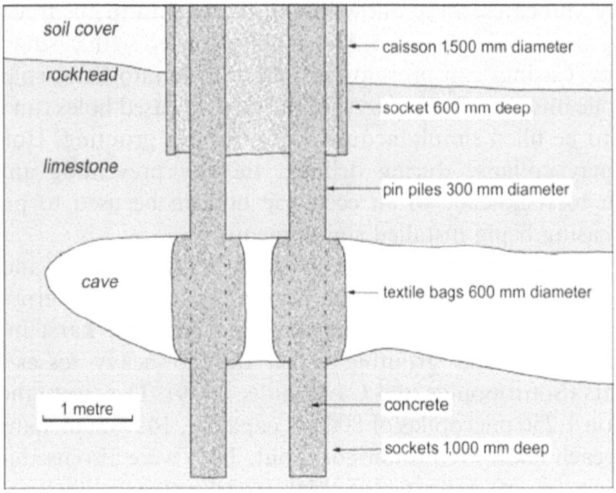

Figure 12.15. Textile bags used to form oversize concrete columns where a cast-in-place pile is bored through an open cave.
After Sotiropoulos and Cavounidis (1979).

longer than the mean rockhead depth determined during ground investigation (Foose and Humphreville, 1979).

12.3.2 Bored piles and caissons

Conventional large-diameter bored piles can be used where heavy structural loads have to be carried. Although drilled shafts may represent the only type of viable deep foundation at some sites on karst, they are expensive. Extra costs may accrue due to additional depths to reach sound rock because of the occurrence of fissured rockhead, buried sinkholes or caves, due to additional rock excavation through weathered zones, due to sloping rockhead surfaces or shale horizons or due to an inability to de-water (Destephen and Wargo, 1992). Also known as caissons or piers, bored piles are normally 1–2 m in diameter and carry loads of 1–5 MN. Most cavernous karst limestone is strong enough to carry the high end loads of bored piles, once the rock is proven to be free of voids immediately beneath the pile. Some of the softer limestones, including most chalks, cannot carry such high point loads, and bored piles are commonly inappropriate on them.

Pile shafts are sunk by conventional methods, normally by large drilling machines, though hand-dug shafts are still used in some difficult ground. Once the shaft is cleaned out to reach solid ground, a rebar cage is lowered into it and the concrete is then cast in place. A casing is normally needed for temporary support through soil, but may be extracted as the concrete is placed. These piles are largely end-bearing. Though part of their load may be carried by skin friction, this is normally minimal where they are through weak soil onto strong limestone bedrock. They are normally socketed about 1 m into sound rock, and rarely need to be belled out on strong karst limestone. With their high end loads, proving of sound rock beneath them is critical in karst, where open caves could diminish or destroy pile integrity (Section 12.3.3).

High-rise buildings commonly stand on large bored piles, because no other pile type has the necessary bearing capacity. Once selected for a project, there is then normally no alternative to boring the pile shafts however deep is required to found them on stable rock. Pinnacled rockhead can therefore provide appalling ground conditions, especially where it is very well developed in tropical karst terrains (Figure 5.3). Adjacent piles on building sites in Kuala Lumpur, Malaysia, have reached rockhead at depths of 5 m and 80 m, where the first pile found a stable pinnacle, but the second passed down a deep soil-filled fissure. Even in less mature karst in Alabama, some pile borings 900 mm in diameter have encountered bedrock on one side of the hole 25 m before it was encountered on the other side, where the site happened to be over the steep wall of a karst fissure (Cooley, 2002). Such unexpectedly deep piles can be required on any site where a deep buried sinkhole breaks the rockhead, and such can occur in almost any karst (Chapter 5).

12.3.3 Proof testing for piles in karst

Cavities at shallow depth constitute a potential hazard on any karst site where structural loads are imposed. However, they are totally critical to the integrity of piles that carry large point loads onto rock, and commonly do so at depths well below the ground in areas of deep rockhead. It is therefore essential to prove the competence of the rock below the level at which any piles are founded.

At any pile site, karst limestone should be assumed to contain dissolution cavities until it has been proven otherwise. In many cases, the position at which a pile is to be sunk should be probed or drilled prior to the pile being emplaced. Where the structural design leaves no options on the locations of bored piles, it may be more economical to drill the pile shaft first, as far as a socket into sound rock, and then probe the floor for the shorter distance to prove intact rock beneath.

The depth of probing required beneath any pile site should depend on the maximum size of any caves that are likely to be present (based upon local knowledge or the class of karst, as in Chapter 2), on the bridging capacity of the rock mass (based upon rock strength and the nature of its fractures and discontinuities) and also on the load and stress to be applied. The first two factors are very difficult to assess within natural ground (Section 7.3), and therefore it is only possible to present the very broadest of guidelines that can be generally applicable (Table 7.1). Engineering practice may be based on local knowledge of specific karst areas, and should therefore produce more useful local specifications, but there are some surprising variations in available published data (Table 7.2). Guidelines that probing should reach a depth of, for example, up to three times the pile diameter below foundation level are based on concepts of the bulb of pressure and take no account of the width of an unsupported cave span; they may prove inadequate in karst terrains with large caves (Section 7.3).

A displacement pile that is driven onto or into limestone may punch through thin layers of intact rock overlying dissolution cavities. In such cases, and assuming that the pile is not damaged, driving must continue until rock of adequate strength provides the resistance for bearing capacity to be satisfied. A bored pile can be continued down through ribs of sound rock and intervening caves, though careful drilling may be needed to re-enter solid rock through any sloping floor of a large cave.

Where any cavities are revealed by a ground investigation beneath the intended site of a building or structure, then relocation to lie over non-cavernous ground is the best solution, if that is possible. If not, further investigation of the cavity or cavities may be required to determine their size and condition, and there may be benefit in using down-hole cameras. These can view either radially or axially, with focusing and rotation of the head controlled from the surface. The heads have their own light source, and some models can work under water, where ultrasonic scanners also have application in flooded caves. Once the location and dimensions of a cavity have been determined, appropriate remediation can be designed. Filling can be achieved by bulk grouting (Chapter 11). However down-hole camera images can be difficult to interpret, and if the caves are large, direct exploration after gaining access is preferable. This is particularly valuable where shuttering can be placed in a cave to prevent grout loss into areas of no structural significance, and where soft cave sediments may require removal before they can diminish the integrity of concrete poured over them.

12.4 TUNNELS THROUGH CAVERNOUS GROUND

Construction of tunnels in karst can encounter caves at any depth beneath the ground surface. Rock collapse may have already occurred or potential collapse may constitute a significant hazard, and remedial measures can match those of surface works that found on cavernous rockhead. Unstable sediment and debris is normal inside caverns, where it may require precautions and treatment comparable to surface sites that are prone to development of subsidence sinkholes. The very varied difficulties that can be posed by large caves have led to recommendations for probing ahead by up to 50 m on tunnel headings in some of Greece's karst (Marinos, 2001).

A large cavern in the path of a tunnel presents a difficult problem, and inevitably delays excavation. Particularly where excavation is by a tunnel boring machine (TBM), the tunnel may have to be relocated or diverted around a large cavern, as it is technically impossible to head a TBM out of a rock wall into an open space. A number of railway tunnels in the well-developed karst of south-west China have had to be re-routed around large caves (Chen, 1994). Some smaller caves can be filled. A cave breached at mid-height by the TBM driving the Trebišnjica hydro-tunnel, in Bosnia Herzegovina, was filled with 386 m^3 of concrete to provide a solid floor over which the TBM could advance (Milanovic, 2000). Tunnels in both China and Herzogovina have been successfully advanced over concrete arches that have been built with spans of up to 15 m inside caves. Concrete filling of caves may only be

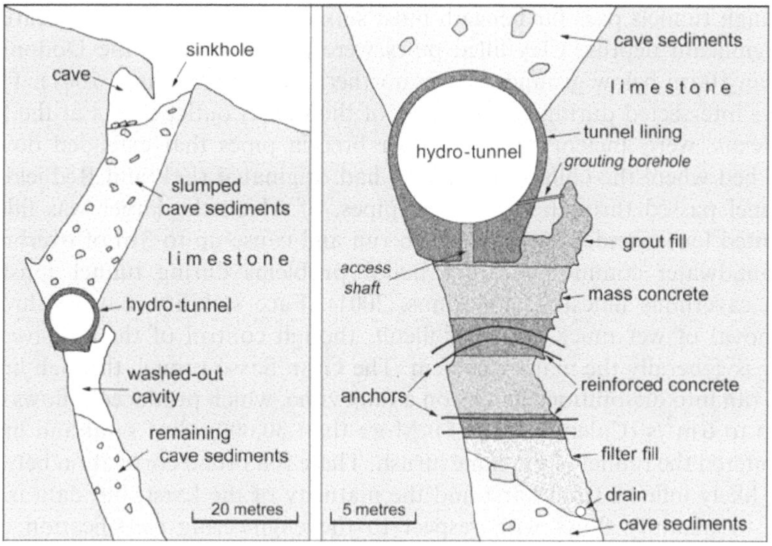

Figure 12.16. Sections through the Capljina hydro-tunnel where it passes through a filled cave in the limestone of Bosnia Herzegovina, with a detail on the right showing remedial works that were required.
After Milanovic (2000).

feasible where shuttering has been built to prevent potentially enormous concrete losses into extensive cave systems and also to prevent concrete blocking active conduits where free drainage must be maintained. Tunnels in both China and Germany have been built through open caves by retaining the tunnel lining as a protective roof, covered with a blanket of granular debris to reduce the impact of breakdown blocks falling from the unsupported cave roof.

Caves that are full of sediment may offer slightly more scope for driving tunnels through them. Advance grouting of the cave fill, either through a cone of drilled holes (spiling), or by site-specific remediation where man-access is created into the cavern, is essentially an extension of the techniques of soft-ground tunnelling. Remedial works may also be required where a tunnel passes close above a large cave with a potentially unstable roof. A tunnel driven through cave sediments may encounter problems after construction, especially where any water leaks from a hydro-tunnel. Water losses from the Capljina tunnel, part of the massive Trebišnjica hydro-scheme in Bosnia Herzegovina, caused erosion of the cave sediments through which it passes (Milanovic, 2000). Sediments washed from below the tunnel threatened its integrity over the emptied cave, and sediments washed from above opened a sinkhole at surface level. Remediation was by sinking a hole through the tunnel lining for man-access into the cave below, where reinforced arches were built inside the cave to support mass concrete placed up to the tunnel floor (Figure 12.16). The sediment above the cave was left untreated, but it stabilised when the water input was eliminated.

Though tunnels pass far beneath most sinkholes, buried and filled features can reach significant depths. Clay-filled pipes were encountered in the Dodoni tunnel more than 100 m below ground level in northern Greece (Marinos, 2001). The filled chimneys intersected during construction of the tunnel outlet works at the Sanford Dam, Texas, were ancient sinkholes over breccia pipes that extended down to a gypsum bed where the collapsing cavities had originated (Eck and Redfield, 1965). The tunnel passed through two of the pipes, of which the larger was filled with uncemented loose sand, which tended to run and cause up to 3 m of overbreak.

Groundwater commonly causes major problems during tunnel construction through cavernous limestone (Marinos, 2001). Face stability may be threatened, and removal of wet muck can be difficult, though control of the massive inflows of water is generally the major problem. The Gran Sasso tunnel, through limestone in Italy, ran into dissolution cavities on a fault zone, which produced inflows claimed to be up to 6 m^3/s (Calembert, 1975). More than 30,000 m^3 of sand and limestone debris entered the tunnel in the same inrush. There is a broad correlation between the scale of likely inflows from karst and the maturity of the karst, but data is not yet accrued to quantify flows with respect to the engineering classification of karst (Chapter 2), and exceptions are always possible where large individual conduits are breached by tunnels. In addition, tunnel construction can lead to de-watering, thereby reactivating old sinkholes and inducing new subsidence sinkholes in any soil cover, as occurred above the Canyon tunnel in Sri Lanka.

12.5 DAM CONSTRUCTION IN SINKHOLE KARST

Sinkholes and caves present numerous problems in the construction of large dams, among which bearing strength and water-tightness are paramount. Long, deep and expensive grout curtains are commonly required to impound water on karst (Bruce, 2003). These become particularly complex where large caves are breached and require individual filling or sealing, but the far-reaching subject of karst hydrogeology lies beyond the scope of these pages. Wide experience has been gained in building numerous dams and sealing their reservoirs in the mature limestone karst of Croatia and Herzegovina (Milanovic, 2000, 2003). There are more than 5,000 dams and reservoirs in the extensive karst terrains of southern China, with examples built, both on the surface and underground, in almost every conceivable situation (Lu, 1986). Most of these dams have been successful, though about one-third of them suffer from serious leakage. Many have failed completely, either collapsing or retaining no water at all, but most of these are the smaller projects built by rural communities and local engineers (Smart and Waltham, 1987). Reservoir impoundment is also significant in inducing new subsidence sinkholes in covered karst that is in China or elsewhere (Section 8.1.1).

Most sound limestone is a strong material with a high bearing capacity, and sufficient bearing strength generally may be obtained within a cavernous rock mass by distributing loads to bear on the more solid rock. For a dam, this may require deeper excavation than otherwise would be necessary. The extent of dissolution

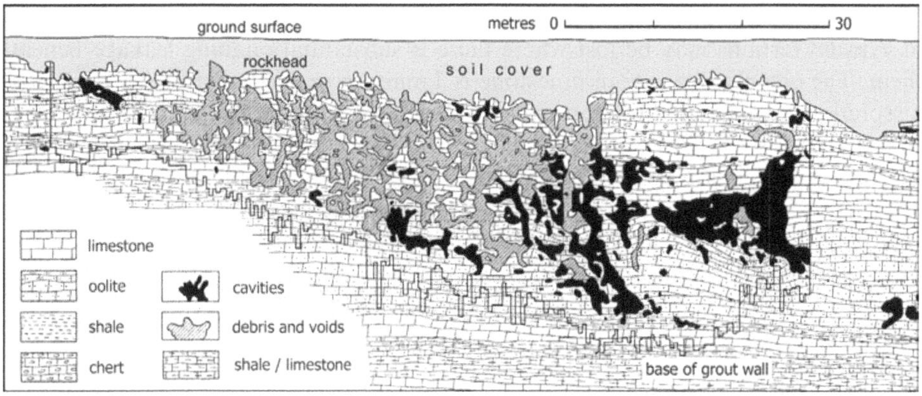

Figure 12.17. Cross section along the grout wall under one part of the Hales Bar Dam, Tennessee, where a zone of extensive, open and soil-filled, fissures and caves lay within the limestone.
After Schmidt (1943).

cavities, both open and sediment-filled, that may lie beneath a river valley in limestone karst, was demonstrated by those revealed during grouting works beneath the Hales Bar Dam in Tennessee (Schmidt, 1943). While much of the dam-site limestone had only a scatter of dissolution cavities, two zones contained networks of dissolution features to depths of more than 30 m below the original bed of the river (Figure 12.17). Extensive grouting programmes, extended in later years, all failed, and the dam was eventually demolished. Besides creating massive reservoir leakage, such cavernous ground can threaten the structural integrity of a dam, but usually only at limited specific points.

Only after construction of Turkey's Keban Dam was completed, a large cave was found under one of its abutments (Bozovic *et al.*, 1981). A whirlpool developed in the new reservoir over where rapidly leaking water scoured open a previously soil-filled sinkhole. Subsequently, the rifts and shafts of the newly opened cave were followed to a depth of 45 m where they opened into a large chamber that extended beneath the dam footprint. The rock was weak (taking 2.1 t/m of grout through boreholes), and the cave roof was unstable. A shaft and 13 boreholes were therefore sunk into the chamber, to pour in 600,000 m^3 of rock, gravel, sand and clay fill, before the sinkhole was plugged with concrete. In contrast, a large, clay-filled cave directly beneath the Grabovica concrete gravity dam in Bosnia Herzegovina was left untouched, though the limestone around it was sealed by grout injection, and the cave exit just downstream of the dam was cleaned out and plugged with concrete to a depth of 5 m (Milanovic, 2000).

As gypsum is more readily soluble than limestone, caves and conduits can develop in gypsum much more rapidly than they can in limestone or dolomite. Under the influence of high pressure gradients, large turbulent flows created by leakage from new reservoirs can widen dissolutional fissures by 10 mm per year, and thereby cause major leakage increases after only 5 years in gypsum

(Dreybrodt *et al.*, 2002). It is therefore possible that structural integrity of dams in gypsum terrains may be lost where there is substantial ongoing leakage beneath them. The comparable rate in limestone is 1 mm per year after 25 years. Limestone dissolution rates of up to 1.02 mm/year, measured at Tennessee's Hales Bar Dam, were up to 10 times the rates in natural drainage systems (Moneymaker, 1968). Though these enhanced dissolution rates are barely significant in limestone, there remains the greater geohazard that piping and removal of sediments can induce rapid and catastrophic sinkhole development where any reservoir is impounded on any karst rock.

13

Case studies

The case studies are intended to demonstrate a wide variety of karst ground conditions and the engineering responses to them in different parts of the world. They demonstrate the processes and principles described in the preceding chapters, but were all written by invited specialists who have contributed material from their own experiences on karst. The subject material crosses the barrier between geology and engineering, and the cases are therefore presented in no particular order.

#1 Remediation of a sinkhole over gypsum at Ripon, U.K.
#2 Collapse sinkhole at Dishman Lane, Kentucky
#3 Caves and sinkholes in motorway construction, Slovenia
#4 Road built over Sung Gul lava tube, Korea
#5 Karst collapse prevention along the Shui-Nan Highway, China
#6 Construction over a cave in Huntsville, Alabama
#7 Sinkhole destruction of Corporate Plaza, Pennsylvania
#8 Subsidence over a chalk pipe at Chalfont St. Peter, U.K.
#9 Geophysical investigations of sinkholes in chalk, U.K.
#10 Detection of caves by microgravity geophysics, Bahamas
#11 Sinkholes and subsidence over salt at Wink, Texas
#12 Subsidence over buried karst at Centrurion, South Africa
#13 Agriculture on sinkhole karst on gypsum, Lithuania
#14 Sinkhole remediation over Weeks Island salt, Louisiana
#15 Hazard assessment on dolomite at Simunye, South Africa
#16 Ground investigation in covered karst at Tournai, Belgium

Case study #1

Remediation of a sinkhole over gypsum at Ripon, U.K.

Anthony Cooper

On Wednesday 23rd April 1997, a large and sudden sinkhole subsidence occurred immediately in front of Field View, a house on Ure Bank Terrace, in the small Yorkshire town of Ripon, U.K. Overnight this hole continued to grow, thereby demolishing two garages, leaving a crater 10.7 m across and 6.2 m deep, and threatening the house that had already sustained minor damage (Figure 13.1.1). The adjacent local road was closed, and photographs of the subsidence appeared on television and in the newspapers. It was the latest collapse on a site that had been subsiding since at least 1856, when the Ordnance Survey map showed a pond at that location (Cooper, 1986). An earlier smaller collapse (subsequently filled in) had already demolished two garages, and it was estimated that the total vertical amount of collapse (ignoring minor subsidence) had been about 21 m over the previous 30 years and at least 24 m in the past 150 years (Cooper, 1999). The site

Figure 13.1.1. The open sinkhole that collapsed in front of the house on Ure Bank Terrace, Ripon.
TW.

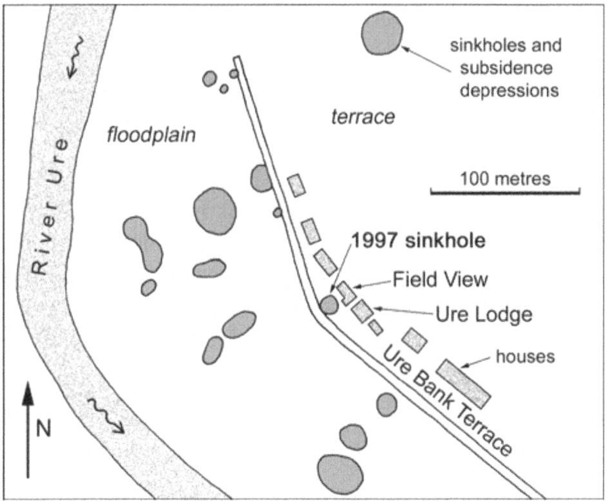

Figure 13.1.2. Location of various sinkholes and subsidence depressions on the east side of the River Ure where it passes through Ripon; the road named Ure Bank Terrace is roughly along the boundary between the floodplain and higher ground to the east.

is in an area of well-documented gypsum karst, where numerous other sinkholes are present nearby and active subsidence has been recorded at many sites.

Both the sinkholes and the many broad and shallow subsidence features have been caused by collapse into cave systems formed in the thick Permian gypsum that underlies this part of Ripon. The gypsum is up to 40 m thick within the marls of the Edlington Formation. Borehole records suggest the gypsum sequence lies directly on the dolomite aquifer in the underlying Cadeby Formation, at about 55 m depth. The juxtaposition of this aquifer with the very soluble gypsum has created a prime environment for the development of maze caves – that are actively evolving with associated collapse and subsidence problems (Cooper, 1986, 1998). Because the gypsum dissolves rapidly, cave systems are believed to evolve rapidly, and catastrophic collapse of the larger cave passages and chambers ultimately cause the dramatic surface subsidence that characterise the area.

Sinkholes in and around Ripon are commonly 10–25 m in diameter and may be up to 20 m deep. These commonly occur in linear belts that may be related to the joint patterns and maze caves within the gypsum (Cooper, 1986). The distribution of the caves, sinkholes and subsidence bowls relates to the paths of water flow through the rock sequence, from the high ground to the east towards the valley floor of the River Ure. Sites of the most severe subsidences are over the sides of the buried valley – where the water escapes from the gypsum into the river gravels (Cooper, 1998). This zone includes the sinkholes along Ure Bank Terrace (Figure 13.1.2).

Because the gypsum dissolution causes such a major geohazard, the local council have a formal planning routine for ground investigation appropriate to gypsum subsidence problems and the approval of new construction designs (Thompson *et al.*, 1996; Paukstys *et al.*, 1999), but the houses at Ure Bank Terrace predated this legislation.

DESIGN PHILOSOPHY FOR REMEDIATION

To remediate the collapse and allow the adjacent road to be re-opened, numerous options were considered – with a background awareness that investigation and remedial measures could trigger further collapse and movement. Re-routing the road was discounted because of cost and access, and because numerous other subsidence features lay along any alternative route. Of the remaining options, the following were considered:

- Leave the sinkhole open and do nothing; discounted because failure of the sides of the open hole would destroy the road and possibly threaten adjacent properties.
- Fill the sinkhole; the simplest option that was eventually undertaken (see below), as it would also support the road and the sides of the hole.
- Line the sinkhole with a high-tensile geogrid and a geotextile membrane, and backfill with broken rock; discounted because it would probably choke the top of the hole and then allow superficial sands and gravels to funnel in beneath the rock and geogrid plug.
- Support the sides of the open sinkhole and then fill it; discounted because of the need to use pile sheeting, with associated vibration and working next to an unstable area.
- Support only the side of the sinkhole adjacent to the road; discounted because shallow support could give an impression of stability while material funnelled away from beneath the road, until the whole support and road failed; deep sheet piles were also discounted because of the associated vibration when working next to an unstable area.
- Reinforce the road to span the subsidence zone; discounted because of cost and the fact that the reinforcing would have to span any potential cone of failure within the soil profile, possibly requiring a span of 30 m or more.

The option accepted was to fill the hole with inert fill of crushed and graded dolomite rock. It was recommended that 37.5 mm, type B, filter drain material (to C1505 of the U.K. Department of Transport specification for highway works) should be used because it was self-compacting, locally available and cheap. This fill would funnel in to the sinkhole, and thereby indicate the progression of any instability. It would also support the sides. Fill of larger size was discounted because it could choke the top of the hole and give a false impression of stability before collapsing catastrophically. Very fine fill (such as sand and fine gravel) was discounted because this could funnel down too easily and then be washed away within the cave system.

It was specified that the fill should be inert, blocky and free-draining. This was to allow water to filter through it as the groundwater level fluctuates. Absorbent fill was

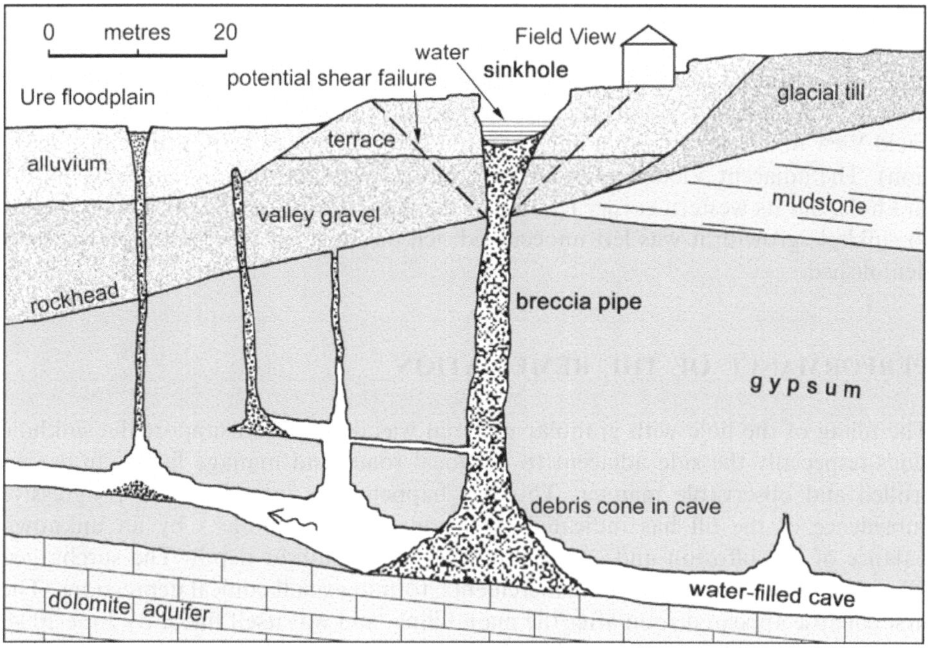

Figure 13.1.3. Stylised cross section through Ure Bank Terrace showing the relationship of the open sinkhole to its underlying cavities. Except for limited surface data from around Field View, the geology shown is either schematic or interpolated from borehole records some distance away, and the structure of the caves is conjectural.

to be avoided as its effective load would increase when it had become saturated before a fall in the groundwater level, thereby possibly causing compaction of the underlying fill or triggering significant movement of meta-stable fill. It was also specified that clay should be avoided since this can turn into a slurry that is easily washed out. Any organic or soluble materials were disallowed as they would degrade and cause settlement of the fill. No toxic material was allowed because water flowing through the cave beneath the sinkhole drains to springs along the Ure Valley.

Another reason to use blocky, free-draining fill is that eventually the material will end up in the cave passage below (Figure 13.1.3). The size of the fill material was selected so that it will form a permeable cone of fill within the cave when it eventually works its way down the entire depth of the sinkhole throat. It is essential that the cave water can pass freely through the fill. Any impermeable clay fill placed in the hole could choke the cave and divert flow through fissures in the surrounding rock. Enhanced dissolution of the gypsum could then create new cavities round the sides of the fill blockage, and thereby promote new collapse and subsidence of ground that was previously stable beside the original sinkhole.

Late in 1999 the sinkhole was filled via a long conveyor belt that was cantilevered over the hole so that no trucks needed to back up close to the opening. The hole was surcharged to a height of 0.5 m. The hole remains unstable, but the collapse of the fill acts as a monitor of the fill performance; a programme of observation and refilling is required to maintain the remediation. The adjacent road was re-opened, the site of the sinkhole was fenced off and remains unused, and the severely damaged Field View house remains standing next to it (with no moves yet towards its demolition). The adjacent Victorian house, Ure Lodge, was not directly damaged by the sinkhole, but its western corner fell within the hazard zone of possible slumping due to sinkhole growth; it was left unoccupied, fell into disrepair and was subsequently demolished.

PERFORMANCE OF THE REMEDIATION

The filling of the hole with granular material was designed to support the sinkhole sides (especially the side adjacent to the local road) and manage failure in a controlled and observable manner. This has happened as intended, and progressive subsidence of the fill has indicated evolution of the sinkhole – by an unknown balance of fill suffosion and continuing rock dissolution at depth. The surcharged fill collapses periodically in small increments, forming small conical depressions. The first collapse appeared soon after the main filling, and was itself filled. By June 2004, further ground collapse had progressed slowly, and another conical depression about 5 m across and 2 m deep had formed near the road over the deepest part of the original collapse. The amount of fill material remaining was just enough to support the side of the road, but any further collapse will require another phase of re-filling with similar granular material.

The current problem revolves around who is legally responsible for the sinkhole, for the filling of the open hole and for support of the local highway. The collapse was on the site of four garages with three owners. These landowners were compensated for loss of their buildings, but home insurance cover does not include the land. The insurance companies claim that the collapse was an "Act of God", and consequently both they and the county authority claim the problem is the responsibility of the landowners. The county authority paid for the initial filling of the sinkhole to enable the highway to be re-opened – but the lawyers are still arguing over who should pay for that and who is responsible for maintaining the fill level.

Case study #2

Collapse sinkhole at Dishman Lane, Kentucky

Patricia Kambesis and Roger Brucker

On a cold afternoon in February 2002, in the southern part of Bowling Green, Kentucky, rush hour traffic on Dishman Lane halted abruptly. A sinkhole 60 m across suddenly collapsed, dropping the road by 5 m. No-one was injured, but four cars were stranded on the subsided road. The cause was a catastrophic collapse of the main passage of State Trooper Cave (Figure 13.2.1), and the subsequent repairs to the road took nine months at a cost of a million dollars.

Back in 1987, the city of Bowling Green had proposed to extend Dishman Lane through the suburbs. The city is located in the Lost River drainage basin of the Sinkhole Plain of southern Kentucky, a part of the state at highest risk from collapse and flooding within the karst, so city engineers commissioned the Center for Cave and Karst Studies (CCKS) at Western Kentucky University to investigate the subsurface of the route of the proposed road.

State Trooper Cave had been mapped several years earlier as part of a regional cave study, prior to any road planning, and the cavers' compass-and-tape survey contained sketched cross sections but no profile. Their traverse extended from an upstream collapse entrance to a flooded zone 1,620 m downstream. The single trunk cave passage is mostly about 10 m wide and 5 m high, containing a river that

Figure 13.2.1. Aerial view of the collapse sinkhole in Dishman Lane, Bowling Green, seen looking towards the south-east
Photo: Centre for Cave and Karst Studies.

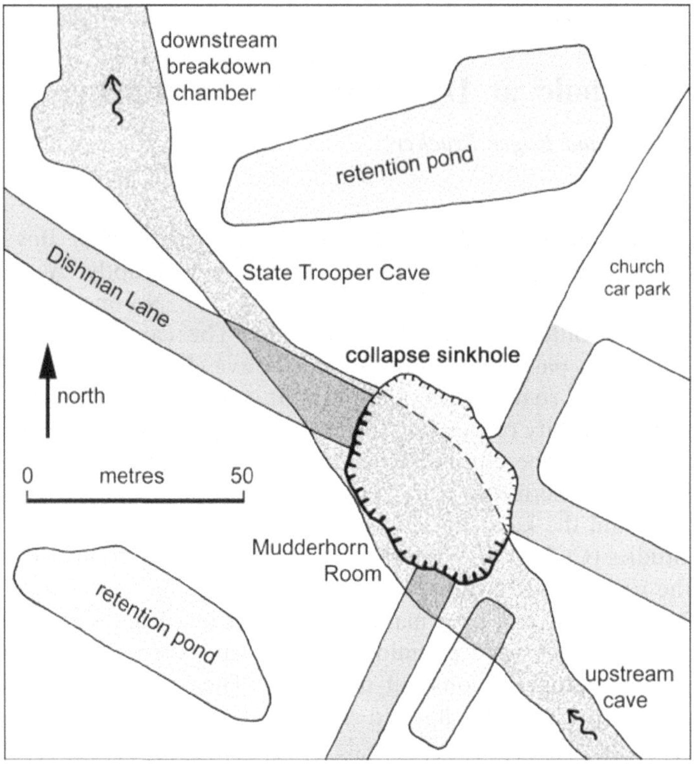

Figure 13.2.2. Outline map of the Dishman Lane sinkhole collapse and the cave beneath.

normally flows 1–3 m deep. It is a segment of the Lost River Cave that drains a basin area of about 92 km^2. Beneath the proposed Dishman Lane, the cave passage is widened and the ceiling rises due to breakdown to form the Mudderhorn Room (Figure 13.2.2); this was named after the thick layers of mud over massive fallen slabs, two of which projected upwards like glaciated horns. The cavers noted that the slightly domed ceiling of the room, several metres above the top of the breakdown debris, was cracked and fissured. Overhead, a condensation cloud could be seen on cold days, rising from the surface 10 m east of the main passage, in line with a tiny side passage draining from a waterfall dome.

In late 1987, the CCKS performed a microgravity survey to confirm the position of caves along the proposed road alignment, and made two low-frequency radio location fixes to confirm the cave room's exact position under the proposed road. The CCKS report recommended modification of the proposed straight road to avoid the Mudderhorn Room. Two curves in the road were suggested to take the road just north of the Mudderhorn, swinging wide of the breakdown and crossing the cave between the chamber and another breakdown area 150 m downstream.

At the request of the area developer, the city decided to place one gentle curve in the road to swing it just south of the downstream breakdown chamber, but elected to pave the road straight over the Mudderhorn Room. It appears that, by the time this decision was made, the original CCKS report had been forgotten or lost, and errors crept into maps that were newly prepared from inadequate data, so that the unstable cave chamber was thought to be away from the road line.

EVENTS OF THE COLLAPSE

In December 2001, cracks began to propagate in the brick walls of a new church being built adjacent to Dishman Lane. Motorists reported a new dip in the road, and protested the city's inaction by alerting the local newspaper. The city then dispatched an engineer to investigate the dip in the road. He concluded that there was no problem. CCKS was neither a party to, nor privy to, this investigation.

After the catastrophic collapse on February 25, investigators from the CCKS returned to the scene. They used a canoe to traverse the cave stream from an upstream entrance as far as the new collapse in Dishman Lane. At the former site of the Mudderhorn Room, they found daylight streaming in from several surface holes above the pile of new and old breakdown. Before the collapse, the Mudderhorn Room had been a wide section in the stream passage with a vaulted ceiling produced by breakaway of the horizontal limestone beds from the arch's tension zone. The mound of breakdown was partially covered by sediment, derived from periodic flooding and perhaps also from soil washed down through the ceiling fissures. After the collapse, loose rock extended from the streambed all the way to the surface along the north-east wall of the old cave room (Figure 13.2.3). The fall of the limestone roof had been followed by an avalanche of debris into the open cave, pushing boulders, cobbles and soil all the way from the north-east wall to the south-west wall. The cave stream flowed between the rocks at the base of this breakdown pile.

The ceiling of the downstream passage could not be seen as the new debris pile sealed off that part of the cave. The roof of the upstream cave passage appeared to be intact, with about 7 m of overburden rock and soil (Figure 13.2.4), and the room's southwestern wall appeared to be the stable original, rather than a freshly broken new surface. The entire rim of the new hole in the cave roof appeared to be along fresh breaks, as did all of the new breakdown blocks that lay over the old debris pile. Some of the new breakdown was coated with soil and sediment from above.

The CCKS investigators concluded that the southwestern wall of the cave room and its ceiling and those of the upstream passage appeared to be stable. The stream surface lay about 16 m below ground level, in a passage about 8.5 m high immediately upstream of the collapse. Additional investigations at the site included gravity surveys and 10 cored holes in and adjacent to the collapse (Kambesis *et al.*, 2003). These revealed voids at several levels above and below the cave passage level, with numerous clean fractures in the rock – consistent with typical cores drilled into the weathered Ste. Genevieve limestone elsewhere in Kentucky. The CCKS team concluded that further upstream collapse was unlikely, since the road alignment avoids the cave between the new collapse and the sink entrance and also the cave

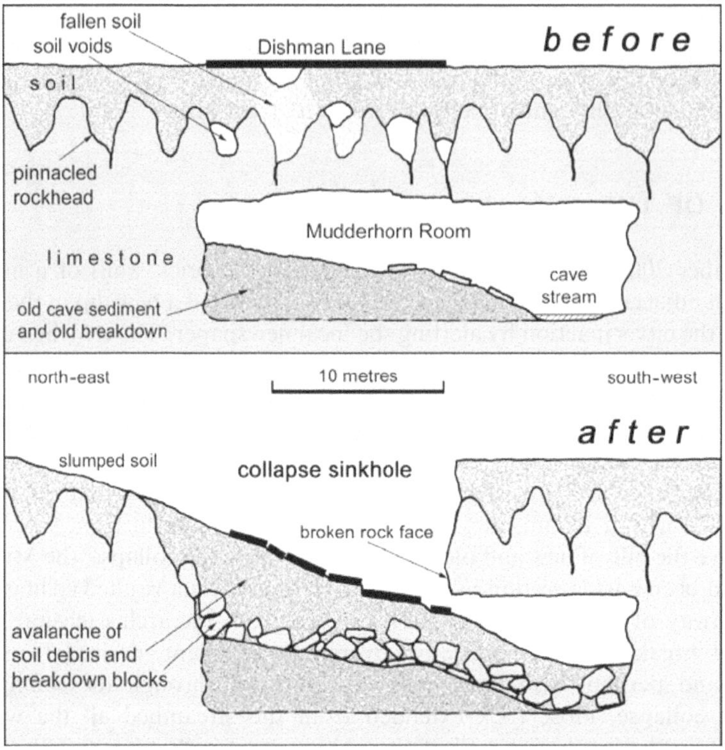

Figure 13.2.3. Cross sections of the cave beneath Dishman Lane before and after the collapse.

Figure 13.2.4. The edge of the collapse zone under Dishman Lane, showing the deeply fissured limestone rockhead beneath the red clay soil cover.
Photo: Hilary Lambert, KEEP.

roof is thicker. Storm-water from the 8,000 m^2 church parking lot, from adjoining properties, and from Dishman Lane was still directed by ditches into the collapse during the nine-month repair period.

The Dishman Lane collapse had features of both bedrock and soil failure. The event was a culmination of factors that included the pre-existing weaknesses in the underlying bedrock and the effects of urban development on those weaknesses. State Trooper Cave is developed along a set of prominent joints that are evident from topographic maps and aerial photographs, and the Mudderhorn Room lies at the intersection of two major joints. Had conditions on the surface remained undisturbed, the low-arch roof profile developed by breakdown across the cave chamber, though very thin, would have remained stable (until it failed through dissolutional thinning over geological time).

However, the Dishman Lane area has undergone 20 years of significant development that involved both construction and the redirection of surface drainage. Placement of parking lots, streets, driveways and retention basins changed the run-off and infiltration characteristics of the soil mantle. Run-off from storm drains and from the edges of the road surface formed flow paths through the soil into open fissures between rockhead pinnacles in the limestone that was already weakened by its many solution channels. Steeper hydraulic gradients and flow velocities accelerated suffosion and removal of the soil, so that regolith arches developed over unstable voids. Because the roadbed acted as a bridging structure, the loss of soil was not apparent until one or more substantial voids had formed beneath it – setting the stage for a major regolith collapse. Since failure of the thin rock roof over the cave chamber was probably caused by a new emplacement of load, it appears likely that a regolith collapse triggered the bedrock collapse of the cave ceiling.

REMEDIATION

To repair the collapse and reinstate the road, three alternative approaches were presented by the CCKS:

- Remove the collapse debris and replace it with stacked large rocks from bedrock to a new road surface; this was the most permanent fix, but also the most costly.
- Find competent rock for foundations on both sides of the breach and completely bridge the sinkhole; in both these cases, the road could continue straight, but any fill in the sinkhole would need to be engineered with large blocks at the bottom to allow the stream to pass through without back-flooding.
- Curve the road around the sinkhole, as originally proposed, and leave the new rubble-filled sinkhole in place; while this was the cheapest option, it would add an unacceptable curve to a high-traffic road, would make access to existing properties awkward, and might still have problems with impeded floodwater.

Further alternatives, of a steel-reinforced roadbed, deep compaction, piled supports and pressure grouting, were considered and rejected.

The city engineers decided to completely excavate the collapsed material down to bedrock, one half at a time, and install a pre-cast concrete culvert 1.2 m in diameter to carry the stream. A vertical drain was also installed above a T-section to carry storm-water down from the repaved road and to allow access for clean-out and repairs. The culvert was covered over with graded aggregate, and the sinkhole was then backfilled with boulders from the excavated collapse, followed by smaller rocks and soil. The road was then rebuilt on its original straight alignment.

Potential future problems do remain. The newly placed rock pile forms a leaky dam in a cave passage that is 30–70 m^2 in cross section and does fill to the roof during storm events. When the much smaller, single, 1.2 m diameter culvert installed at stream level runs pipe-full, water could back up in the cave to cause upstream sinkhole flooding. The culvert size was chosen on the basis of incomplete data on the cave hydrology, and a double or triple culvert would have been more appropriate. Future flood events could destabilize the new fill by carrying trailing edge boulders downstream and thereby cause renewed subsidence of the road.

Six months after the completion of the repair, the church parking lot still drains onto the new road, where curbs carry the water to storm drains on each side that connect to the vertical culvert, as in any other injection well in the city. Adjacent drainage basins retain standing water. A dip in the road over the reconstructed area is being monitored regularly for any signs of renewed collapse.

Loss of the road was due to inadequate design procedures, where professional hydrogeological expertise was sought but then not applied. Designing the road on the basis of incorrect maps, when an earlier correct map had been lost or forgotten, was a catastrophic error in management of the ground investigation. It appears that the Dishman Lane collapse was totally avoidable.

Case study #3

Caves and sinkholes in motorway construction, Slovenia

Martin Knez and Tadej Slabe

One of the larger ongoing projects in Slovenia is the construction of major new highways, and for many years karst scientists have been participating in the planning and construction of motorways across the karst. The value of karst science, which comprehensively considers the characteristics and protection of various types of karst and understands the development and hydrology of the caves, is increasingly appreciated by the construction industry. As almost half of Slovenia is karst terrain, and more than half of its water supply is from karst aquifers, the wisdom of cooperation between builders and karst scientists has been well proven (Knez and Slabe, 2001). During road construction in the last decade, experience has shown that too little attention had been devoted to certain recognised karst phenomena.

Most work has been on examples from the Kras, the karst region often known as the classical karst that gave its name to landscapes on soluble rock. The Kras plateau lies between the northern Adriatic Sea and an extensive flysch upland to the south-east. Surface waters flow from the flysch to create an extensive contact karst along the margin of the carbonate outcrop. With altitudes of 200–500 m, and a surface area of 440 km^2, the Kras plateau is formed of Cretaceous and Paleogene limestone and dolomite with considerable lithological diversity; it belongs in a wider sense to the area of the Outer Dinarides.

On the Kras plateau there are no remains of the surface waters involved in the early development of the plateau. Original drainage flowed at or just below the surface and vertical percolation was minimal, but the water table later dropped by several hundred metres. Today, all the rivers in the Kras region sink where they flow from the flysch onto the limestone outcrop, and then flow underground toward the source of the Timavo River in Italy. The largest stream is the Reka River, which sinks in the Škocjan Caves, while about 65% of the total karst water sinks by percolation through the limestone outcrop. The ecology of Slovenia's Kras warrants protection as it is both valuable and vulnerable.

THE KARST GEOHAZARD

Construction of motorways on the Kras plateau is greatly influenced by the cavernous nature of the karst. Along 50 km of motorway built in recent years, more than 300 caves were newly discovered (Figure 13.3.1). The karst has a complex history of development, and is also marked by active tectonics and lithological diversity, so that the locations of inaccessible caves are difficult to predict in advance, though caves are more numerous along the contact of flysch

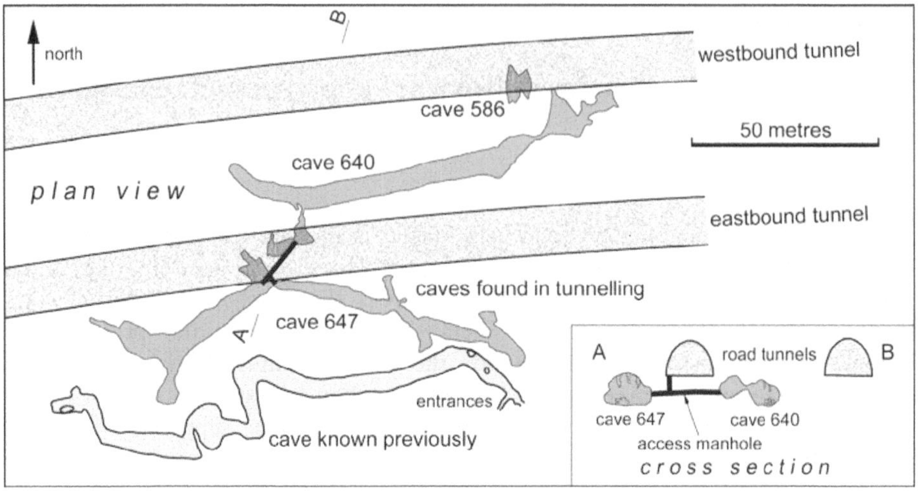

Figure 13.3.1. Outline map of caves intersected by twin tunnels for a motorway near Kozina.

with limestone. The number and extent of the caves (the cavernosity of the karst) is therefore determined on the basis of a thorough knowledge of the Kras combined with meticulous fieldwork during the planning and construction of the motorways.

The surface of the Kras plateau is dotted with solution sinkholes and unroofed caves. Some sinkholes are very distinct, while others are less recognisable and filled with soil. The epikarst is criss-crossed by fissures that are especially large in the Cretaceous limestone, and are mostly filled with soil. The caves include many old passages and chambers, of various shapes and sizes, that are either empty or filled with alluvium; almost two thirds of caves are filled with sediment. There are also numerous shafts formed where drainage sank vertically from the permeable karst surface towards the deep water table; the deepest shaft found in road construction was 110 m deep.

Almost one-third of the caves newly discovered on road projects are unroofed caves (Knez and Slabe, 2002a). These are old cave passages that became exposed by lowering of the karst surface, so that they have now lost their roofs. Shapes of unroofed caves relate to the types and shapes of the original caves and to the development of the karst surface within its geological, geomorphological, climatic and hydrological conditions. The surface appearance of an unroofed cave is dictated by the speed at which its sediments were washed out compared to the rate of surface lowering. Where sediment removal was slow, areas of alluvium and flowstone may be visible on the surface where soil and vegetation develop. Where it was faster, unroofed caves appear as sinkholes, strings of sinkholes or oblong depressions across the karst surface. Sinkhole-like features occur where the lowering surface breached cave passages filled with alluvium and flowstone, and subsidence sinkholes can also develop in the cave sediments that fill a large uncovered passage. Strings of such features may reveal the shape and size of the half-destroyed cave. Long and thin

Figure 13.3.2. Large unroofed cave passage that lay parallel to the karst surface, near Divača, seen after its sediment has been removed in preparation for road construction.
Photo: Martin Knez.

depressions develop from larger unroofed cave passages that lay parallel to the new surface (Figure 13.3.2). Typically, the modern topography is a complex interweaving of various old caves with later surface karst features and the modern epikarst (Knez and Slabe, 1999).

ENGINEERING RESPONSES

Soil and fine-grained sediments must be removed from all sinkholes and unroofed caves, so that fissures and shafts in their floors and sides are all exposed. Any floor fissures are choked and spanned by large blocks of rock and concrete to prevent

Figure 13.3.3. Compacting a graded fill placed inside a large sinkhole that had previously been filled with loose soil, near Divača.
Photo: Tadej Slabe.

water gradually washing away loose sediment. The sinkholes and caves are then filled with rock and layers of gravel (Figure 13.3.3).

Most caves, both shafts and parts of old passages, are opened during excavation of cuttings and tunnels, either beneath the intended road or in the sides of the new cuts. Many cave roofs collapse due to blasting or during subsequent work – notably during final roadbed compaction with large vibrating rollers. In practice, the compaction work is a good test of the road's future stability. Many caves are also opened in the course of building embankments. Caves can quickly be concealed during blasting, either as they collapse or as their entrances are covered with rock and rubble. Slower excavation and more careful groundwork can be worthwhile.

Before construction begins on any project, the available data on caves is checked and augmented by new fieldwork and generic interpretations. With the better information, a model of the cavernous karst is prepared, in order to produce prognostic subterranean maps with special emphasis on the anticipated lithologic and tectonic contrasts in the rock. The potential degree of cave development is examined as thoroughly as possible before site work begins. Locations of suspected caves are then checked by drilling, which also determines the nature of their sediment fills. The shape, type and frequency of caves in adjacent areas can be partially predicted by interpolation of known surface and underground features. All caves uncovered during construction are mapped to determine their morphology and genesis, and samples of alluvium and flowstone are taken for paleomagnetic, mineralogical and palynological research and dating. This continually increases the database on the karst of the Kras plateau (Bosak *et al.*, 2000).

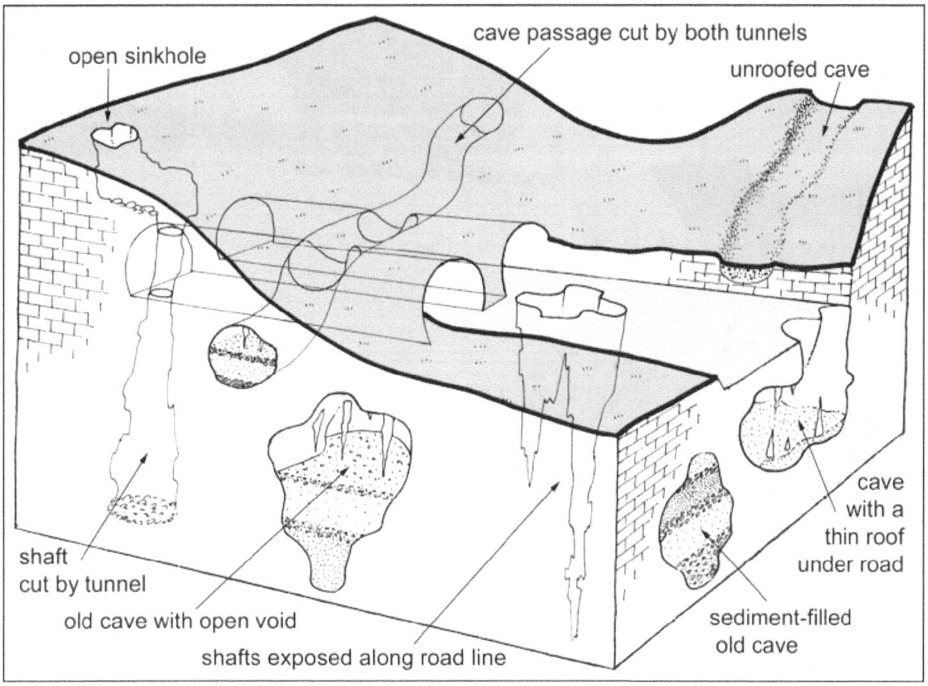

Figure 13.3.4. Various forms of caves, sinkholes and shafts that can be encountered during construction of cuttings and tunnels for motorways through the classical karst.

Thorough surveys of the caves have proved indispensable for engineers and builders. Possibilities of rock falls are assessed, and attempts are made to establish the location of inaccessible parts of the caves. Designs of structures, that bridge over and preserve any caves, are adapted to each individual site and its conditions, in particular to the extent of fracturing and fissuring in the rock. The work is largely dictated by the shapes of individual caves that lie beneath and beside the highway alignments. Many small entrances open out below into much wider and larger caves, and it is not unusual that their roofs are thin enough to create a significant danger of collapse during further stages of construction or during the subsequent use of the roads (Figure 13.3.4).

At most sites, road builders place concrete slabs to reinforce the ground over caves with thin rock roofs, though this is not possible where the cave or shaft walls have been fractured by blasting and are mechanically weakened across a wider zone. Research is in progress on various types of blasting around caves, which will help in future construction and also in the preservation of karst features. Monitoring of motorways that bridge over caves is regarded as necessary, and both ground radar and electric geophysics are employed. A depression has developed many times under a Postojna motorway due to renewed suffosion over a cave in limestone close to the flysch margin.

Figure 13.3.5. Construction of an impermeable roadbed through a cutting where a breached cave has been walled off, along a motorway to Italy.
Photo: Tadej Slabe.

As many caves as possible are preserved. Shafts are the easiest to preserve since their small entrances are simply closed with concrete slabs. Many old caves that have sound rock walls are preserved in a similar fashion. Caves bisected by road cuttings are left with entrances closed with rock walls in both sides of the cuttings (Figure 13.3.5), and some caves are left open for viewing. The most interesting and well-preserved caves are protected completely, and even though they are under the motorway or behind the walls of tunnels, they are made accessible through concrete pipes behind locked grids beside the road or through doors in the tunnel walls. A key factor in the selection of road alignments is preservation of the integrity of the karst landscape, by application of guidelines to avoid the more important surface landforms (including sinkholes and poljes) and any known caves. Special attention is given to the impacts of the motorway construction and its subsequent use on the karst hydrology. Roadbeds across the karst are impermeable, and run-off from them is collected in oil traps and cleaned before being returned to the natural aquifer.

The cooperation of karst scientists during all stages of motorway construction across karst has broadened basic knowledge on the origins and developments of the karst and on road-building techniques in such regions. There are many different types of karst, and each demands a unique approach; therefore, there must be regular cooperation with road builders in every case. In the last 10 years, this fact has been recognised successfully in Slovenia, and the cooperation between planners, road builders and karst scientists serves as a model for planning and executing construction projects in karst regions.

Case study #4

Road built over Sung Gul lava tube, Korea

Hyeong-Dong Park and Tony Waltham

Cheju Island lies 80 km south of mainland Korea and consists entirely of a single shield volcano. Nearly a million people thrive on farming and tourism on a coastal strip formed of old basaltic lavas – which generally provide stable ground, except over the old lava tubes. Near the western tip of Cheju, Sung Gul is a lava tube nearly 1 km long that passes beneath the village of Shinchang; it also extends beneath the island's perimeter road, which was recently enlarged and rebuilt over the cave.

The entrance to Sung Gul is an ancient collapse hole that now lies between the houses of Shinchang, and the main tube passage, 5–14 m wide, extends down-flow towards and beneath the perimeter road (Figure 13.4.1). Much of the roof of the Sung Gul tube is a broad, low-profile arch, with small glaze structures that indicate it is in its original state. Five short sections of the tube have more extensive roof collapse, where blocks up to 2 m across have fallen away to create large piles of breakdown under higher, domed sections of roof. Some of these dome roofs appear to have developed into stable arches within the jointed basalt, and there are no signs of ongoing collapse on any of them.

The collapse section furthest down-flow extends into an area of total roof failure (Figure 13.4.2). On the surface this has created a rocky basin at least 15 m long, which was partly backfilled during road construction. Part of the tube survives round the north side of the collapsed blocks (Figure 13.4.3), and it appears that the roof failure occurred in a very wide section of the tube or over a loop passage beside the main tube. A concrete drain 600 mm in diameter passes from the open lava tube, through the blockfall and backfill, into a concrete culvert that has been built across the collapse depression.

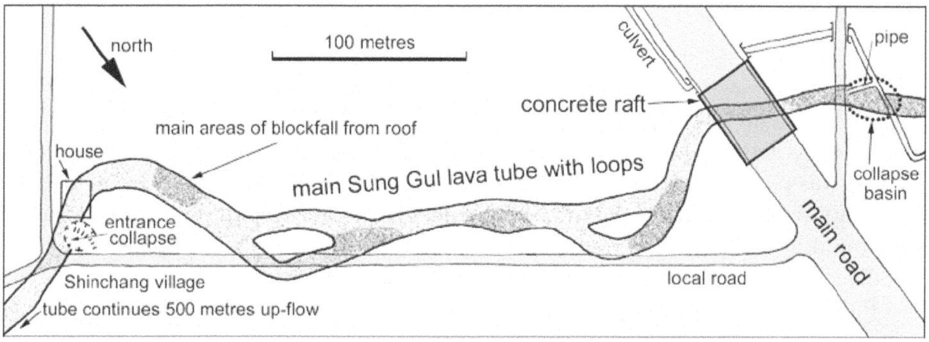

Figure 13.4.1. Outline plan survey of the Sung Gul lava tube where it crosses beneath the island's perimeter road.

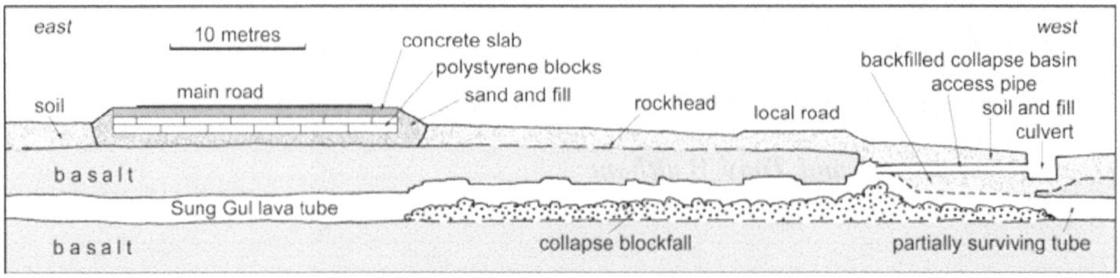

Figure 13.4.2. Sketch profile along the lower part of Sung Gul; the tube has an unbroken roof where it passes beneath the road, but has extensive breakdown down-flow of the road as far as the zone of total collapse, through which lies the access pipe.

Figure 13.4.3. The collapse zone down-flow of the road, with part of the original tube roof surviving on the right.
TW.

Cheju Island's perimeter road crosses over the Sung Gul tube between its two down-flow collapse sections (Figure 13.4.1). The road was a rough track until 1961, when it was paved. This was widened to two lanes in 1974, and to four lanes in 1998. It was not realised that the cave in Shinchang village extended under the road until it was found in 1994 during ground investigation for the last phase of road widening. No diversion of the road was then considered, as the line was already established.

Purely by good fortune, the line of the road lies over the most stable part of the cave, where a smooth arch barely 7–8 m wide rises 2 m above a clean floor of ponded lava (Figure 13.4.4). An incomplete shell of chilled lava lines part of the walls, but there are no fallen blocks on the floor. Assessment of the cave in 1994 found that the

Figure 13.4.4. The stable arch roof in sound basalt where the Sung Gul lava tube lies directly beneath the main road.
TW.

roof consisted of 3.5 m of rock and 2 m of soil. The internal report (unpublished) concluded that "the rock seems to be strong enough, but jointing and fractures could be a potential problem for stability of the road." Lack of data on the rock structure precluded any quantitative analysis, but it was deemed necessary to strengthen the ground.

PRECAUTIONARY ENGINEERING

The cave position was determined only by survey from the adjacent collapse entrance, and no check probes were needed. Soil was removed, and a flat surface was prepared by placing clean sand at least 100 mm deep over the uneven rockhead. For a 20 m length of the road over the cave, two layers of expanded polystyrene blocks (each 500 mm thick) were placed on the sand for the full 23 m width of the road (Figure 13.4.2). On this, two rafts of reinforced concrete 400 mm thick and 30 m long, each 11.2 m wide, completely span the cave (Figure 13.4.5). The road pavement is 50 mm thick over the concrete. The reinforced rafts can act as single-span bridge

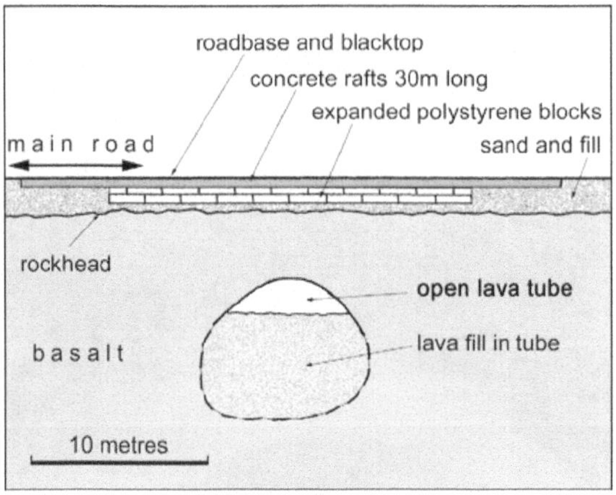

Figure 13.4.5. Cross section through the Sung Gul lava tube and the concrete slab that carries the main road above it.

elements, eliminating any hazard from even a complete failure of the lava tube roof. The rafts' span of 30 m was appropriate as the cave passes obliquely beneath the road. The minimum overlap onto solid rock, of about 3 m at two of the corners was adequate to retain integrity, especially as the rafts' widths extend further onto more sound rock at the opposite corners.

Most of the roof rock that is visible within the Sung Gul tube appears to be in good condition, with few major structural weaknesses. Fractures within the basalt are mostly tight and irregular, and should therefore have no abnormal impact on the rock mass strength within a compression arch over the tube. Stability of this rock arch was crudely assessed by Finite Difference Analysis (Waltham and Park, 2002). A worst-case scenario was created for the road over the cave with a width of 12 m and a rock-roof thickness of 3 m. The basalt unconfined compressive strength (UCS) was taken as 160 MPa, and a rock mass factor of 0.1 was applied as a reduced strength on the ubiquitous joints. Failure of the tube roof occurs under a vertical loading of 4,000 kN applied on a square pad of 0.5 m edge length. This is far in excess of any possible highway loading or design criteria, but the figure relies on estimates of the fracture weakness within the basalt, and individual fractures could not be adequately represented in the computer model.

The available data suggest that the concrete slab may not have been entirely necessary to ensure integrity of the road, but the cost of the concrete slab was small as a proportion of the project costs, and its inclusion was a reasonable precaution. The alternative options of collapsing the tube and building over the stabilised breakdown, or filling the tube with concrete, both represented higher costs than the simple bridging slab as sound rock was available on each side. Various houses and side roads in the village of Shinchang do stand over the Sung Gull lava tube (Figure 13.4.1), some over sections with roof much thinner than under the main road (Waltham and Park, 2002), but none shows any sign of distress or imminent failure.

Case study #5

Karst collapse prevention along Shui-Nan Highway, China

Mingtang Lei and Junlin Liang

More than one-third of China's land area has outcrops of soluble rocks. With the unparalleled urban, agricultural and infrastructure expansion in recent years, more and more sinkhole collapses have occurred and have become the main geohazard in the karst terrains – with significant influence on the economics of construction and land development. From only incomplete statistics, more than 30,000 sinkhole collapses have occurred within 1,400 karst sites spread across 22 of China's provinces (but mostly in Guangxi, Guizhou, Jiangxi, Yunnan, Hunan, Hubei and Sichuan, all in the south of the country). Of the 101,354 km of modern highway, about one-seventh (more than 13,084 km) are across ground zoned as high-risk or very-high-risk with respect to karst collapse by sinkhole development. Remedial treatment of sinkhole collapses along highway, and preventive measures within new designs, are the main problems of highway construction in China's extensive karst terrains.

An important section of the new road from Chongqing south to Zhanjiang, is the Shui-Nan Highway, between Hechi and Nanning (both in Guangxi). Its total length is 236 km, and this includes more than 10 km through a karst valley within the catchment of the famous Disu underground river system. This is a basin of mature karst with numerous well-developed stream sinks, large solutional dolines, open shafts, sinkholes of all types and dry valleys. Potential karst collapse (by development of new sinkholes) creates a serious risk to the highway construction. Consequently, measures to prevent collapses have been incorporated during construction of the 10 km of highway at greatest risk. This section of the Shui-Nan Highway site lies along a valley whose floor width is generally only 300–1,200 m, with tall conical hills of mature fengcong karst on each side (Figure 13.5.1). Karst features developed within the limestone of the road corridor include solution sinkholes, collapse sinkholes, dry and blind valleys, open fissures, open shafts, flooded shafts (cenotes) and stream sinks (ponors). The valley floor is distinguished by a karst rockhead that is deeply fretted between tall pinnacles, some of which project through the soil surface as stone teeth (Figure 13.5.2). The extensive cover of alluvial soils contains numerous subsidence sinkholes.

Devonian, Carboniferous and Permian limestones, dolomitic limestones and dolomites reach a total thickness of 4,000 m in the Disu karst basin. The main karst is developed on the very massive and very pure Maokou limestones within the Nanjiang double syncline. The whole rock sequence is strongly folded and fractured, and limestones dip at 30–60° beneath the valley that carries the Shui-Nan Highway. Quaternary soils are mostly clays and silty clays, generally 0–20 m thick with a patchy and discontinuous distribution. Perennial and seasonal streams flow within the valley, but most flow to sinks where they drop underground. The

Figure 13.5.1. The deck of a long viaduct with piled piers carrying the Shui-Nan Highway over difficult karst ground between steep conical hills in the fengcong karst.
Photo: Mingtang Lei.

Figure 13.5.2. Stone teeth – the tops of bedrock pinnacles projecting through the soil cover – are widespread in the Disu karst of Guangxi.
Photo: Mingtang Lei.

caves are unmapped as most lie submerged beneath the water table. Though main cave conduits are 15–20 m wide and 3–5 m high, they cannot take high wet season flows, and the valley floors are frequently flooded for a week at a time. Maximum flood level is 154 m a.s.l., less than 3 m below the design altitude of the highway surface.

Along the valley floor, the available corridor for the new highway was zoned with respect to potential sinkhole collapses, by a combination of field mapping and geophysics that included ground penetrating radar, seismic reflection and high-density electrical surveys. These investigations recorded 317 caves, 18 karst shafts, 25 solution dolines, 20 stream sinks and 24 open fissures within the valley along the 10 km of the highway route, making the whole area one of high-risk with respect to potential karst collapse. When engineering works started, excavation and stripping of the soil cover exposed 9 stream sinks, 2 caves and 46 solutional sinkholes, all within the footprint of the roadbed. The solutional sinkholes were all considered as hazards because they swallowed drainage after rainfall and were prone to development of subsidence sinkholes (inside the broader solution depressions). This catalogue of individual landforms was additional to the fissured and pinnacled rockhead that provided difficulties along almost the entire route across the karst.

CONSTRUCTION MEASURES TO PREVENT KARST COLLAPSE

Extensive experience within, and modelling of, China's karst terrains indicates that the most important process of karst collapse is piping and suffosion to create subsidence sinkholes in the cover of Quaternary soils deposits; this is caused by seepage flows of infiltrating drainage, and is especially induced by fluctuation of the groundwater level. Measures of efficient drainage control, which effectively reduce the influence of groundwater on the roadbed, are therefore the key to preventing karst collapse along the new highway.

Rock filling was selected as the main subgrade type for the sections of the highway over karst (Figure 13.5.3). After excavating the soils, fissures and depres-

Figure 13.5.3. A rock fill embankment that required considerable ground treatment to carry the Shui-Nan Highway through the Disu karst.
Photo: Mingtang Lei.

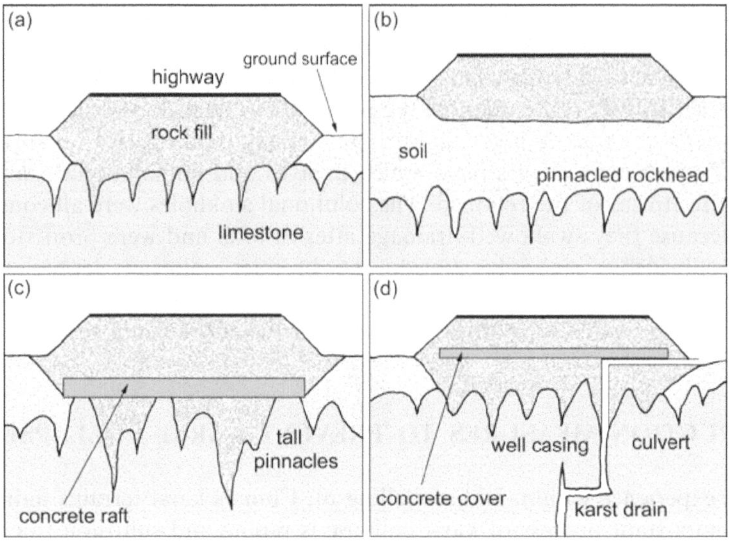

Figure 13.5.4. Diagrammatic design concepts (not to scale) for the Shui-Nan Highway: (a) rock fill placed on bedrock where soil is <3 m thick; (b) mattress of rock fill placed on a thick soil cover; (c) concrete raft spanning wide soil-filled hollows between footings on sound bedrock pinnacles; (d) culvert installed to maintain natural drainage into a karst sink.

sions between the rockhead pinnacles are backfilled with coarse rock rubble that provides better mechanical properties and also has higher permeability. Different construction designs are applied to different sections along the road, each with respect to soil cover thickness and rockhead fissure width (Figure 13.5.4). Soil thickness is immensely variable over the pinnacled limestone, so is measured down to the tops of the majority of the pinnacles. Most rockhead fissures taper rapidly downwards to widths of 200 mm or less, but the widths measured for construction design are those across their tops between adjacent pinnacles or bedrock blocks.

Where soil thickness is less than 3 m and karst fissures are generally less than 1 m wide, the soil cover is entirely removed (Figure 13.5.4(a)). However, where the soil thickness is greater than 3 m, only the surface soil is excavated (Figure 13.5.4(b)). After excavation, coarse fill is placed up to roadbed level, with a grain size of 300 mm at the base, decreasing gradually upwards. Minimum acceptable strength of the filling material is 15 MPa. Voids between the blocks are filled with smaller aggregate and gravel, and the entire fill is compacted by road rollers weighing 15 tonnes. The uppermost 800 mm to the roadbed are formed by a mixed soil of silts and gravels with maximum grain size of up to 100 mm. Soils are compacted with heavy vibrating rollers, so that the top 1.5 m below the roadbed is compacted to 95% of the maximum possible, with compaction reaching 93% below that.

Water levels vary from a dry season water table that lies below rockhead up to flood level that rises half way up the embankment. Consequently suffosion into

Figure 13.5.5. Excavating the soil fill from fissures and hollows between elongate, bladed pinnacles in the karst rockhead, before building the Shui-Nan Highway.
Photo: Mingtang Lei.

bedrock fissures from the lower parts of the soil and fill cover is a natural and almost inevitable process with an inherent risk of soil cavitation and piping failure. It is intended that the potential for suffosion has been virtually eliminated by careful grading of the placed rock fill where rockhead has been exposed (Figure 13.5.4(a)). However, the possibility of soil movement is recognised where thick soils are not entirely replaced (Figure 13.5.4(b)), and a programme of monitoring and research has been started to ensure that the highway is not catastrophically undermined.

Where soil thickness is less than 3 m and karst fissures are commonly wider than 1 m, the soil is first cleaned out from all the fissures and hollows between the bedrock pinnacles (Figure 13.5.5), and the spaces are backfilled up to the main rockhead level, using coarse rubble with grain size decreasing gradually upwards. A raft of reinforced concrete 1 m thick is then placed across the pinnacles and the intervening fill (Figure 13.5.4(c)), and fill is placed over the raft to reach roadbed level (Figure 13.5.6).

Where soil less than 3 m thick is removed and an open shaft or stream sink is discovered, a culvert pipe is installed to carry drainage from the natural catchment

Figure 13.5.6. A concrete slab spanning wide and deep fissures in exposed limestone, with hand-placed stone facing on the road embankment above.
Photo: Mingtang Lei.

Figure 13.5.7. Piers stand on piles founded on sound bedrock to carry the Shui-Nan Highway over a section of difficult karst ground.
Photo: Mingtang Lei.

(usually a broad solution doline next to the road) through to a vertical pipe seated over the natural karst drain (Figure 13.5.4(d)).

Where soil thickness is less than 3 m over a very complex rockhead consisting of wide fissures between pinnacles commonly taller than 10 m, bridges with piled foundations set onto sound bedrock are used to span the karstic openings (Figure 13.5.7). Such conditions have required construction of a number of long low viaducts in place of rock-fill embankments (Figure 13.5.1).

The section of the Shui-Nan Highway through the karst was still in the early stages of construction at the time of writing this case study. Monitoring over the previous year had shown that the highway had not suffered from any collapses or significant subsidences. The design philosophy for ground treatment in the karst terrain appears to be successful.

Case study #6

Construction over a cave in Huntsville, Alabama

Warren Campbell and Tony Waltham

The city of Huntsville, in northern Alabama, U.S.A., grew up around the prolific water supply that is offered by Big Spring. However, this is a karstic rising, with implications that were not appreciated until later. It is developed in the Mississippian Tuscumbia Limestone. This strong limestone, with its intermittent chert beds, tends to be cavernous (and several wells produce more than 200 m^3/ha from it today). Just behind the rising, the original main square of Huntsville was laid out on higher ground. In its centre stood the old courthouse, directly above the cave passage that feeds Big Spring, but fortunately with sufficient rock cover to support the modest structure. In the 1960s, this was replaced by a new and larger Madison County Courthouse – which was designed and built with full consideration of the karst ground conditions.

The footprint of the new building lies right across Big Spring Cave (Figure 13.6.1), whose existence was already known because the passage could be entered just above the rising, which lies only 50 m south-west of the edge of the construction site. An accurate line survey fixed the position of the cave, but the complex nature of the passage hindered its full investigation. The cave beneath and upstream of the roof dome is up to 12 m wide and is extensively modified by roof collapse, so that the accessible part appears as a low wide passage over the top of breakdown blocks that are stacked 4–6 m deep (Figure 13.6.2). The survey (Figure 13.6.1) shows that the cave roof has about 5 m of rock above it, but it has already migrated upwards by about the same amount due to progressive roof failure of the original dissolutional passage (which was at floor level). The roof dome and the passages downstream reach higher, but are less modified by collapse.

An initial phase of ground investigation drilled 19 boreholes across the site, except in the centre where the old court house still stood. Eight boreholes in the southern half of the site found solid limestone with only a few, very small dissolution cavities beneath soil cover 1–4 m thick. Five more holes reached Big Spring Cave. North of the known cave, all six boreholes encountered extensive karstic cavities. Borehole #19 found 8 m of soil, above only 3 m of rock and over 10 m of cave void, before reaching solid rock nearly 23 m down (Figure 13.6.3). Rockhead drops to deeper levels in the north-east corner of the site (Figure 13.6.1), and borehole #7 found soil 12 m deep in what appears to be a buried and filled sinkhole.

The cavernous nature of all the ground north of the cave prompted design modification for the new Courthouse. An office block eleven floors high would stand on the sound limestone south of the cave, while the building above the cave and onto the northern cavernous ground would rise to only three floors. Conventional foundations were used for the high block on good ground. Pads 1,500 mm

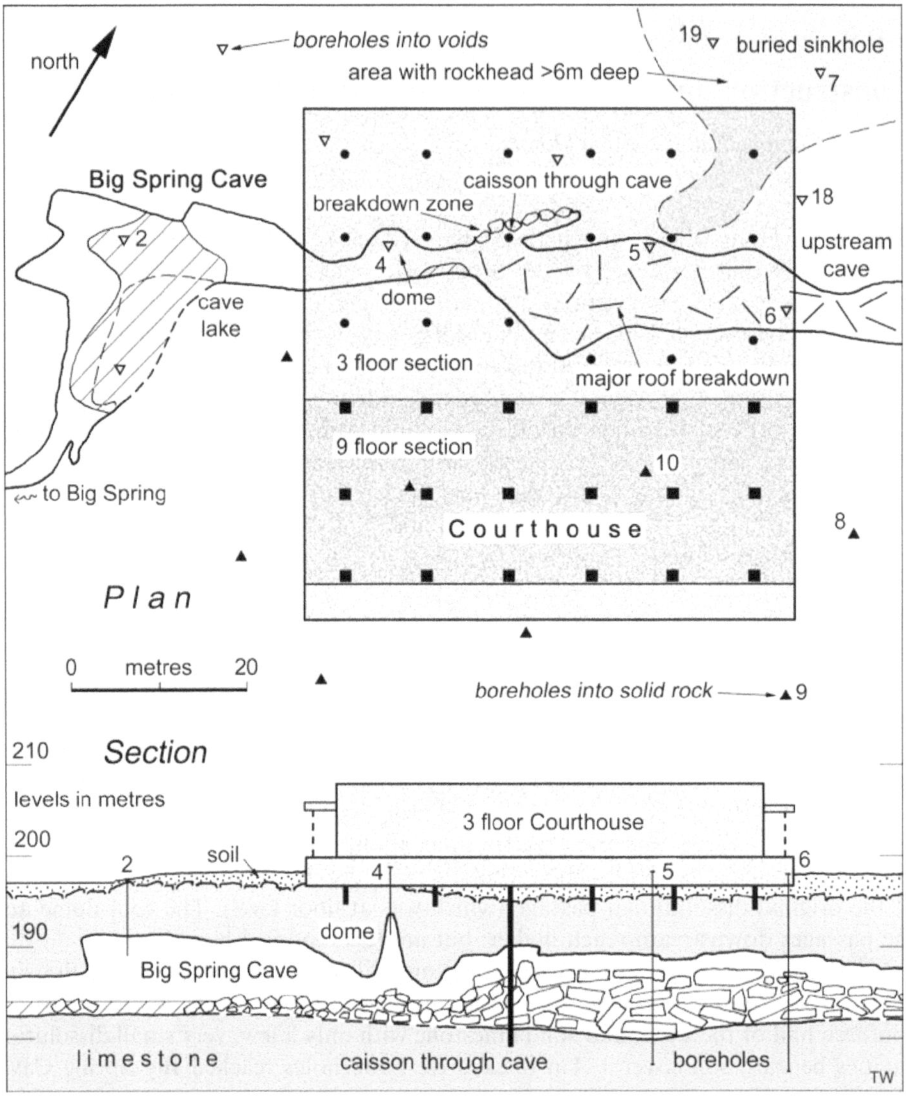

Figure 13.6.1. Footprint of the Madison County Courthouse in Huntsville with its column bases and founding caissons, locations of the ground investigation boreholes and the position of Big Spring Cave beneath. Boreholes are only numbered where they are identified in the cross sections or in the text. The cross section is drawn along the line of the cave, beneath the lower part of the Courthouse. Profiles of both the rockhead and also the cave floor beneath the breakdown are conjectured between points observed or reached by boreholes. The only caisson marked to depth is that seen in the cave; the other caissons are in broken rock beyond the cave in this view.

Drawn largely from data in unpublished plans by Worthington, Smith and Kranert, and cave survey by Huntsville Grotto.

Figure 13.6.2. The main passage of Big Spring Cave directly beneath the Courthouse in Huntsville; the roof has broken to a collapse profile by the progressive breakdown that has left the block pile, more than 5 m deep, that forms the "floor" in this view.
TW.

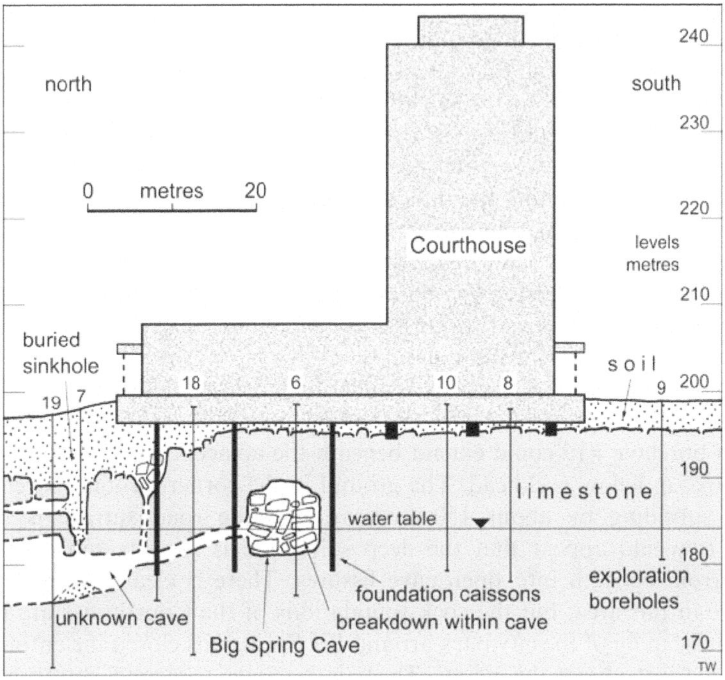

Figure 13.6.3. Cross section, drawn along the eastern edge of the Madison County Courthouse building, showing its asymmetrical profile designed in response to the cavernous ground beneath the northern half of the site.

square, or short bored piles under-reamed to the same footprint, were founded on solid rock with a design safe bearing pressure of 4 MN.

The low extension to the north was designed to rest on 18 steel-cased concrete caissons (bored piles) each 750 mm in diameter and drilled into solid rock. A second phase of ground investigation placed another 13 boreholes at or close to the sites of all caissons north of the cave: these confirmed the presence of numerous cavities, though nearly all proved 5 m of solid limestone below the 180-m level. Though the limestone is almost horizontal, dissolutional cavitation appears to be distinctly asymmetrical around Big Spring Cave. In contrast to the mostly solid rock south of the accessible cave, the north side has numerous voids. Whether these represent earlier positions of the main conduit (whose catchment area is known to be largely on the north side of the known cave), inlet caves or distributary passages is unclear. This cavernous zone extends both above and below the water table, whose level is controlled by the conduit outlet at Big Spring. Subsequent collapse has been extensive, and the zone of breakdown extends well to the north of what can be seen in the accessible cave, notably in the central area where the cave's northern "wall" is just impenetrable breakdown.

Each caisson in this northern half of the site was drilled into the rock until the caisson could rest on at least 3 m of solid limestone, below the level of the known nearby caves. The plan of the caissons was modified from a simple square grid where three of them were displaced south to avoid the cave (Figure 13.6.1). The middle line of caissons was initially intended to lie just north of the open cave, but one is exposed within the breakdown zone that forms the northern "wall". The steel caisson shell can be seen emerging from the slightly over-size hole in the roof, passing down through the open cave and into a drilled hole through a very large breakdown block (Figure 13.6.4), before entering solid rock out of sight. The caisson stands just out of vertical, reflecting the difficulty of drilling through such broken ground, but is large enough to retain integrity under its modest loading. All the caissons are linked north–south (and therefore spanning the cave) by reinforced concrete girders 1,500 mm deep, and east–west by reinforced ground beams 600 mm deep. Failure of any one caisson, due to loss of footing or damage by rock collapse, would therefore not threaten the integrity of the building.

The buried sinkhole and large cave under the north-east corner of the site remain untreated beneath a perimeter strip of footway and gardens. The high cave known only from borehole #19 could extend beneath the adjacent road, but the large void has its roof 5 m below rockhead. The ground in the corner, around borehole #7, is currently subsiding by about 150 mm/year, and the road surface is repeatedly repaired. It would appear that the deep soil cover is steadily settling while it is ravelling from beneath into open cave fissures. There is clearly scope for further soil failure in this area, but the rock foundations of the Courthouse are not threatened. Landscaping of the city park around Big Spring has closed the entrance to Big Spring Cave just above the rising. There is however a second entrance that was originally an open sinkhole 5 m deep into a small tributary passage, 150 m upstream of the Courthouse. This now lies beneath a city street that stands on a

Figure 13.6.4. A caisson that supports the Courthouse in Huntsville, passing through the major zone of karstic breakdown along the northern side of Big Spring Cave.
TW.

large reinforced concrete slab lying on rockhead, but a manhole in its centre allows a ladder descent into the cave.

The Huntsville Courthouse is a model of sound construction on very broken and cavernous ground. It has stood the test of time, and there are no signs of rock distress or renewed collapse either in the ground around it or within the accessible cave beneath it.

Case study #7

Sinkhole destruction of Corporate Plaza, Pennsylvania

Percy Dougherty

In February 1994, the redevelopment efforts of the City of Allentown came to an abrupt halt when a large sinkhole opened in Seventh Street and undermined the Corporate Plaza office building and an adjacent multi-story car park. Corporate Plaza, the crown jewel of downtown revitalisation, was a seven-story, modern office building with over 10,000 m^2 of office space occupied by over 400 employees. The building was a complete loss, resulting in an insurance claim of $8 million and several legal actions (Figure 13.7.1).

An old industrial city that has experienced an extended period of urban decline, Allentown has yet to re-emerge from the sinkhole disaster that struck on that cold

Figure 13.7.1. The newly collapsed sinkhole in Seventh Street, with the conspicuous sag in the Corporate Plaza building above it.
Photo: Robert Seel.

winter day. It is the major city within a metropolitan region of 750,000 people in the Lehigh Valley. Downtown Allentown contains a concentration of government buildings, cultural venues, and office buildings that reflect its importance as the regional centre.

The sinkhole-prone Lehigh Valley is a NE–SW structural valley better known as a part of the Great Valley of the Appalachians, continuing as the Cumberland Valley through southern Pennsylvania and the Shenandoah Valley in Virginia. It is bounded on the north-west by the Silurian and Devonian rocks of the folded Appalachian Mountains and on the south-east by the Cambrian and pre-Cambrian rocks of the Reading Prong of the Blue Ridge Province. The Valley floor, some 30 km across, is composed of rolling hills of Ordovician shale on the north-west half and a flat to gently rolling plain on Ordovician and Cambrian limestones to the south-east. With its fertile soils on the limestone and its natural transportation routes, the Valley houses a string of cities, including Allentown, Bethlehem, and Easton. This places a large population at risk to sinkholes. Each year, the Lehigh Valley suffers several millions of dollars in damage by sinkholes, a figure only exceeded by Central Florida for karst-related damage in the U.S.A.

Most of the city of Allentown is built on the Cambrian Allentown Formation composed of medium-grey, thick-bedded, dolomitic limestone with dark grey chert stringers and nodules, interspersed with laminated oolitic and stromatolitic beds. The rockhead surface is a pinnacled karst with depths ranging from surface outcrops to more than 30 m. A combination of glacial, alluvial and colluvial deposits and weathered bedrock cover the sound rock. The limestone areas of downtown Allentown and the Lehigh Valley have a long history of karst collapses (Myers and Perlow, 1984; Dougherty and Perlow, 1988; Dougherty, 1991). Across open country, the Allentown Formation has an average sinkhole density of 3.4 per km^2, but ground disturbance within the city has led to over 150 sinkhole sites being recorded by subsidence movements, pipeline breaks or structural damage within the 4 km^2 occupied by downtown Allentown.

Although the land is relatively flat, thus encouraging the development of the city, the subsurface is complex both lithologically and structurally. The Allentown Formation is composed of near-shore dolomitic limestone beds each 2–5 m thick. These break into thinner laminae upon weathering and are interspersed with thin shaly beds. Magnesium content varies greatly in the carbonates, and becomes more conspicuous by color differences upon weathering. Because of the near-shore origins of the rock, there is rapid change in characteristics from place to place. Folding and faulting have created complex structures that concentrate groundwater flow and influence the formation of sinkholes. This complexity means that the weaker beds and structural weaknesses cannot be followed, so forecasting the potential location of sinkholes is rarely possible.

Indications of an impeding disaster at the Corporate Plaza surfaced at midnight on 23 February, 1994, when workers at the Allentown Water Plant noticed a drop of 600 mm in the water level at the city reservoir within 30 minutes. At 4.00 am, the leak was found on Seventh Street, in front of the Corporate Plaza, where water was shooting 7 m into the air and a sinkhole had already formed in the street. It is

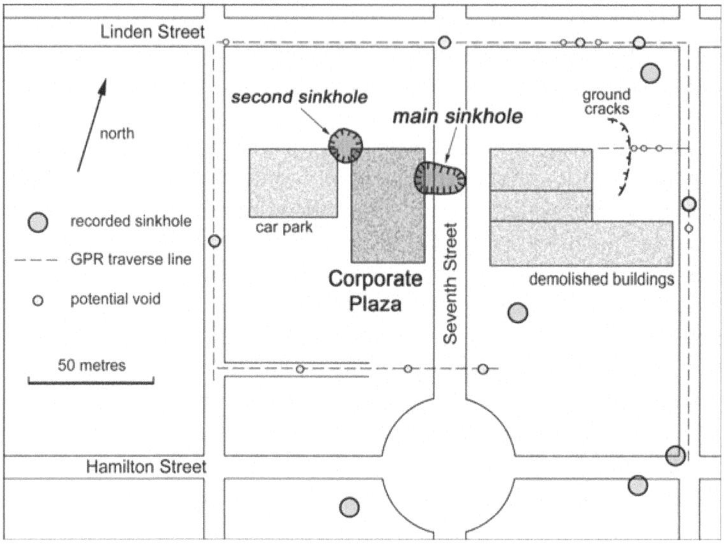

Figure 13.7.2. Outline map of the Corporate Plaza site. Locations of earlier sinkholes (those that are either suspected, reported or unverified) are from the records of the City of Allentown. Potential soil voids were recorded by ground probing radar along the street traverses indicated.

estimated that about 30 million litres of water were lost before the three ruptured water mains (one 300 mm in diameter, and two of 100 mm) were shut off. All the water disappeared underground into the thin bedding planes of the dolomitic limestone, and a light dusting of snow on the city street remained intact to show that there was no overland flow.

The mid-street sinkhole continued to enlarge, so that it extended beneath the supports of the Corporate Plaza building, causing extensive damage to its Seventh Street facade. By daybreak the sinkhole was 5 m deep and 15 m across (Figure 13.7.2). One of the support columns for the building was undermined and a V-shaped bow had developed in the lower floors of the building facade. By noon, three columns along the Seventh Street frontage were engulfed by the sinkhole. Each of these had stood on spread footers that had been totally undermined; the brickwork columns had disintegrated and the internal steel stanchions were left suspended from the building's frame (Figure 13.7.3). In addition, the northwestern corner of the building sagged 3 m into a second sinkhole, which also damaged an adjacent multi-story car park (Figure 13.7.4). Damage was so extensive that the City of Allentown forced the evacuation of the Corporate Plaza and issued a demolition order. The entire building was demolished 23 days later in a spectacular implosion (partly funded by the sale of souvenir bricks and a lottery to trigger the detonation). Three more buildings across Seventh Street, and on the alignment of the two new sinkholes, were also damaged and were subsequently demolished.

Figure 13.7.3. The Seventh Street facade of Corporate Plaza damaged by the collapsing sinkhole. The column on the left is damaged over its subsided column base, exposed except for a thin cover of snow and debris. The column in the centre has its steel stanchion hanging from the building's frame after all its footing has dropped into the sinkhole and its brick cladding has fallen away. The corner column, on the right is almost undisturbed.
Photo: Percy Dougherty.

Figure 13.7.4. Subsidence of the northwestern corner of the Corporate Plaza where the second sinkhole developed.
Photo: Percy Dougherty.

DEVELOPMENT OF THE SUBSIDENCE SINKHOLE

In hindsight, it is apparent that the extensive damage resulted from the building having been placed on spread footings when it was constructed in 1986. Prior to construction, 21 test borings had been taken to depths of 9–13 m on a grid at 15-m centres. No bedrock or voids were encountered along the northern side of the site, so the engineering report recommended a flexible steel-frame structure bearing on spread footings (Leddin, 1994). Concrete column bases 2,500 mm across were founded on soils 1,500–1,800 mm below the ground surface. Had the columns been founded on bedrock, the building would have survived the sinkholes that developed within the soil.

The severity of the damage and the size of the sinkholes are similar to other sinkhole events in the Lehigh Valley. Rather than any collapse of bedrock into a cave, these are classic subsidence sinkholes, formed by suffosion and failure of regolith arches in the unconsolidated soils that cover the bedrock. As the unconsolidated sediments are entrained into fissures along the bedding planes and joints, a void forms within the soil, and this continues to enlarge until the imposed stresses of buildings and surface structures cause a collapse. The process is particularly effective in urban areas where buildings and paved streets cover the surface and result in an "urban subjacent karst".

The hazard was exacerbated in Allentown by the abnormally cold and wet winter of 1993–1994. Breaches in the protective cover formed by paved streets, combined with an early thaw, concentrated the flow of water into the soil and rock beneath the Corporate Plaza site. Reports from workers in the building indicate that there had been trouble with doors not closing correctly and that there were instances where closet doors opened on their own; both features were undoubtedly caused by a slow subsidence of the building into the developing void within the soil profile. A threshold was passed on 23 February, 1994, when the cast-iron water pipes fractured and burst due to deformation of the sagging soil under Seventh Street. The escaping water washed unconsolidated sediment into the bedrock fissures, creating the large void into which the street and the building then collapsed.

Lessons were clear from the collapse of the Corporate Plaza. The sinkhole was a result of the suffosion of the soil into karstic fissures in the underlying dolomitic limestone, a situation that is common in the Lehigh Valley. In an urban area with an ageing infrastructure, slow subsidence can result in the rupture of water pipes that in turn enhances the removal of unconsolidated soils and causes the collapse of the surface over wide soil voids. This event demonstrated the importance of thorough geotechnical investigations of karst ground conditions before construction of large structures. In areas of soil-covered mature karst, large structures should be founded on stable bedrock, and not on spread footings within potentially unstable soil profiles, as was the case at the Corporate Plaza.

Case study #8

Subsidence over a chalk pipe at Chalfont St. Peter, U.K.

Clive Edmonds

In the chalk karst west of London, a rural home near Chalfont St. Peter, Buckinghamshire, was damaged by subsidence due to reactivation of a sinkhole feature. The house had a large, two-storey, brick extension, built in the 1960s with a tiled roof and standing on conventional shallow strip footings. The surrounding ground is fairly level and is underlain by Quaternary fluvioglacial terrace sands and gravels, overlying Cretaceous Upper Chalk Formation. Chalk underlies much of southern and eastern England and is prone to the karstic development of solution features. Thousands of solution features are recorded within a computerised natural cavities database, maintained by Peter Brett Associates (Lord *et al.*, 2002), including large numbers of swallow holes, solution pipes and subsidence sinkholes in the area around Chalfont St. Peter. The karstic geology and geomorphology of the site, together with the potential for subsidence hazard, was therefore well known at the time of the subsidence event.

During 1993 a swimming pool in the large gardens of the house developed problems and appeared to be losing water. In October it was decided to inspect the pool more closely, after pumping out the water. This was achieved over a period of a few days, and the water was disposed of onto a grass lawn beside the house, which was regarded as a convenient soakaway (Figure 13.8.1). The day after water pumping was completed, the ground subsided. This caused structural damage to the 1960s extension on the house, with significant cracking of the walls across its south-west corner, in response to considerable subsidence of the foundations, the floors and the adjacent external veranda (Figure 13.8.2).

The structural damage resulting from the subsidence led to an insurance claim being made. A phased ground investigation therefore followed, in order to define the cause of the ground movement and the extent of the problem. Initially an electromagnetic survey within the garden area around the south-west side of the house suggested the presence of a buried solution feature at the position of the subsidence, and also indicated other features in the garden area beyond.

The position and engineering characteristics of the solution feature at the subsidence location were investigated by heavyweight dynamic probing (DPH; in accordance with British Standard BS1377 Part 9). This probing technique consists of percussively driving a sacrificial cone into the ground. The driving energy is provided by a 50 kg weight that is dropped for 500 mm onto an anvil on rods that connect into the back of the cone. The hammer operation is automated using a chain drive mechanism to control the rise and fall of the hammer weight. The probe provides a quantitative profile through the ground by recording the number of blow counts taken for the cone to penetrate the ground for each increment of 100 mm. This is referred to as the N100 value. No samples are obtained, and the probe records were interpreted by reference to logged profiles from two boreholes (Figure 13.8.1).

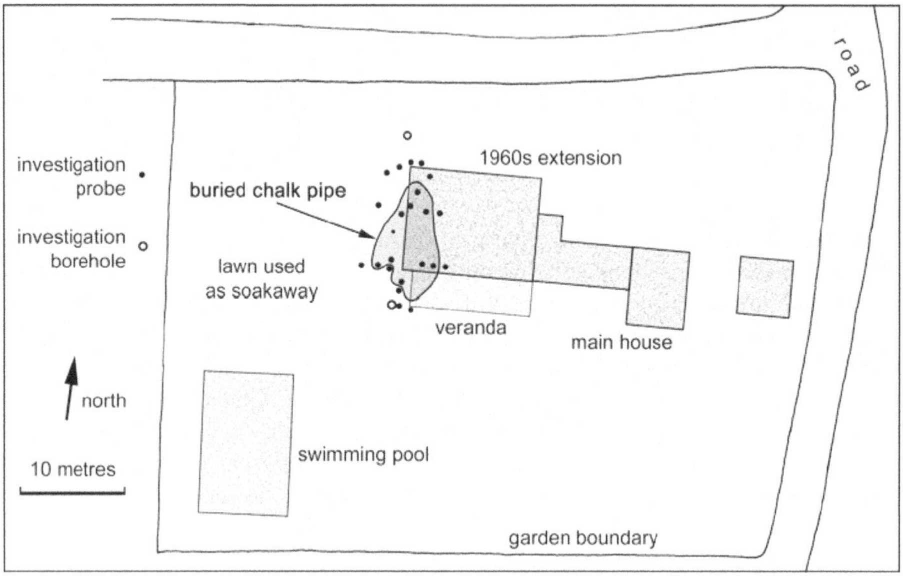

Figure 13.8.1. The site near Chalfont St. Peter, with the house over the buried chalk pipe and the location of the investigation boreholes and probes. The boundary of the pipe is drawn at rockhead, and is approximate as it is based only on the probe and borehole logs.

Figure 13.8.2. The west end of the house extension, showing the new open break where the floor has dropped away from the wall, and also cracks in the higher part of the wall where it is failing in cantilever.
Photo: Clive Edmonds.

The probing profiles are plotted as N300 values (by summing three consecutive N100 values) and classed according to numerical ranges in a classification of blow count values developed empirically from experience with many thousands of dynamic probing profiles within and around solution features in England's chalk. Certain characteristics are typical of buried solution pipes in the chalk. Generally N300 values of <7 tend to occur within the highly disturbed central portions of solution features where the infills are essentially loose to very loose, structureless and highly porous, the original bedding characteristics having been lost. On a macro-scale, the fills form sag-synclinal structures as they undergo subsidence into the developing solution features. During this process, air-filled voids can develop and progressively migrate upwards to produce subsidence sinkholes at the surface. N300 values of 8–14 tend to indicate a transitional change from disturbed to undisturbed ground, where the deposits, still largely intact but undergoing subsidence, tend to dip towards the centre of the solution feature. During this change of dip, the deposits weaken as the soil fabric breaks and shears, and porosity increases as a result. N300 values of >14 indicate bedded deposits beyond the solution feature, and are usually found to be undisturbed.

Boreholes at the Chalfont St. Peter subsidence site revealed a geological sequence, outside the solution feature, comprising fluvioglacial interbedded sands and gravels extending to depths of 9–10 m. Intact chalk was penetrated for about 5 m below rockhead. Dynamic probes through the undisturbed ground either side of the solution feature revealed typical N100 blow counts of 10–40 within the sands and gravels. Below this the N100 blow counts fell within the range 5–15 within the chalk, with scattered higher blow counts where flints were encountered. Within the solution feature, the highly disturbed infills are typically loose to very loose, possessing significant microvoid space and locally upward-migrating, air-filled voids beneath arching soils; N100 values for the disturbed infills were typically in the range of 0–3. Transitional zones of partially disturbed ground could also be seen in the probing profiles where N100 values lay in the range of 3–5. It was clear that the strengths of the disturbed infills within the buried solution feature are greatly reduced compared with the adjacent, undisturbed, normally bedded soils.

Interpretation of the results showed that the buried feature is a soil-filled pipe developed by dissolution of the chalk. It is elongated north–south with a maximum diameter of about 10 m and a width of about 6 m (Figure 13.8.1). The probes showed that, within the solution feature, rockhead levels on the chalk fall steeply to more than 20 m depth (Figure 13.8.3). Given the plan size of the pipe, it was considered possible that the deepest portions of the soil infills extend to depths of 30 m or more, but the lower reaches of the pipe were not reached by the probing. The investigation clearly demonstrated that the ground containing the solution pipe, partially beneath the house, could not be relied upon to provide satisfactory support to the built structure. It was considered that, without remedial action, the south-west part of the house would deteriorate and possibly collapse in time.

In order to stabilise the structure it was therefore recommended that either the property be underpinned, or the fill within the pipe be grouted. The underpinning scheme required piles located beyond the influence of the solution pipe, using beams

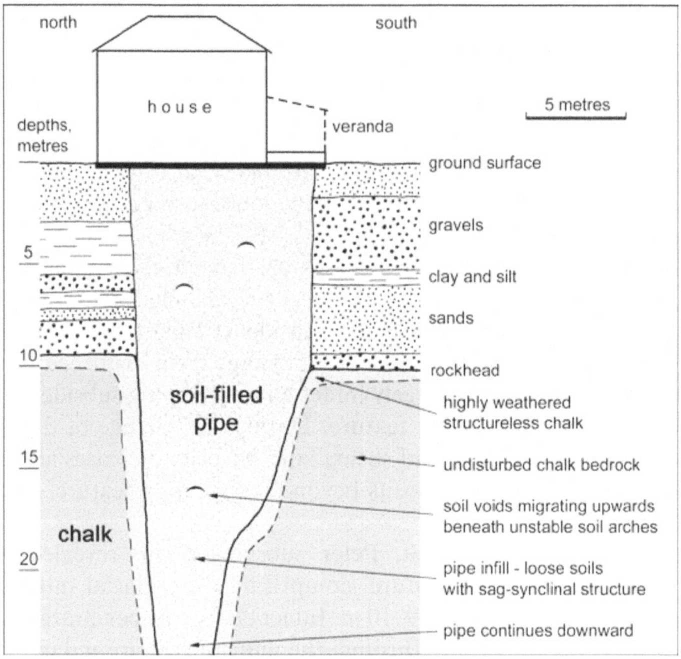

Figure 13.8.3. Interpreted profile of the buried pipe, compiled from logs of the various probes and the two boreholes, and drawn along the line of the west wall of the house.

spanning between pile caps to support the foundations. However, it was noted that partial underpinning of buildings can lead to differential movement and subsequent cracking of the structure. Also, a piled scheme does not prevent the solution feature from deteriorating further, possibly resulting in future subsidence and a risk to the house and its owner. Alternatively, the highly disturbed, loose soil fills within the solution pipe could be treated by compaction grouting to help reinstate stable ground support below the structure. A remedial grouting programme was preferred as it treats the cause of the instability directly and removes the potential for further movement. In this way both owner and structure are protected. The degree of ground treatment can also be tested *in situ* and its success validated.

The costs of either ground treatment approached the value of the house, and compensation was therefore paid by the insurance company. The entire building was demolished, to be replaced by a lightweight timber-frame house founded on piles into the gravel, but on almost the same footprint. The buried solution pipe remains untreated beneath the house and the garden, which now lacks a swimming pool.

Before the subsidence event, there was no surface indication of the chalk pipe, the buried sinkhole, at this site. However, comparable features were known in the area, and due precautions with respect to drainage should have been exercised. Emptying the swimming pool onto the lawn was a classic case of self-induced subsidence.

Case study #9

Geophysical investigations of sinkholes in chalk, U.K.

Peter McDowell

A small area between Chichester and Arundel in West Sussex, southern England, provides several examples of ground subsidence related to dissolution of chalk (McDowell, 1989). Roads and buildings in this area have already been damaged, and the risk of subsidence damage to future development is considered to be high. Consequently ground investigations have included a wide range of methods, both at and adjacent to sinkholes and other subsidence features.

At all the investigated sites, the fine-grained, marine Slindon Sand rests on an irregular chalk rockhead, which is part of a raised beach cut into the Upper Chalk (Figure 13.9.1). The sands are overlain by well-graded, silty and sandy, head gravels, and locally also by a thin layer of silty-clay Brickearth, both of Quaternary age. The regional water table lies deep within the chalk, but perched water can exist within the sands where there is a residual clay over the chalk rockhead.

Localised ingress of water and sand into fissures and solution pipes in the chalk produces voids in the sands that ultimately collapse to form subsidence sinkholes (McDowell and Poulsom, 1996). The most numerous subsidences are sudden dropout failures that create circular, steep-sided sinkholes up to 4 m wide and deep. A cluster of new dropout sinkholes formed at Fontwell, near Arundel, in 1985, when ground collapses followed the burst of a water pipe. A circular area about 90 m across developed more than 70 steep-sided dropout sinkholes in the soil cover (Figure 13.9.2). There were also various groups of circular soil cracks that represented sites of incipient sinkholes or potential renewed subsidence. The shape and size of these features indicated that dome-shaped voids up to 4 m in diameter could develop within the sands (about 3 m thick at the site) and into the overlying head gravels. A section of road and several houses were damaged. A second sinkhole type common in the area is represented by wider depressions with lower side slopes. These are classic suffosion sinkholes developed by void migration through the sands

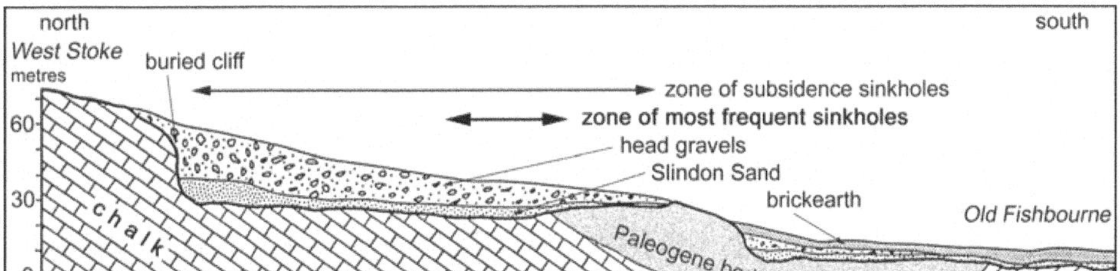

Figure 13.9.1. Geological cross section through the main zone of subsidence sinkholes on the chalk of West Sussex; the profile is 5 km long with a greatly exaggerated vertical scale.

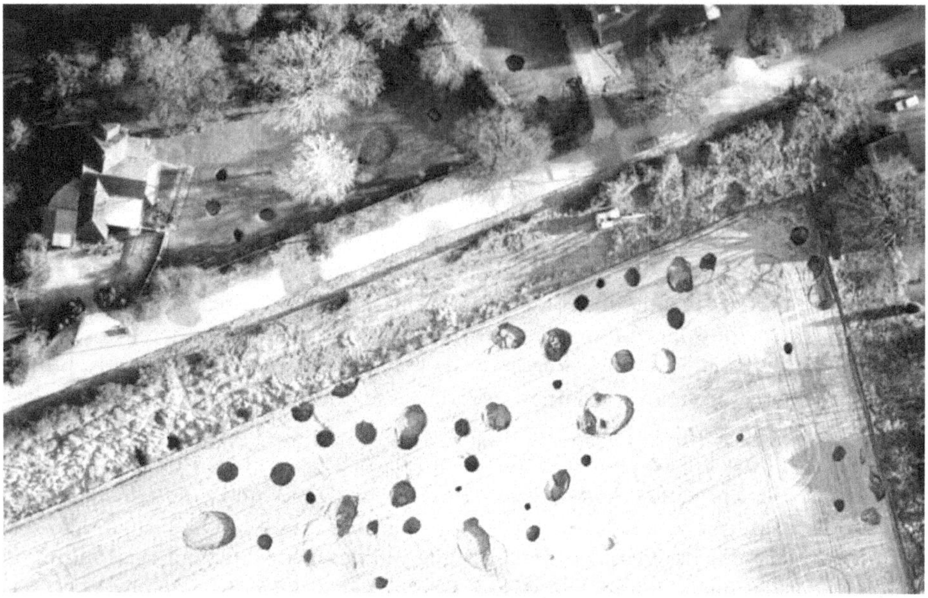

Figure 13.9.2. The cluster of subsidence sinkholes over chalk that developed near Fontwell when a water main burst (in the top right corner of this view).
Photo: Sealand Aerial, Chichester.

and gravels over the chalk. In 1976, a wide and shallow depression caused damage to a road and houses in Chichester.

At both these sites, borehole investigations confirmed the anticipated stratigraphy and groundwater conditions, and also provided relevant geotechnical information. Identification of voids and loose ground, and the presence of sand within fissures in chalk, confirmed the soil suffosion mechanism of sinkhole formation. However, it became evident that direct investigation by drilling, even when supplemented by trenching and dynamic probing, could not be relied upon to locate all voids and loose ground within a reasonable budget. A combination of indirect investigation, by geophysical methods, and direct investigations was considered to be more appropriate.

A variety of geophysical methods were considered, particularly for investigation at Fontwell, which was adopted as a test site for research projects. Seismic methods were not used, although it was recognised that acoustic tomography, utilising shear waves or measuring P- and S-wave attenuation, could potentially be suitable in this type of ground. These techniques would require closely spaced boreholes and probably would only be cost effective for investigation of the ground beneath existing structures. Magnetic methods, although shown to be effective elsewhere for the rapid location and delineation of clay filled dissolution hollows in limestone, would not identify open voids and loose ground in these granular soils. Most consideration therefore was given to electrical and electromagnetic methods.

The electrical resistivities of the soils and rocks were established by calibrating the resistivity sounding against borehole logs. Disturbed sand and gravel was shown to have values of >300 ohm-m, significantly greater than undisturbed sands and gravels (100–200 ohm-m) and the underlying chalk (100 ohm-m) in this area. Voids were expected to have much higher resistivity values, particularly during the dryer months of the year. This was considered a sound basis for using electromagnetic ground conductivity surveying to cover large areas quickly, with the Geonics EM-31 and EM-34 instruments, and also for more detailed investigations by electrical resistivity imaging. However, the EM non-contact surveying produced confusing results, possibly due to the very rapid variations in ground conditions both laterally and vertically (and the proximity to power cables at some sites).

Initial results of resistivity imaging, using ground electrodes, were more encouraging, and this technique was adopted for evaluation at the Fontwell site (Rigby-Jones *et al.*, 1993). A traverse adjacent to the main cluster of sinkholes used a Wenner array, with an electrode spacing of 2 m, to show the variation in electrical resistivity of the ground to a depth of about 8 m. This was then compared to the section subsequently exposed in a trench excavated along part of the traverse (Figure 13.9.3).The anomalously high resistivity values of >300 ohm-m (at chainage 134–157 m and depth 3–6 m) correspond well with the two voids subsequently exposed by the trench. However, this fails to identify the voids individually, and the resistivity survey does not differentiate between the open void spaces and the

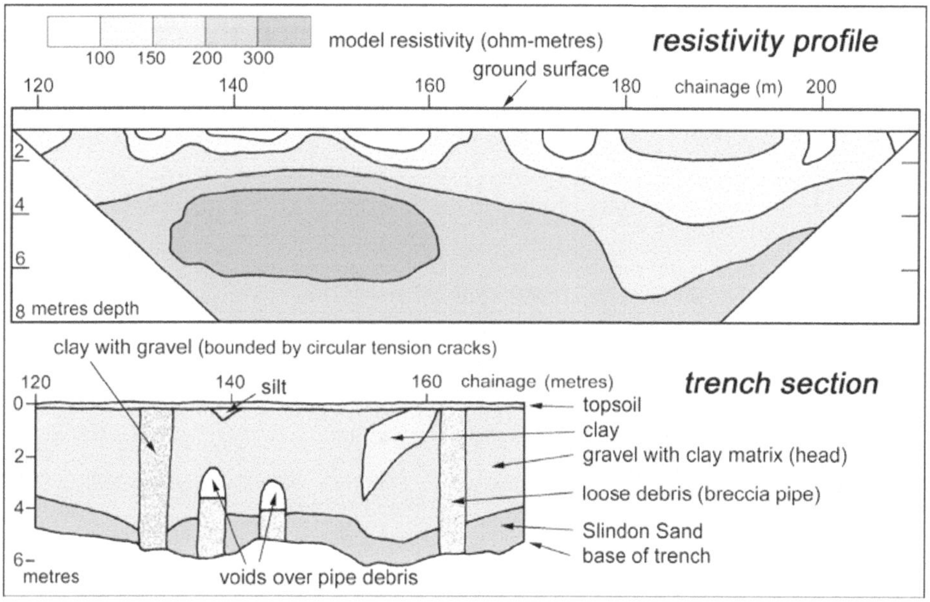

Figure 13.9.3. True resistivity along the Fontwell traverse, using a Wenner array with 2 m electrode spacing (above), compared to the section subsequently exposed along the trench. After Matthews *et al.* (2000).

pipes of loose gravel that have formed beneath the migrating voids. Pipes of loose sand and gravel associated with previous collapses were not clearly identified, probably because they were only 2–3 m diameter and were of lower resistivity due to their clay content.

Improved definition was achieved by using a pole–pole array with a 1 m electrode spacing (Rigby-Jones *et al.*, 1993). Again, a three-layer case was shown, with a high resistivity middle layer, at depths 3–7 m, corresponding to the expected depth and thickness of the Slindon Sand. A zone of particularly high resistivity (400–500 ohm-m) was shown by trenching and drilling to represent a small air filled void at 3 m depth, within an extended column of loose, dry sand and gravel. This survey also indicated the slightly greater depth to the chalk rockhead at this site.

Attempts were made to improve accuracy in the location and definition of voids by the use of ground probing radar (GPR). A GPR survey using a 500-MHz antenna successfully located a void at 1 m depth, but was inadequate for deeper investigation because of the high antenna frequency and the relatively low resistivity of the surface layer. The results of further surveys using a 100-MHz antenna were also disappointing, though it was considered that inclined, planar reflected structures could have been caused by shear planes related to subsidence. Better results may have been obtained if the surveys had been carried out during a dryer period, when the surface gravels had lower moisture content. It is significant that any choice of geophysical methods is site-specific, and resistivity was found to be useful in West Sussex whereas it may not be applicable elsewhere.

Case study #10

Detection of caves by microgravity geophysics, Bahamas

Peter Styles

Cavities were encountered in karstic limestone during investigation for construction of a container terminal at Freeport, on Grand Bahama. The Bahamas are formed on a carbonate platform composed of a variety of carbonate sediment types including coralgal and oolitic limestones. They contain significant cave systems that can be extensive at shallow depths, and these cause major problems for engineering projects on the widespread limestone. Caves fall into four main types (Wilson *et al.*, 1995). Banana holes are oval chambers, generally less than 12 m across and 4 m deep, formed at the levels of past water tables on the islands' freshwater lenses. Flank margin caves can be large chambers or extensive passage networks formed in the zone of water mixing and enhanced dissolution at the edge of the freshwater lenses. Pit caves are youthful, circular to elliptical, vadose shafts up to 7 m wide and 10 m deep. Blue holes are deep water-filled shafts that were formed in Pleistocene environments of low sea levels and were drowned by the deglaciation sea level rise; some lie inland but many connect with lagoons or the open ocean.

The site was originally a karst plain just a few metres above sea level with a thin soil cover and scrub vegetation. A microgravity survey was selected as a rapid means of estimating the extent of caves that were correctly anticipated beneath the site. With the development of modern high-resolution equipment, careful field acquisition techniques and sophisticated reduction and analysis, gravity anomalies as small as 10 microgals can be detected and interpreted. Not only do the anomalies reveal the presence of density variations associated with the location of caves and voids, but they also provide information on cavity depths and shapes from their spectral content, characteristic gradient signatures and from modelling studies (Bishop *et al.*, 1997; Patterson *et al.*, 1995; Daniel and Styles, 1997; Styles and Thomas, 2001; Branston and Styles, 2003).

GRAVITY DATA COLLECTION AND ANALYSIS

Microgravity surveying enabled the detailed delineation and characterisation of an extensive network of below-sea-level flooded caves beneath the site of the proposed container terminal, which occupies a square about 600 m across. Data were collected using four Scintrex CG3-M gravimeters on a 5-m grid. Altitudes were determined using a grid network of stations established with Topcon Total Stations, and the heights of individual points were established using a rotating laser level to an accuracy of 5 mm. Several gravity base stations were used and the meters were returned to these at hourly intervals to determine Earth tides and drift correction, which were applied in addition to the internal Scintrex corrections. Such practice is

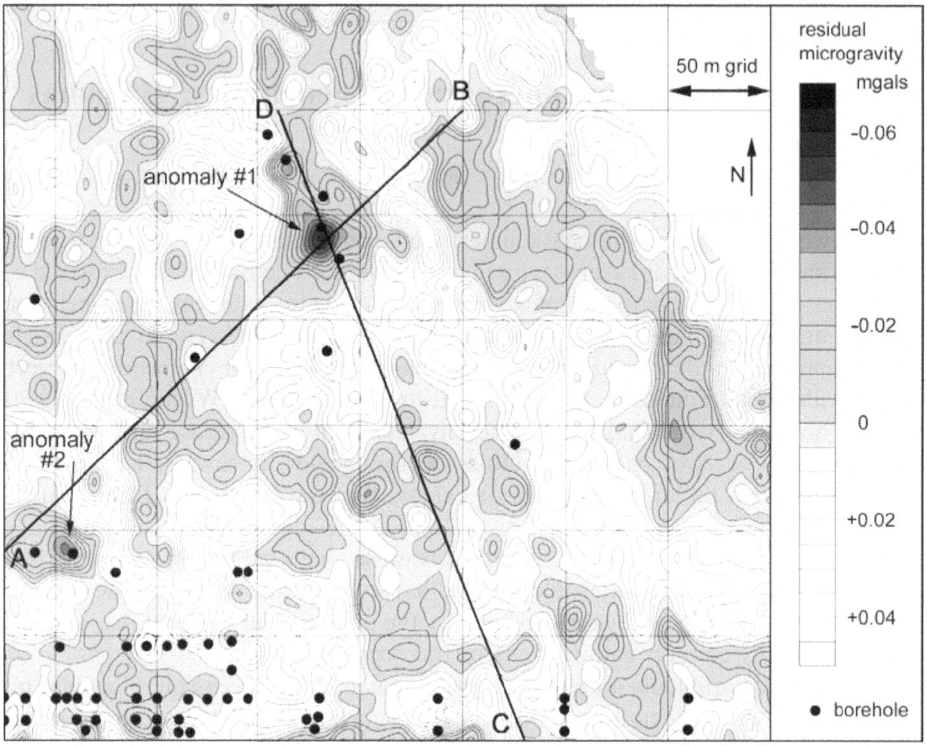

Figure 13.10.1. Microgravity residual map of the Freeport site, with the locations of boreholes and the analysed cross sections.

essential to obtain adequate accuracy for this type of survey. The data were corrected for base station offset and Earth tides, and free-air Bouguer corrections were applied to give a raw Bouguer microgravity map (Figure 13.10.1).

The position of the bodies that cause the microgravity anomalies were determined by a process known as Euler Deconvolution. For gravity (and magnetic) fields, it is possible to determine the position of the causative body, based on an analysis of the gravity field and its gradients and some constraint on the geometry of the body. The profiles selected for automated depth analysis cross a significant negative anomaly (#1 on Figure 13.10.1). Profile A–B crosses the anomaly from south-west to north-east, and analysis gives the calculated position of the top surface of the causative body as a window of calculation is moved along the profile. The main cluster of interpreted caves occurs at 10–12 m beneath the ground surface over a width of about 15 m. This is the short axis of this feature, which appears to extend 30 m or more in an east–west direction. An additional Euler Deconvolution analysis was made along Profile C–D, which crosses anomaly #1 in a nearly north–south direction. This confirmed the depth of the body that causes the anomaly to be at about 10 m, with a shallower body near by.

DATA INVERSION AND GROUND MODELLING

The technique of Euler Deconvolution, together with confirmatory depths from the probe drilling, was then used as an independent constraint that allowed inversion of the microgravity map to give thickness dimensions to the identified cave passages and chambers. Inversion required assumptions to be made concerning the likely depth of the median plane of the cavity and the density contrast between the surrounding rock mass and the cavity. The depth could be constrained at about 10–12 m below surface (by the Euler Deconvolution solutions and the drilling results) and the density contrast was assumed to be −1.25 g/cc for a water-filled cavity in the low-density, porous limestone that typifies Bahamian ground. Inversion made a first guess of cavity thickness based on a simple Bouguer slab formula, and then iteratively adjusted the thickness at each grid node by comparing calculated gravity with observed gravity and scaling the thickness accordingly. This generally converged within a few iterations to a very good fit between the observed and calculated anomalies.

Constrained 2-D cross sections and their gravity models have been created for each profile (Figure 13.10.2). Profiles A–B and C–D cross the site diagonally, coinciding at negative anomaly #1. Both profiles show a major cavern at the site of this anomaly, with a thickness of about 6 m extending from about 11–12 m below grade to about 17–18 m deep. A smaller, high-level cave is located at a depth of about 4 m, lying above the major cave, and this is seen best on profile C–D.

Profile A–B also shows the northern edge of a second major cave at anomaly #2 (Figure 13.10.1), which at a 2–3-m thickness does not appear at its full extent on this profile. The main cave system appears to be continuous across the surveyed area and to have several sub-branches. The cavity at anomaly #2 has an entrance accessible from the harbour (along the southern edge of the map, Figure 13.10.1) and cave divers found a very good correlation between the gravity-derived map and their underwater mapping of a sub-horizontal network of flooded caves. Another high-level cave was revealed at about 4 m depth by drilling a probe hole into the centre of anomaly #2 (Figure 13.10.1). This was also identified by the Euler Deconvolution analysis on profile A–B, which indicated a target at this depth. Using Gauss's theorem the mass deficit for this anomaly was estimated to be almost 1,600 tonnes.

An extensive system of caves with passage heights of 1–2 m appears to lie at a depth of about 13.5 m below grade (10 m below sea level) along both profiles. This agrees with the positions of cave entrances mapped from the diving survey of the cut face offered by the harbour wall. The network may be preferentially located at a specific stratigraphic horizon, and is likely to have originated at or just below a past water table or along a past freshwater/seawater interface – both locations where dissolution processes are concentrated. The investigated site shows significant negative microgravity anomalies that are associated with large amounts of missing mass identified from this survey. Missing mass may be caused either by discrete caves or by distributed void space in the form of micro-caves, fissure networks or features of primary porosity, but the shape and size of the anomalies suggest that open caves of significant size are present beneath many of these features. The cavity maps and

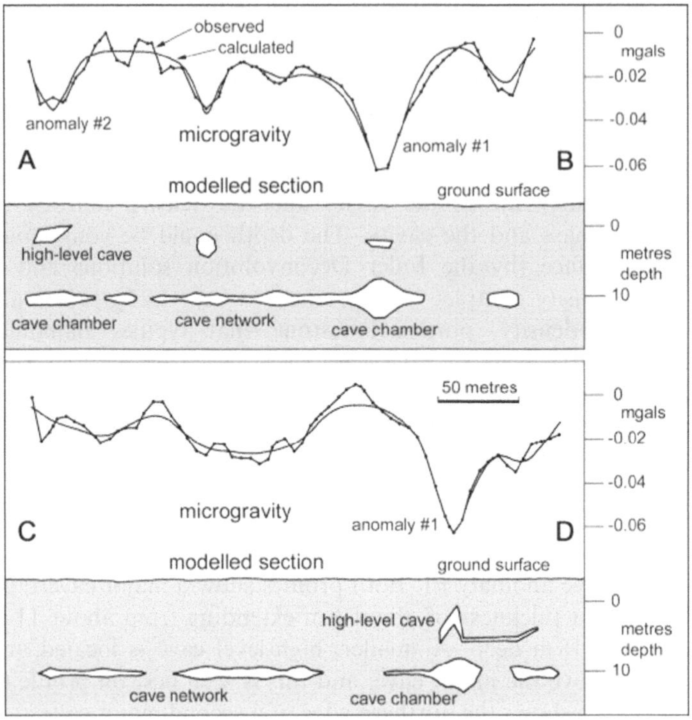

Figure 13.10.2. Profiles A–B and C–D with the calculated and observed gravity in the upper plots and their gravity modelling plotted as 2-D cross sections below.

cross sections were constrained by microgravity, available drilling and cave diving inspection, and represent the best estimate from this dataset.

It was calculated that some 32% of the site could be expected to be at risk from significant ground voids. The major anomalies appear to be linked by sinuous features, which may be interpreted as a network of flank margin cave systems with small passages linking larger chambers and with conduit connections to the sea, probably to both south and north of the island. A network cave appears to underlie the western part of the site with a large feature at anomaly #2 (Figure 13.13.1), which is probably a cave of significant size. This feature alone has an estimated volume of 1,300–1,400 m^3. This cave appears to connect to the sea to the south, and it probably connects north to an even larger underwater cave chamber at anomaly #1, and to more cave passages extending further north and east. Continuity of the anomalies suggests the presence of a linked cave system beneath the whole site, with probable extensions further inland, to the north, and also to the east. The estimate of mass deficiency beneath the whole site is 120,000 tonnes, which represents nearly 100,000 m^3 of water-filled void. Microgravity surveying proved cost-effective and invaluable for investigating and characterising the cavernous limestone beneath an area of 36 ha without enormous numbers of probe holes.

Case study #11

Sinkholes and subsidence over salt at Wink, Texas

Kenneth S. Johnson

A small area near the town of Wink, in Winkler County, Texas, has developed two catastrophic sinkholes and two zones of gentle land subsidence since 1980. All these features occurred within an unpopulated area of about 2.5 km^2 near the middle of the Hendricks field, a giant oil field that was discovered in 1926. Wink Sink #1 and Wink Sink #2 are large caprock sinkholes, respectively 110 m and 240 m wide, each developed around a plugged borehole. The two nearby subsidence zones contain several producing or abandoned oil and water wells.

It appears that these sinkholes and subsidences formed by ground failure into underground dissolution cavities developed in salt beds of the Permian Salado Formation, which is 260–320 m thick in the area and is about 400 m below ground level (Figure 13.11.1). Natural dissolution of Salado salt is well known in western Texas, but the sinkhole collapses and the subsidence sites near Wink appear to have resulted from, or to have been accelerated by, oilfield activities.

THE FIRST SINKHOLE

Wink Sink #1 formed on 3 June, 1980, and within 24 hours had expanded to a maximum width of 110 m (Figure 13.11.2); two days later, the sinkhole was 34 m

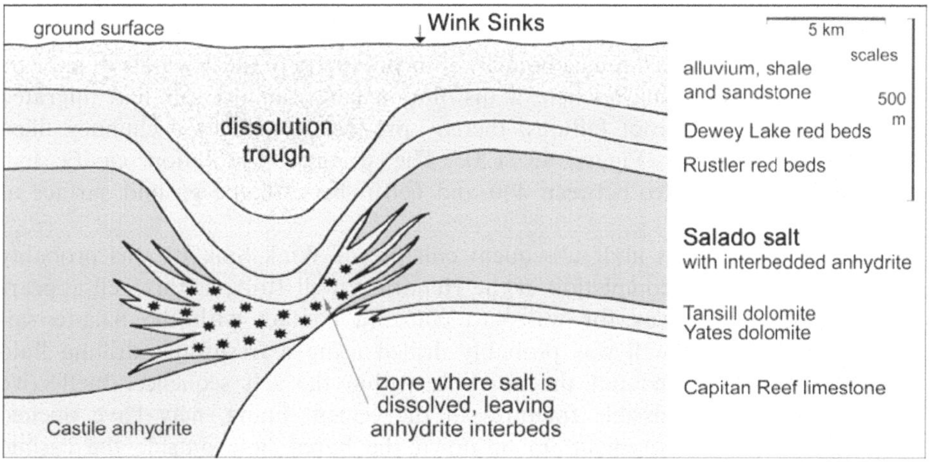

Figure 13.11.1. Cross section showing location of the Wink Sinks close to a dissolution trough formed by natural salt dissolution in Winkler County; the Dewey Lake Formation and all lower beds are of Permian age.
After Baumgardner *et al.* (1982).

Figure 13.11.2. Aerial view of Wink Sink #1 in June 1980, when it was 110 m across. Photo: Robert Baumgardner, Jr.

deep, with a volume estimated at about 159,000 m^3 (Baumgardner *et al.*, 1982; Johnson, 1989b, 1998; Johnson *et al.*, 2003). The abandoned Hendricks oil well 10-A (drilled in 1928) was incorporated within the sinkhole, and a second oil well was later plugged and abandoned because of its proximity to the new hole. It appears that Wink Sink #1 formed where a dissolution cavity in the salt had migrated upward by successive roof failures, thereby producing a collapse chimney filled with brecciated rock (Figure 13.11.3). The original dissolution cavity had developed in the Salado, between 400 and 660 m beneath the ground surface at the sinkhole.

Cavity development and subsequent collapse of Wink Sink #1 was probably related to drilling and completion of the Hendricks well 10-A, as this well appears to have been a pathway for water to come in contact with the Salado salt (Figure 13.11.3). The well was probably drilled using a freshwater drilling fluid that enlarged or washed out the borehole within the salt sequence. Ineffective cement sealing, and possible fractures in the cement lining, may have opened pathways for water movement up or down the bored hole outside the casing. Because of undoubted borehole enlargement during drilling in the Salado salts, the small amount of cement reportedly used to set the casing in the hole was probably enough to cement only the lower part of the hole; this would have left most of the salt exposed behind the uncemented casing.

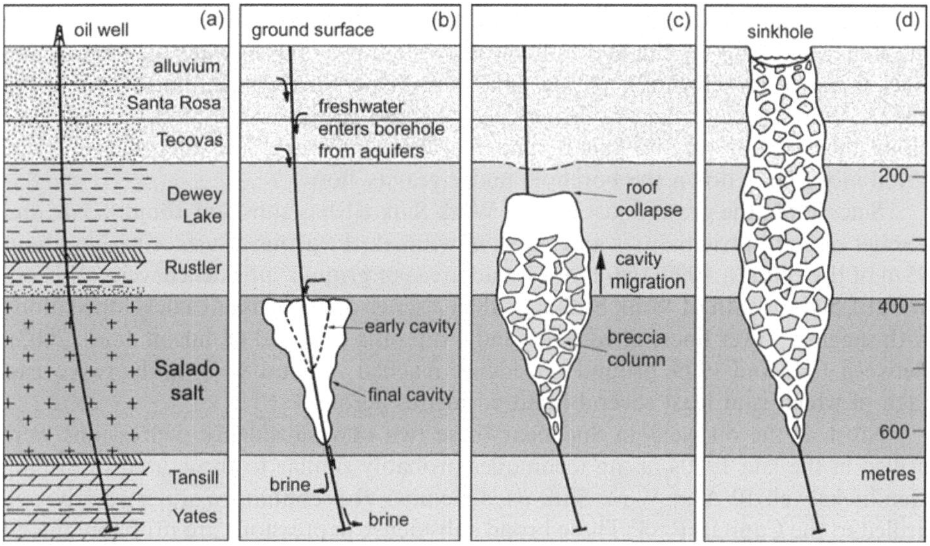

Figure 13.11.3. Schematic cross sections showing the interpreted relationship between Hendricks oil well 10-A and the development of Wink Sink #1: (a) well drilled in 1928; (b) dissolution cavity grows in the salt during and after oil production; (c) cavity migration through beds above the salt; (d) surface collapse to form the sinkhole in 1980.

The well casing was probably punctured by corrosion due to the abstraction of large quantities of brine along with the oil, in the production period of 1928–1951; this would parallel the casing corrosion observed in the nearby Hendricks well 3-A, which had a similar history of drilling, completion and production. Use of explosives to realign well 10-A while drilling in the underlying Tansill Formation not only fractured the rock and increased its local permeability, but also may have fractured the cement lining higher in the borehole. Final removal of the casing from well 10-A in 1964 left an unlined borehole from the base of the Santa Rosa aquifer to the top of the Rustler Formation for 16 years, until formation of Wink Sink #1.

All these oilfield activities, although consistent with standard industry practices during the life of Hendricks well 10-A, could have aided in conducting freshwater from shallow aquifers down the borehole to the salt beds (Figure 13.11.3(b)). Available outlets for the high-salinity brine, which was formed by dissolution of salt around the borehole, included the permeable beds underlying the Salado Formation, as well as any possible dissolution channels already existing within the Salado. It therefore appears that a large dissolution cavity developed around well 10-A, probably in the upper part of the salt sequence (Figure 13.11.3(b)), and this eventually became large enough to permit collapse of its roof (Figure 13.11.3(c)). With successive roof failures, the cavity then migrated upward until it finally reached the land surface and created Wink Sink #1 (Figure 13.11.3(d)).

Waters capable of dissolving the salt may also have ascended from below, with the same end result, as the hydraulic head of slightly brackish water in the Capitan Reef is above the elevation of the Salado at Wink Sink #1 (Baumgardner *et al.*, 1982). A flow-cycle driven by brine-density could cause unsaturated water to rise along the borehole to the Salado under artesian pressure, and then denser brine could move back down the borehole under gravity flow.

Since 1980, the ground just east of Wink Sink #1 has subsided about 8.5 m, and a series of concentric fissures and faults now break the ground surface within about 85 m of the sink. In 1999, two more broad areas of ground subsidence were observed about 1,500 m south of Wink Sink #1, when a series of earth fissures developed along with sagging power lines, tilted poles and a dip on a well pad (Johnson *et al.*, 2003). Between 1970 and 1999, ground subsidence reached 7 m and 8.5 m in the two areas, each of which is at least several hundred metres across.

Most of the oil wells in and near these two new subsidence depressions were drilled in the late 1920s, using techniques probably similar to those used in drilling Hendricks well 10-A at Wink Sink #1. One area also contains a water-supply well drilled to the Capitan Reef. These broad subsidence depressions are probably due to subsurface dissolution of the Salado salt, and they may well be partly related to oilfield activity in the area. As at Wink Sink #1, oil test holes and other boreholes may be conduits whereby unsaturated water has flowed against the salt beds and caused development of solution cavities. If deep-seated cavities do exist under the subsidence depressions, they have not yet breached the ground surface to form discrete sinkholes.

THE SECOND SINKHOLE

On 21 May, 2002, a second major sinkhole developed 400 m west of the existing subsidence depressions, and 1,500 m south of Wink Sink #1 (Johnson *et al.*, 2003). By the end of the first day, Wink Sink #2 was about 140 m long, 90 m across and 30 m deep to water level. The sink continued to enlarge as its vertical banks sloughed material into the water-filled hole, so that it was about 240 m by 185 m across in March 2003 (Figure 13.11.4). The sinkhole is still expanding and the pond has not yet filled with sediment, in spite of the large volume of soil that has slumped into the hole.

Wink Sink #2 is centered on the site of a former water-supply well, the Gulf WS-8 (Johnson *et al.*, 2003). This well was completed in September 1960, drilled into the Capitan Reef, to a total depth of 1,092 m. It met the top and base of the Salado Formation at depths of 412 m and 686 m, so the Salado salt is 274 m thick. There are not yet sufficient data to assess the specific relationship of Wink Sink #2 to the Gulf WS-8. However, it may be similar to Hendricks 10-A in that the well could have been a conduit for unsaturated water to reach the Salado salt beds, where a solution cavity was created and then migrated upward to cause collapse of the land surface.

Remediation engineering of major sinkholes, such as the Wink Sinks, is difficult. The sinkholes generally are too large and unstable for backfilling or for other

Figure 13.11.4. Aerial view of Wink Sink #2 in March 2003; the sinkhole is about 185 m across, with its 240 m length foreshortened, and each circular earth berm is about 100 m in diameter.
Photo: Jim Newman.

standard engineering techniques, and remediation costs could be excessive. Most of these large sinks are fenced off to control access, and are then allowed to partly fill or stabilise on their own. Further enlargement of dissolution cavities in the salt can be arrested by injection of salt-saturated brine.

To prevent formation of such catastrophic sinkholes, modern techniques of drilling, completion and plugging practices are designed to minimise accidental access of unsaturated water to salt deposits in boreholes. This is accomplished by drilling with salt-saturated fluids, injecting cement to completely fill the borehole annulus between casing and salt, setting cement plugs in the borehole to isolate zones of unsaturated waters, testing the integrity of casing and tubing to assure that corrosion does not allow fluid access to the salt, completely cementing the borehole through the salt beds on abandoning the well and possibly removing the casing and tubing.

Case study #12

Subsidence over buried karst at Centurion, South Africa

Fred Calitz

Site investigations were conducted in 1994 to ascertain the cause of ground movement beneath a state-owned warehouse that was damaged by subsidence in the eastern suburbs of Centurion, near Pretoria, South Africa. The warehouse was built in 1974 as a single-storey U-shaped brick structure around a paved courtyard. It is significant that many sinkholes, some with diameters greater than 20 m, have occurred in this area during the period 1974–1993, causing the destruction of another warehouse, a cafeteria complex and a large swimming pool that all lay close to the warehouse investigated in 1994.

The warehouse was built along the crest of a low ridge with very gentle side slopes. The site is underlain by chert-rich dolomitic rocks of the Eccles Formation of the Proterozoic Transvaal Supergroup. The bedrock exhibits typical karst topography with hard-rock dolomite pinnacles separated by deep fissures that have developed along fractures and zones of weakness; these are filled with a weathering residuum of highly compressible, manganese-rich soil known as wad. Gravity surveys and air percussion drilling subsequently confirmed that the rockhead topography varies considerably across the site, with a prominent subsurface depression elongated NW–SE that underlies the northern part of the warehouse. This is one of a series of linear features traced over some kilometres in length; it appears to be a buried karst valley, comparable to a large buried sinkhole as a cause of ground subsidence. Depths to rockhead were 12–39 m. Scattered floaters of hard-rock dolomite within the residual material appear to represent detached or fallen pinnacles.

Prior to construction of the warehouse, a detailed geotechnical investigation was conducted in 1974 to assess the risk of development of karstic instability features. This investigation utilised a regional gravity survey (on a grid spacing of 50 m) and the drilling of five air-percussion boreholes. Based on the subsidence history of the area and the results of these surveys, it was concluded at the time that the area exhibited a high risk from new sinkholes and/or subsidences. It was decided that the development of the warehouse may continue, provided that an efficient surface drainage system be implemented, that surface water be removed to at least 10 m away from the warehouse and that no sewerage lines or water mains be placed beneath the structure. However, the implications of improper development on dolomite karst terrains were not well understood by the developer at the time, and the recommended precautionary measures were not implemented during the design and construction of the warehouse. This oversight led to the eventual destruction of the built structure.

In November 1993, numerous cracks began showing in external and internal walls, as well as across concrete floors, in the northern part of the warehouse. The

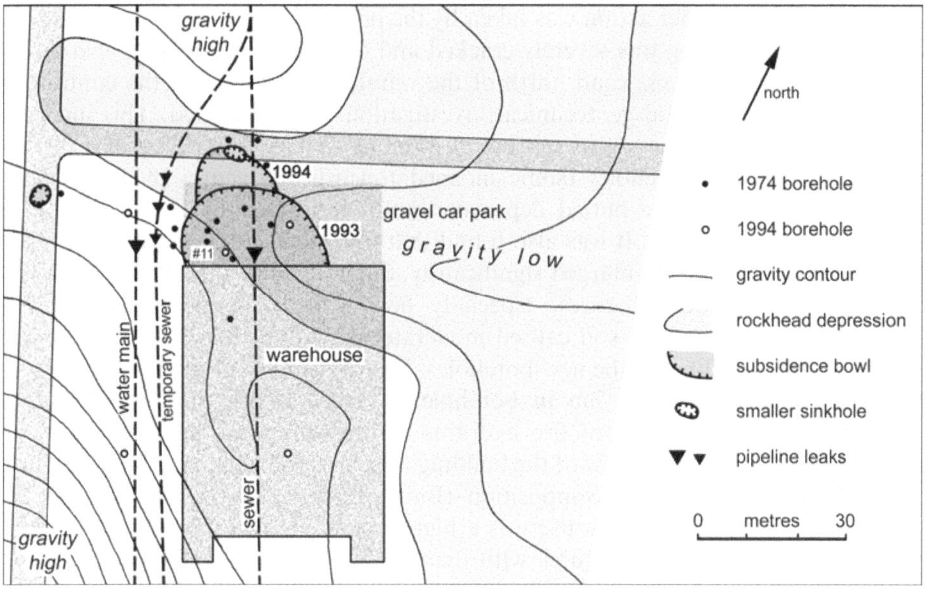

Figure 13.12.1. Outline map of the subsidence features and geotechnical investigations at the Centurion warehouse; the rockhead depression is delineated only broadly, on the basis of the gravity contours.

ground outside developed a series of roughly concentric surface cracks and fissures that extended right up to the northern walls of the warehouse. Some of these fissures opened to 300 mm wide and 4 m deep, and they described the perimeter of a subsidence feature with a diameter of about 20 m that was forming partly beneath the northern end of the building (Figure 13.12.1). A site visit revealed the presence of both a sewer and a water main beneath the warehouse, and also showed that the precautionary measures recommended during the 1974 investigation had not been implemented. Evidence of poor surface drainage around the site was also noted. The building was subsequently evacuated and the pipeline tests showed that a sewer was losing its flow through leakage just north of the warehouse.

A geotechnical investigation was immediately launched to determine the cause of the subsidence feature. This study incorporated another gravity survey (again on a grid spacing of 50 m), and the drilling of seven new air-percussion boreholes. This confirmed the presence of the elongated rockhead depression beneath the northern portion of the warehouse. It was found that this depression is filled by residual soils composed of chert rubble interbedded with thick layers of low-density, highly compressible and easily eroded wad (manganese oxides), capped by a thin layer of alluvial gravel and clayey silt. Bedrock lies at depths of at least 30 m but was not reached in all boreholes; drilling was hampered by poor sample recovery and air loss that were deemed to be indicative of the presence of interconnected voids within the soil. It was recommended that the northern portion of the building was demolished, the surface drainage improved and the subsidence feature backfilled.

However, no further action was taken by the property owner until May 1994, by which time the building was severely cracked and the subsidence feature also threatened an important access road north of the warehouse. In light of the continuing subsidence, an additional geotechnical investigation was conducted. This incorporated a gravity survey on a grid spacing of 30 m, as well as the drilling of a further eight air-percussion boreholes (some inclined towards the centre of subsidence), accurately delineating the buried depression and confirming the karstic nature of the rockhead topography. It was also found that the voids within the soil inside the rockhead depression had enlarged significantly, due to compaction and suffosion of the wad into rockhead fissures, especially beside the dolomite pinnacles. The cavernous condition of the soil caused moderate to total air loss and poor sample recovery during drilling all the new boreholes. Fast-flowing, foul-smelling water was encountered at a depth of 7 m in borehole #11 west of the subsidence feature (Figure 13.12.1). Inspection of fire hydrants at the warehouse and further north revealed that a water main west of the building appeared to be damaged. Application of the Method of Scenario Supposition (Buttrick *et al.*, 2001) showed that the northern portion of the warehouse was a high-hazard site (class VIII) for development of karstic instability features with diameters greater than 10 m (Table 10.5). This assessment was based on the gravity survey results, on drilling records from the sites of both the damaged warehouse and several nearby structures and on the known subsidence history of the area.

Figure 13.12.2. Ground fissures around the subsidence feature adjacent to the northern end of the Centurion warehouse after it finally collapsed in November 1994.
Photo: Council for Geoscience.

Heavy rain during November 1994 led to further ground settlement over the rockhead depression, and the northern part of the warehouse finally collapsed (Figure 13.12.2). The zone of subsidence extended to the north-west and a small subsidence sinkhole developed on its margin, both the result of further consolidation and suffosion losses of the wad, destroying another access road (Figure 13.12.1). Demolition of the remainder of the warehouse revealed a cavity 1.5 m deep beneath its concrete floor. The site was subsequently cleared of all buildings and landscaped, with no further development allowed.

SITE ASSESSMENT AFTER THE EVENT

Evaluation of the observed ground movements and the geotechnical data suggested that a series of events had led to the destruction of the warehouse (all features are located on Figure 13.12.1).

- Water from the roof of the warehouse had drained via several down-pipes, and then ponded in a gravel-lined hollow directly north of the building.
- Surface water regularly infiltrated into the soils over and in the rockhead depression, causing the material to consolidate and also be lost by suffosion.
- A main sewer, at a depth of about 2 m directly north of the warehouse, was ruptured by ongoing settlement of the soils, leading to sustained leakage and therefore to further settlement.
- Soil consolidation was eventually expressed at the surface by a large subsidence feature that caused initial minor damage to the warehouse.
- Storm-water from roads to the north and west of the warehouse flowed unimpeded into fissures around the new subsidence feature, leading to further consolidation of the soils within the rockhead depression.
- Water leaking from joints in a temporary sewer line, placed on the surface to the west and north of the warehouse, also seeped into the soils.
- Ongoing settlement finally caused the rupture of a water main feeding fire hydrants west of the warehouse, introducing an additional large flow of water that may have led to considerable suffosion of the soils into fissures between the dolomite pinnacles flanking the rockhead depression.

With the benefit of hindsight, it appears that removal of the water and sewage pipelines from the site, together with implementation of proper surface drainage measures, would have safeguarded the warehouse from the inherent geohazard created by wad-rich residuum over the dolomite. Furthermore, removal of the active pipelines beneath the entire structure, directly after the initial damage was reported, may have prevented the loss of the warehouse. Although the damage to the structure was caused by infiltration from ponded surface water and from leaking pipes, subsequent collapse of the warehouse was the direct result of failure to implement suitable precautionary measures. These had been proposed after the initial geotechnical investigation, prior to construction. This failure indicates an insufficient knowledge on the part of the developer and land user regarding development on dolomite karst, especially those areas with a soil cover of unstable wad, and proves again that the education of engineers, town planners, developers and local authorities with regard to the geotechnical characteristics of dolomite terrain is essential.

Case study #13

Agriculture on sinkhole karst on gypsum, Lithuania

Bernardas Paukštys

The gypsum karst of northern Lithuania is characterised by high densities of sinkholes, subsidences and karst lakes. The Late Devonian gypsum is overlain by Quaternary deposits that are 0–10 m thick. Karstification began in Devonian times and is still active today, causing damage to buildings and roads, and affecting the development of urban infrastructure. In rural areas it complicates agricultural practices. The karstic region includes 1,187 villages and 5 towns, and has an average population density of 29 km^2. In the past, most of the karst was given over to intensive farming activities, which have involved the use of inorganic and organic fertilisers, herbicides and pesticides. Although this intensive farming gave rise to higher crop yields, it also led to regional groundwater pollution and more active karst development – on a scale now regarded as a national problem.

The karst is highly vulnerable to pollution by agricultural activities, and the level of nitrogen and organic contaminants in the groundwater often exceeds the maximum allowable concentrations (MAC) for drinking water (Paukstys, 1999). In some wells, nitrate concentrations reach 200 mg/l (MAC = 50 mg/l), and the chemical oxygen demand is 19 mg/l (MAC = 6.5 mg/l). In addition, human activities interfere with the natural equilibrium of the vulnerable karst ecosystem and accelerate the rate of karst landform development. A recent survey of aerial photographs revealed the appearance of 61 new sinkholes during 23 years within an area of 3 km^2 of the active karst.

More than 8,500 sinkholes of different shapes, sizes and diameters occur in the active gypsum karst, which has an area of about 400 km^2 (Paukstys and Karise, 1998). Some 87% of the sinkholes have diameters of 10–50 m, and 93% are less than 5 m in depth. Some sinkholes are open to depths of about 40 m. Most sinkholes are dry, but some open to the water table, where they are filled with water (Figure 13.13.1). Some clusters of sinkholes have coalesced to form subsidence basins now occupied by small lakes; the most notable is Lake Ilgasis (Long Lake) 1,100 m long and 200 m wide across a belt of 30 sinkholes. Karst springs occur along valley sides, and were focal points for settlements in the past. Only one cave is known in the gypsum karst region, but other old caves have collapsed.

The density of sinkholes has an inverse correlation with the thickness of overlying deposits (Figure 13.13.2). Most of the karst is covered by less than 5 m of Quaternary till, loam and sand. Zones with most sinkholes tend to occur along river valleys, with up to 200 sinkholes per km^2. The distribution of sinkholes is also related to the fracture patterns within the gypsiferous strata. Rock fracturing has guided the development of river valleys, and is also the primary focus for cave development in the soluble gypsum.

Figure 13.13.1. A large, old, flooded sinkhole in woods surrounded by farmland near Birzai on the soil-covered gypsum karst.
Photo: Anthony Cooper.

PROBLEMS FOR AGRICULTURE ON THE KARST

The development of sinkholes and areas of subsidence in the karst represents a continuing problem for the local agriculture and infrastructure. Some sinkholes and smaller depressions have been filled by local farmers but it is not practicable to fill them all. The presence of fissures and voids in the ground facilitates rapid infiltration of precipitation that, in turn, aids soil erosion. As a consequence, the karst region (that was formerly a zone of intensive agriculture) has been designated by the Lithuanian government as a protected area in which human activities, notably agriculture and groundwater extraction, are limited. Much effort by Lithuanian geologists and environmentalists has been put into demonstrating to local farmers and to the government that the limitation of human activities is the only way to reduce groundwater pollution and karst development. As a result, a governmental decree in 1991 declared the karst lands as a protected region with a special developmental regime. Land has not been taken out of production, but limitations have been imposed on the use of fertilisers, which has led to production of crops that demand less fertiliser and can grow in natural, gypsum-rich soils. Government subsidies to

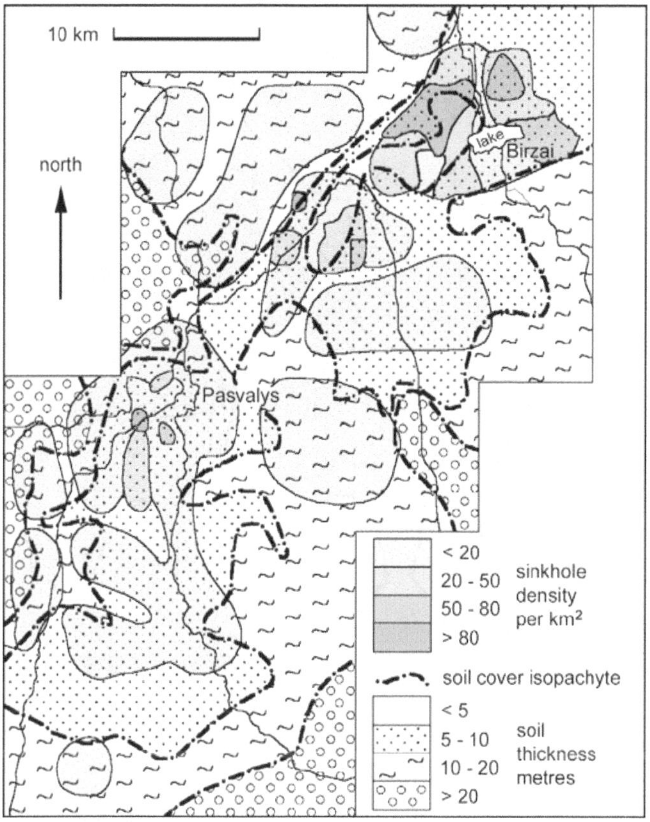

Figure 13.13.2. Spatial correlation between the density of sinkholes and the thickness of Quaternary soil cover in the Lithuanian gypsum karst; sinkholes are very rare or absent in the unshaded areas.

help farmers to start the new, less-polluting types of agriculture were made available from the time when the special status was imposed.

Karst protection measures have been introduced on the basis of the results of investigations of the natural and human-induced factors that have contributed to the karst development. These factors are land cultivation, application of chemical and organic fertilizers and also human activities (i.e., extraction of sand, gravel and gypsum, and groundwater abstraction) that cause water table decline with development of cones of depression and acceleration of groundwater flows in the karst aquifers. In the first instance, the karst region was divided into two zones. The first zone, of 276 km^2, comprises that of active karst development. The second zone, of 1,660 km^2, constitutes the karst protection zone. In addition, the abstraction of groundwater is restricted to the main Devonian aquifer (sandstones and dolostones) that lies below the karstic gypsum. Groundwater abstraction from the karst aquifer is forbidden, as water level decline accelerates removal of fill material from

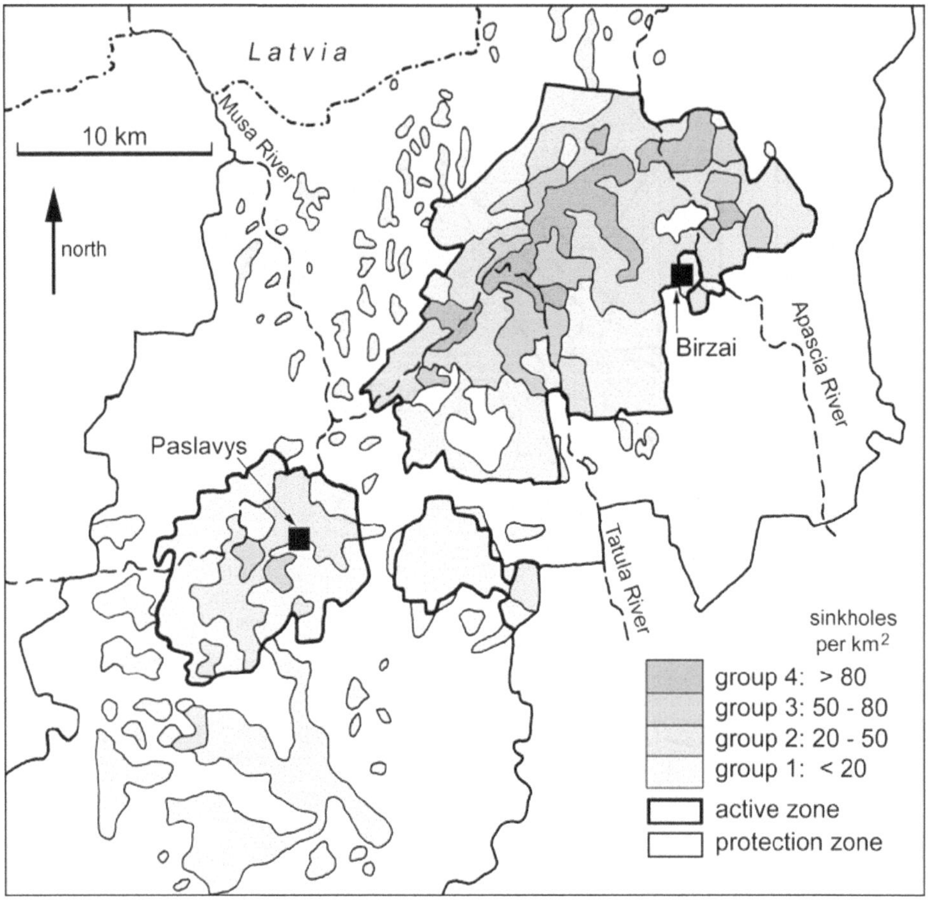

Figure 13.3.3. Karst zones and land groups on the Lithuanian gypsum.

karstic cavities and accelerates the development of new sinkholes. With the current low rate of groundwater abstraction from the Devonian aquifer, there is no danger that such exploitation will lead to any lowering of the karst water table, with the associated consequences. Any increased groundwater consumption in the future should not exceed the safe yield.

The active karst zone was further subdivided into four land groups (Figure 13.13.3), largely on the basis of the spatial density of sinkholes. Each land group has its own levels of restriction imposed on any agricultural activity that can continue within it (Table 13.13.1). These limitations have been based on scientific assessments of natural soil fertility, proposing particular types of crops that need less chemicals and also reduce nitrogen leaching into the groundwater. Maximum applications of fertilisers are defined by their nitrogen/phosphorus/potassium (NPK) active ingredients, and triasinic herbicides and chloroganic insecticides are

Table 13.13.1. Land groups established for the Lithuanian gypsum karst, with their defining sinkhole densities related to their agricultural permitted practices and restrictions.

Land group	Sinkhole density per 100 ha	Agricultural practices	Restrictions
1	< 20	>50% grain crops, 40% pasture, <10% root crops	NPK maximum – 90 kg/ha Manure maximum – 80 Mg/ha
2	20–50	Grain crops and pasture Root crops banned No new orchards or gardens	NPK maximum – 60 kg/ha Manure maximum – 60 Mg/ha
3	50–80	Only pasture farming allowed	NPK maximum – 60 kg/ha No mineral nitrogen fertilizers No pesticides, except fungicides
4	>80	Meadows and forests only	All fertilizers and pesticides banned

NPK = nitrogen/phosphorus/potassium.

prohibited in the entire region of active karst. A protection zone with a radius of 25 m is required around any sinkhole in any of the land groups, within which only grass may be grown without fertilizers or pesticides. Protection measures also include the building of waste-water treatment plants and manure storage facilities, along with ecologically sound agricultural plans for each land group, and these involve the introduction of organic farming into the karst region.

Birzai regional park has been established to include three karst landscape reserves with an aggregate area of 3,586 ha, together with several notable geological features at its core. In this way, it is hoped that rural tourism will become an alternative, more attractive and more beneficial way of life for some rural inhabitants of the karst region.

The 12 years that have passed since the implementation of protection measures in the karst is too short a period to notice any but the obvious environmental improvements. Nonetheless, the results of groundwater monitoring indicate a decrease in groundwater pollution by chemicals. The limitation of karst agriculture has had no significant impact on national agricultural production as the area of active karst is only 1,660 km^2, representing less than 5% of agricultural land in Lithuania. However, these measures are vital to environmental protection, and they demonstrate how agriculture and other human activities can be developed in a sustainable and environmentally friendly manner. The additional problems caused by sinkhole collapses within the karst, including damage to roads and buildings, are being reduced as the region gradually becomes more focused on tourism instead of intensive agriculture.

Case study #14

Sinkhole remediation over Weeks Island salt, Louisiana

James Neal

Weeks Island is formed over a salt dome in the coastal marshland of Louisiana. Within it, a pillar-and-stall salt mine was adapted during 1975–1981 to store 73 million barrels of crude oil as a part of the United States Strategic Petroleum Reserve.

In May 1992, a sinkhole was noted over the edge of the mine (Figure 13.14.1). It lay hidden in woodland and appeared to have developed incrementally over about a year before it was found (Neal and Myers, 1995). When first seen, the sinkhole was 11 m across and 9 m deep, with nearly vertical sides in the Pleistocene loess that covers Weeks Island (Figure 13.14.2). It formed an obvious hazard, as it lay just 25 m from the main access road to an operating mine. The initial reaction was surprise rather than alarm, and it was even thought to be due to collapse of an old house cellar, while an unofficial policy of "wait and watch" was adopted. But knowledge of the sinkhole spread, and local geologists and miners suggested not only the exact cause, but also that more occurrences could be expected. Meanwhile, the

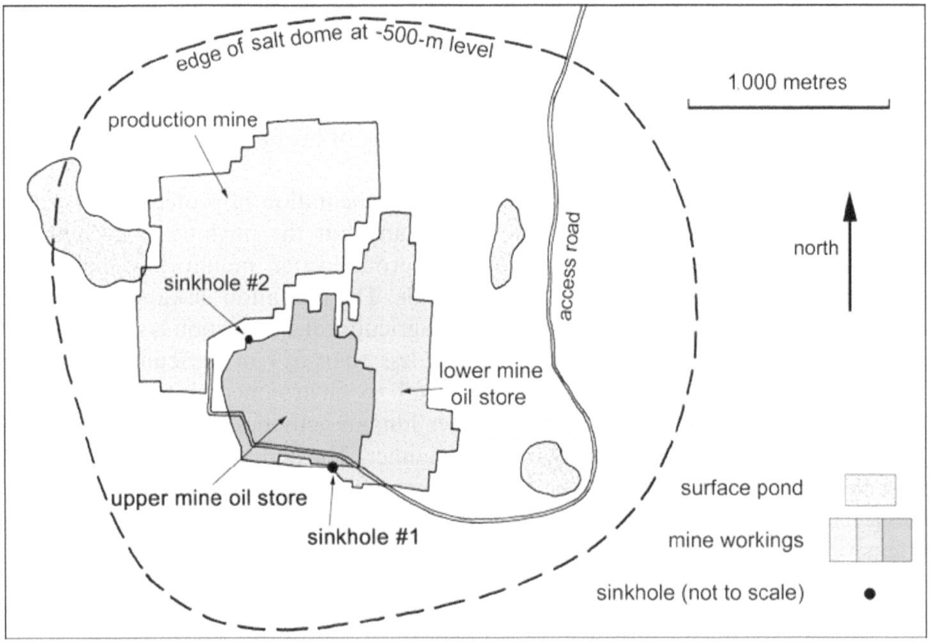

Figure 13.14.1. Location of the two sinkholes over the edge of the upper oil storage mine within the Weeks Island salt dome.

Figure 13.14.2. The main sinkhole soon after its discovery in the woodland of Weeks Island. Photo: Jim Neal.

sinkhole gradually deepened and was actively enlarging. It was recognised that the site should be closed, with the oil transferred elsewhere, and decommissioning began in 1994. Anxiety over the threat of an oil spill from the mine increased when a second sinkhole was shortly noticed nearby; but this one was only about 4 m wide and 3 m deep.

The main sinkhole occurred over the southern perimeter of the upper level of the two-tiered pillar-and-stall mine that held the oil store from 1981 through to 1995. The mine had opened in 1902 and salt was extracted commercially until 1977, when a new mine was developed just to the north-west (Figure 13.14.1). Minor leaks of water were noted at various times during the 75 years of active mining, but inflows were controlled by in-mine grouting. With the benefit of hindsight, one small brine seep, noted during studies to convert the mine to oil storage, appeared to have anomalous isotope geochemistry indicative of meteoric water; this lay near the later sinkhole, and also near an earlier exploratory borehole, but evidence for causal connection was not available. Attention then focused on remediation and removal of stored oil.

SINKHOLE DIAGNOSTICS

Groundwater inflow into the mine was indicated by increasing quantities of brine caught in the sumps that were used during oil-fill operations. Flows increased from 5 to about 14 litres per minute by early 1994, enough to establish a distinct, ongoing change. At the same time, the main sinkhole was filled with sand because it had deepened to 13 m. As soon as the fill was placed, slumping proceeded at a rate of about $2\,m^3$ a day, requiring new fill weekly. This showed that dissolution was ongoing, and there was a rough correlation between the increased amount of brine recorded in the mine sumps and the amount of unsaturated water necessary to leach a volume of salt equivalent to that of the added sand fill. The second sinkhole did not deepen or change, and was never considered a threat.

Rock mechanics modelling (Neal *et al.*, 1998) showed that the mine perimeter would have been in tension and that fractures in the salt could have formed as early as 1970 as a result of the mining. Such cracks could be exposed to unsaturated groundwater, and would gradually enlarge while further propagating under mining-induced tension. The modelling was substantiated by survey data showing subsidence over the mine.

Inclined boreholes provided direct information on the sediment and salt geometry and also on the hydrogeological environment. Cross-hole seismic tomography then showed clear evidence of low-velocity material below the top of the salt, verifying that a sediment-filled sinkhole occurred between the higher velocity salt on either side. This also showed that the area of dissolution below the sinkhole did not extend. Seismic reflection profiles identified the water table within the loess cover with a cone of depression 4 m deep around the sinkhole. One borehole was aimed at the throat of the sinkhole, reaching the salt at the expected depth and continuing through salt into a major void at least 22 m deep. An installed flowmeter then indicated essentially vertical groundwater movement down the sinkhole at a rate of >1 m per day, and its displacement indicated that sediment was moving down the throat at a rate of 25 mm per day, presumably in response to dissolution of salt by unsaturated groundwater at some point below. Rhodamine dye injected into the sinkhole fill was not traced into the mine, but fluorescein dye injected through a borehole into the sinkhole at depth (after the freeze-curtain was completed – see below) was recorded 21 days later in the mine sumps, proving the genetic link between the sinkhole and the mine.

REMEDIATION

Once the geometry of the sinkhole was available, saturated brine was injected directly down its throat, at a rate of about 12 litres per minute under gravity feed. The result was that subsidence at the sinkhole was arrested, and virtually no downward movement was measurable. At the same time, the water table recovered to a normal position around the sinkhole as the cone of depression disappeared. Evidently the brine had stopped the dissolution of salt, but whether this could be a permanent remedy was debatable. Injection continued from August 1994 until the stored oil was removed, though the near-saturated brine was sensitive to temperature fluctuation, and halite crystallisation often led to line clogging and flow cessation.

Grouting was considered as a permanent remedy, but was not instituted because of the lack of urgency and uncertainty about how and where to place grout. Instead, an engineered freeze curtain was constructed around the sinkhole in 1995, as a further control to ensure containment during withdrawal of oil from the mine. This was formed by chilling calcium chloride refrigerant to an average temperature of −38°C, and circulating it through 54 wells drilled in three concentric rings in and around the sinkhole. An outer ring of 22 wells was drilled 3 m into the salt to anchor the freeze wall into the salt dome, and a middle ring of 22 wells was drilled to or slightly into the salt. An inner ring of 10 wells was drilled to the level of the salt, but 5 of these did not reach salt and appeared to end within the sinkhole fill. The end-product was a cylindrical freeze wall 6 m thick with an external diameter of 21 m, reaching a depth of 67 m to seal it to solid impermeable salt (Figure 13.14.3). Brine levels in the wells were modified to concentrate freezing at the lower depths near the top of the salt, creating an ice plug that sealed off the soil fill in the upper part of the sinkhole. Boreholes that reached the lower sinkhole were used to position a flow-meter, and also to inject brine in order to inhibit further salt dissolution during the time it took to freeze the overburden.

Once the freeze wall around the sinkhole solidified and the flow-meter indicated groundwater inflow had stopped, the oil was evacuated from its underground store. The mine was subsequently filled with 85% saturated brine so that stabilisation would be achieved rapidly and subsequent subsidence would be reduced (Molecke, 2000).

LESSONS LEARNED

The sinkholes at Weeks Island formed over the edge of the old mine as a result of geological, hydrological and mine-induced factors that were poorly understood at the time the oil repository was established. They were solution sinkholes formed rapidly within the salt, then expressed at the surface by suffosion and collapse of the loessic soil into the growing voids beneath. While the first sinkhole continued to mature, the second smaller one appeared to have sealed itself by the inflow of sediment. The main sinkhole was located towards the edge of the salt dome

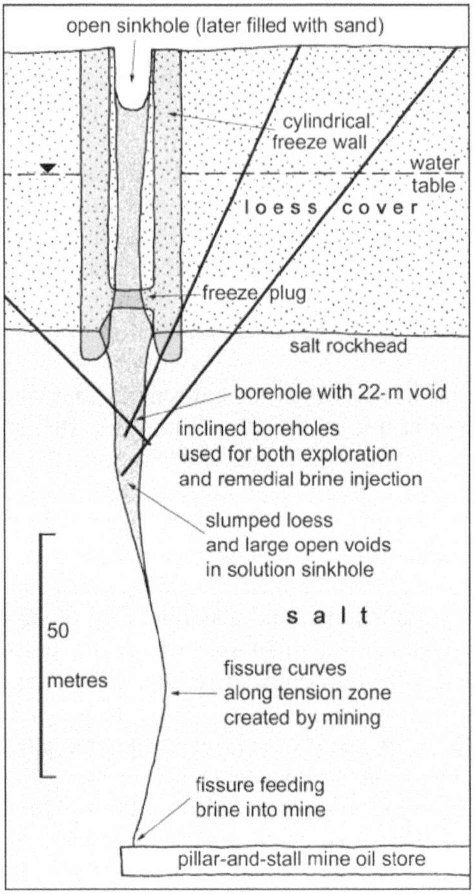

Figure 13.14.3. Diagrammatic section through the main Weeks Island sinkholes and its remedial freeze curtain.

footprint, probably astride a shear zone that had developed during diapiric rise of the salt and was itself capped by a shallow rockhead valley. Underground openings in the mine placed the salt over its periphery in tension, thereby favouring fracture development, probably as early as 1970. Eventually, an incursion of unsaturated groundwater traversed the fractures for 107 m of salt from the rockhead down to the mine, where it emerged as brine. Gradually increasing dissolution enlarged a void at the top of the salt, creating the collapse environment for the sinkhole that formed c.1991.

Exploratory drilling and geophysics defined the void beneath the sinkhole, enabling the introduction of saturated brine directly into the throat, which completely arrested the subsidence at the sinkhole by eliminating ongoing dissolution. The construction of a freeze curtain around the sinkhole then enabled the oil to be removed from its mine store and transferred to other sites with environmental surety.

Case study #15

Hazard assessment on dolomite at Simunye, South Africa

David Buttrick

One-fifth of Gauteng Province, the mining, commercial and industrial heartland of South Africa, is underlain by dolomite. In the west of Gauteng, the West Rand has an infamous history of catastrophic sinkhole development induced by de-watering of the dolomite aquifers to permit safe mining of the gold bearing quartzite reefs beneath (Box 8.1). In June 1986, de-watering commenced in the West Gemsbokfontein groundwater compartment, and in the early 1990s the Simunye site was identified within it for the emergency relocation of a settlement of 5,000 families, which at that stage was threatened by ground instability (Figure 8.1.1). This was the first occasion in South Africa that consideration was given to placing a new residential area in a compartment currently being de-watered.

The geological, geophysical, borehole and geohydrological information gathered during an extensive investigation was analysed to assess the stability of the delineated site. The main evaluation factors to assess the hazards of sinkhole development were: extent of cavities in both the soil and bedrock, groundwater flow in response to the pumped drawdown and ingress of water, potential sinkhole size, nature of the soils and bedrock morphology (Buttrick *et al.*, 2001). The site was then characterised in zones, each of which was defined in terms of the likelihood that a certain hazard may manifest itself; using the method that is standard in South Africa (described in Section 10.3.2). The ground was characterised primarily in terms of eight Inherent Risk Classes (Table 10.5), of which three were recognised in the Simunye site.

GEOLOGY AND GEOHYDROLOGY OF THE SITE

The Simunye site is directly underlain by dolomite and chert of the Monte Christo Formation, within the Transvaal Supergroup. It lies within the West Gemsbokfontein groundwater compartment, between syenite dykes that have intruded the dolomite and isolated the hydrological compartments in the karst. Karoo rocks have been deposited over the dolomite, and are also preserved in paleokarst depressions. Tertiary to Recent colluvial soils blanket the terrain. Ground elevation is around 1,610 m. The original water table was at 1,559 m, but was falling at a rate of 1–10 m per year in response to the mine de-watering. Records showed a major elongate zone of preferential groundwater drawdown associated with a deeply leached paleokarst valley along the eastern boundary of the site.

A gravity survey, combined with borehole information, strongly guides appraisal of the sinkhole hazard in the karst of the West Rand (Kleywegt and Enslin, 1973), and was therefore a basis for the eventual zoning map (Figure 13.15.1). Much of the site is underlain by a plateau structure where the dolomite

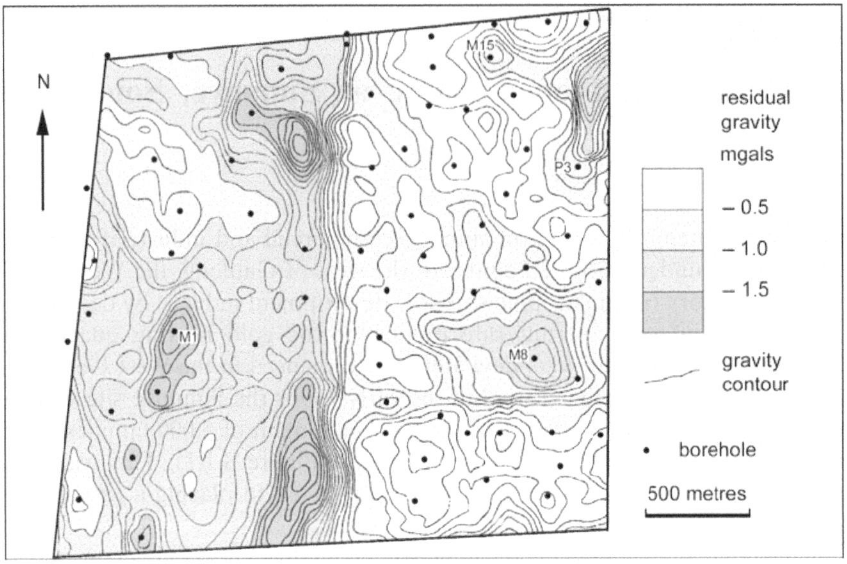

Figure 13.15.1. Residual gravity map of the Simunye site; numbered boreholes are referred to in the text.

rockhead lies well above the original water table elevation of 1,559 m. Boreholes indicated rockhead generally at depths of 22–50 m, though there are isolated fluctuations outside this range. A deeply eroded paleokarst valley, related to the Panvlakte fault zone, was delineated by a north–south step beside two elongate negative anomalies in the gravity values. Other east–west and north–south lineaments were also noted in the gravity data. On either side of the main paleovalley, two paleokarst depressions (large buried sinkholes), each more than 85 m deep, have their floors well below the original water table (boreholes M1 and M8, Figure 13.15.1). Paleosinkholes were also interpreted at the sites of boreholes P3 and M15.

The gravity data show a good correlation with the borehole records, and this allowed compilation of a second map showing overburden thicknesses (Figure 13.15.2). Potentially compressible soils, notably residual manganese wad, clayey silts, fine chert residuum and some Karoo sediments extended below the original water table under parts of the site. Under some further areas of the site, compressible materials were recorded below the water table but are so thin that they are regarded as insignificant. Karoo shales enhance ground stability by restricting ingress of water, and they cover most of the eastern part of the site, but are absent in the northwestern sector (Figure 13.15.2).

HAZARD CHARACTERISATION OF THE SITE

The Simunye site was broadly divided into areas of distinctive karst morphology (Figure 13.15.3) that were largely reflected in the maps of gravity contours and

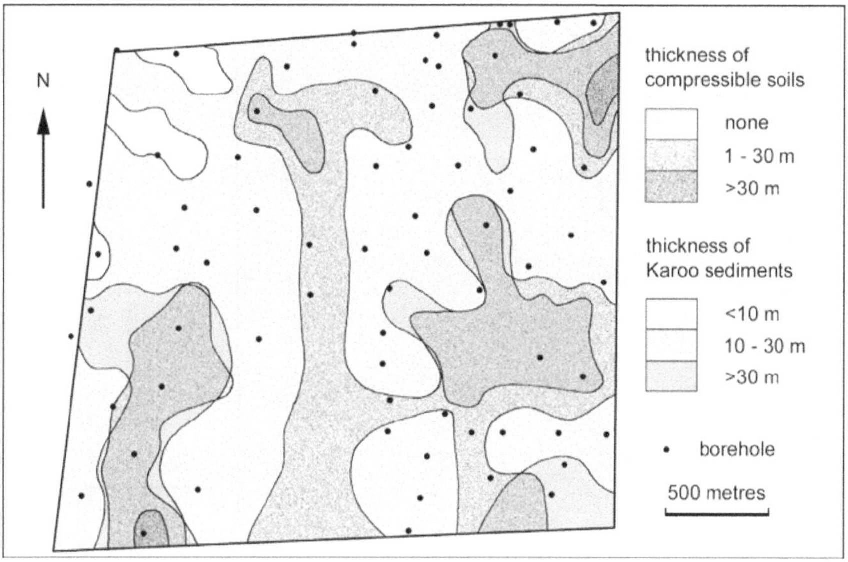

Figure 13.15.2. Soil thicknesses under the Simunye site, identifying both the compressible residual soils and the stable Karoo sediments that influence the hazard zoning.

overburden isopachytes. These fell into three Dolomite Stability Zones that could each be ascribed to classes within the hazard classification (Table 10.5); within South Africa, these are known as classes of inherent risk.

Zone 1: Hazard Class 1

This zone is identified by the areas of high gravity and their immediate surrounds that have gentle gravity gradients. Areas with steep gravity gradients are eliminated from the zone due to the propensity for sinkhole formation on steep rockhead gradients. The dolomite rockhead lies mainly above the original water table. The zone includes small sub-areas where rockhead is below the water table in areas of gentle gravity gradients, where minimal potentially compressible soil materials are recorded below the water table and where Karoo shale overlies the dolomite. Karoo shale, colluvium and alluvium constitute a large proportion of the karst cover, and these residual and reworked materials typically have permeability values of the order of 1×10^{-7} to 1×10^{-9} m/s.

Voids in dolomite bedrock lie at depths in excess of 30 m. Very large sinkholes could therefore be formed if the soil cover was fully mobilised, but these are considered as very rare, and none has been observed in this setting. The water table decline occurs within the dolomite bedrock, and therefore has minimal negative impact on the overlying soils; soil mobilisation potential due to groundwater drawdown is assessed as low (NSH < 0.5, where NSH = number of new sinkhole occurrences per km^2 per year). Ingress of water to initiate subsurface erosion or

Figure 13.15.3. Division of the Simunye site into three zones with respect to stability and potential sinkhole hazards over the karst dolomite.

piping is largely precluded by the low-permeability soils with poor internal drainage; so their mobilisation potential with respect to ingress of water is also low.

Characterisation of the zone as land of Hazard Class 1 reflects the low levels of hazard from the formation of both subsidence sinkholes and compaction sinkholes, with respect to both water ingress and water table decline.

Zone 2: Hazard Class 4

This zone is also identified by high gravity values on areas of shallow dolomite rockhead that lies above the original water table. Any cover of Karoo rocks or reworked sediment is generally absent or too thin to make any significant positive contribution to ground stability. Soils are primarily composed of colluvium, or chert and wad that are residuals from dolomite dissolution.

Water table decline is within bedrock as rockhead is located above its original level at 1,559 m. Mobilisation potential is therefore low with respect to groundwater drawdown. The soil cover typically is characterised by materials with medium permeabilities (1×10^{-2} m/s to 1×10^{-4} m/s), so good internal drainage readily permits subsurface erosion where sustained and concentrated infiltration of surface water occurs. Mobilisation potential of the soil is therefore moderate with respect to any ingress of water. With anticipated bedrock voids at depths around 30 m, and with a soil cover composed of two or three horizons, potential sinkhole sizes could be large to very large.

Subsurface conditions in this zone characterise the land as Hazard Class 4, reflecting a medium hazard (NSH 0.5–5.0) from the formation of all sizes of subsidence sinkholes and compaction sinkholes with respect to ingress of water, and a low hazard (NSH < 0.5) with respect to water table decline.

Zone 3: Hazard Class 8

This zone is distinguished by negative gravity anomalies, wider areas of lower gravity, areas of preferential groundwater drawdown along fault zones that are directly linked to the main cone of depression where the de-watering is driven, and/or any areas where thick sequences of potentially compressible soil materials (notably the residual wad) are recorded below the original water table.

Typical geotechnical characteristics of the residual soil are dry density of 0.8–1.6 Mg/m3, void ratio of 1.3–6.0, collapse potential of 1–5%, compression index (C_C) of 0.6–1.5 and coefficient of consolidation (C_V) of 9–10 m^2/year (Brink, 1979). The cover of Karoo rocks is highly variable, from a very thin cover to substantial beds over 100 m thick. Paleo-sinkholes indicate historical instability. These features are typically filled with permeable, aeolian, silty, fine sands with good internal drainage, and may be reactivated either by water table decline or by infiltration of surface water.

Mobilisation potential of the soil materials below the original water table is high with respect to water table decline. Any such change in the moisture regime may lead to the formation of subsidence sinkholes or to soil consolidation and the generation of compaction sinkholes. Superimposed on this are potential stability problems related to infiltration of water. Where Karoo shale is present, it may act as an aquitard, and therefore preclude subsurface erosion. However, where the protective Karoo horizon is absent, the residual soils are exposed to the erosive effects of any concentrated flows of infiltrating water. Soil mobilisation potential is therefore high with respect to water table decline, and with respect to ingress of surface water is low on Karoo cover to medium where there is no Karoo cover.

Land in this zone is therefore characterised as Hazard Class 8. Though there is only a low to medium hazard (NSH <5) from the formation of subsidence sinkholes and compaction sinkholes with respect to ingress of water, there is a high hazard (NSH >5) with respect to water table decline, and the higher hazard establishes the class.

RISK MANAGEMENT STRATEGY

Hazard assessments of both Zone 1 and Zone 2 areas permit housing development. This is the direct result of land characterisation in terms of hazard within the Method of Dolomite Land Hazard and Assessment that has evolved from the original Method of Scenario Supposition (Buttrick *et al.*, 2001). Water precautionary measures, involving appropriate services design and maintenance, are applied in both zones, more stringently in Zone 2 than in Zone 1. The land in Zone 3 may not be urbanised, and is retained as open land or used for agricultural activities.

Construction of Simunye commenced in 1992, and 5,000 housing units now occupy the site.

This development was permitted in the context of a proactive Dolomite Risk Management Strategy. Monthly monitoring of the township is undertaken, and includes visual inspections to detect evidence of structural distress in buildings, visual recording of ground cracks and precise levelling of various strategically placed survey beacons. The data is then all analysed for evidence of ground settlement.

By January 2004, eleven sinkholes (some compaction and some subsidence) had developed in the high-risk Zone 3 areas. These features have ranged in size from large to very large, so that the majority are >15 m across. All the subsided ground in the sinkholes has been immediately rehabilitated in accord with the Dolomite Risk Management Strategy for the area. No ground stability problems have been recorded in the land areas of Zone 1 and Zone 2. However, the vigilance and monitoring continue.

Case study #16

Ground investigation in covered karst at Tournai, Belgium

Olivier Kaufmann and John Deceuster

Since the beginning of the 20th century, localised and sudden ground collapses have been reported in the Tournaisis, the region of covered karst around the city of Tournai, in southwestern Belgium. Most of these subsidence sinkholes are circular or elliptical in plan and cylindrical or conical in profile, while they can exceed 10 m in diameter and depth.

In the Tournaisis, the bedrock of impure Carboniferous limestone is overlain unconformably by a cover that mainly consists of Cretaceous marls and sandy or clayey Tertiary sediments. Though this cover is geologically a caprock, it is poorly consolidated and the sinkholes through it are best regarded as subsidence sinkholes. The topography of the Tournaisis has few landforms typical of karst terrains, but quarry faces show that paleokarst features are common in the underlying limestone. These features developed in localised zones of jointing where progressive dissolution of the carbonates leaves a soft porous weathering residue. As the host limestone is siliceous, this residuum is also siliceous, but there is still a significant content of carbonate in some of the intermediate weathering products, and the porosity is commonly very high. It is widely recorded that most new sinkholes develop directly above these paleokarst features, mainly in areas of water table decline in the karst aquifer (Kaufmann and Quinif, 2001, 2002).

In an area where collapses had already occurred, there was serious concern over the structural stability of a public building nearly 100 years old and four floors high (Figure 13.16.1). It was standing on conventional foundations less than 5 m deep on the Cretaceous marls. A number of sinkhole collapses, each 2–3 m across, have occurred under and around the building in the last five years. One of these, 2.0 m wide and 1.7 m deep, occurred below the lift shaft during former remediation works, while others caused floor failures between the foundation elements. An investigation was therefore conducted beneath the building foundations, in order to establish the ground conditions and map the complex topography of the buried rockhead. This was essential to identify problematic zones, and then to design proper and effective remediation.

At the site, the limestone bedrock is overlain by no more than 10 m of clays and marls. Underpinning the foundations onto solid limestone is therefore an option to secure the building. However, installation of micropiles under the northern wing of the building during an earlier phase of remedial works, along with a few boreholes around the site, showed that soft weathered limestone and residual soils are more than 15 m thick under some parts of the site.

To improve knowledge of underground conditions at the site, several cross-borehole resistivity panels have been determined beneath the foundations. The zones of weak, weathered limestone and its residuum were already known to have

Figure 13.16.1. The building under investigation near Tournai, with a fence surrounding a small subsidence sinkhole in the foreground.
Photo: Olivier Kaufmann.

a good contrast in resistivity values with those of sound limestone rock (Kaufmann and Quinif, 2001). Twelve boreholes were drilled around the building and three more were placed inside the cellars in order to provide resistivity panels that gave the best assessment of the subsurface conditions. Borehole depths range between 12 and 24 m, with at least 1 m into sound limestone wherever possible.

The boreholes revealed the presence of backfill and sandy soils for the first 1–3 m of depth, and the underlying clays and marls are generally 5–10 m thick. Where the limestone is weathered, the weathering residuum varies from 1 to more than 20 m thick. Thus, depths to sound bedrock ranges from 6 to more than 25 m below ground surface. Each borehole was equipped with a system of permanent electrodes (Daily and Ramirez, 1999) with one electrode every metre in order to measure the underground electrical resistivities.

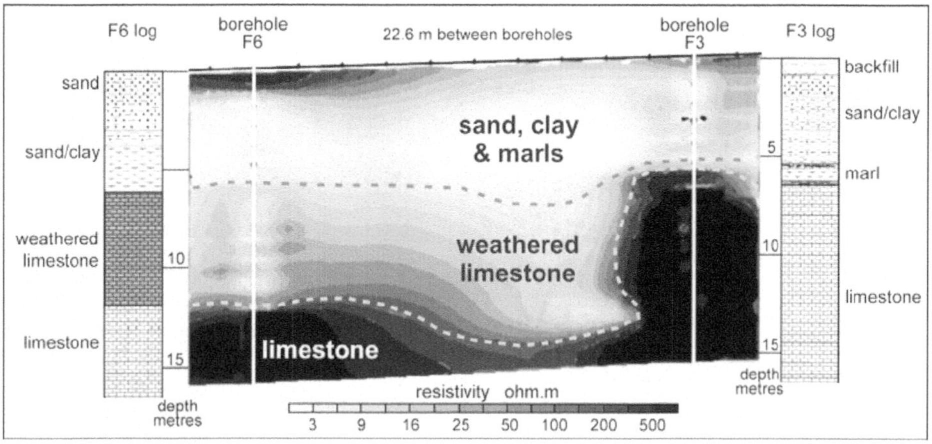

Figure 13.16.2. 2-D cross-borehole tomography of ground resistivities within a single panel, correlated with logged data from the two boreholes F6 and F3.

On-site experiments led to selection of both dipole–dipole (AM-BN) and also pole–dipole (AM-B and A-BN) arrays (Bing *et al.*, 2000). For this investigation, 23 cross-borehole resistivity panels were constructed between pairs of boreholes, and at least 2,500 different apparent resistivities were measured for each panel. Inversion was solved by using Res2DInv software based on a smoothness-constrained, least-squares method (Loke *et al.*, 1996). Each panel was then plotted as a 2-D tomograph to show a ground cross section by its interpreted resistivity values. Figure 13.16.2 shows the panel between boreholes F6 and F3 (which are located on Figure 13.16.3). Resistivities below 20 ohm-m (shown in lighter tones) correspond to the clays and marls, while the high resistivities (>300 ohm-m, shown in darker tones) are correlated with limestone bedrock at depth and with granular backfill near the ground surface. Intermediate resistivities are interpreted as the weathered limestone and dissolutional residuum. Resistivity variations within the weathered zones can be explained by changes in water content and by differences in the weathering intensity.

On Figure 13.16.2, the cover of clay and marl appears to be quite regular and is about 5 m thick. However, the depths to sound bedrock show important variations. Around borehole F3 the depth to the limestone rockhead is about 6.5 m, whereas it is about 12 m around borehole F6. From borehole F6 to about 10 m away, a layer of weathered limestone about 7 m thick overlays the sound limestone rock. Beyond this, a trough of highly weathered limestone is clearly distinguished, between 10 and 17.5 m from F6, and this feature is also detected on parallel panels. A map of sound rockhead elevation was then synthesised by computed interpretations of the 23 tomographic panels together with available data from other boreholes and micro-piles (Figure 13.16.3). Weathered zones, where sound rock lies at greater depths, are clearly delineated, and bedrock joint directions are also distinguishable.

The geophysical survey provided valuable information that enabled selection of appropriate remediation techniques, and also minimised remediation costs. A

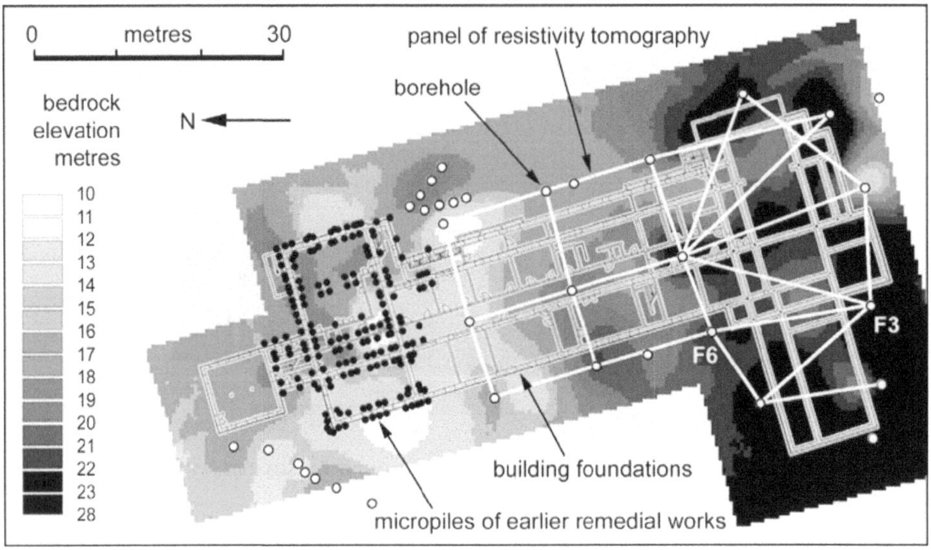

Figure 13.16.3. Elevations of sound limestone rockhead beneath the foundations of the building, with locations of the investigation boreholes and the tomographic panels prepared between them.

programme of grouting and micropiling is planned to ensure continued integrity of the building. Two new boreholes already drilled have confirmed rockhead at depths within less than 1 m of the positions indicated by the geophysics, and the depth data from the earlier phase of micropiling also conform with the interpretations of the resistivity tomography. The geophysical data appears to be validated.

References

Abdullah, W.A. and Mollah, M.A. (1999) Detection and treatment of karst cavities in Kuwait. In: Beck, B.F., Pettit, A.J. and Herring, J.G. (eds), *Hydrogeology and Engineering Geology of Sinkholes and Karst*, pp. 123–127. Balkema: Rotterdam.

Abkemeier, T.J. and Stephenson, R.W. (2003) Remediation of a sinkhole induced by quarrying. *American Society Civil Engineers Geotechnical Special Publication* 122, 605–614.

Alexander, E.C. and Book, P.R. (1984) Altura Minnesota lagoon collapses. In: Beck, B.F. (ed.), *Sinkholes: Their Geology, Engineering and Environmental Impact*, pp. 311–318. Balkema: Rotterdam.

Aley, T.J., Williams, J.H. and Massello, J.W. (1972) Groundwater contamination and sinkhole collapse induced by leaky impoundments in soluble rock terrain. *Missouri Geological Survey Engineering Geology Series* 5, 32pp.

Allred, K. and Allred, C. (1997) Development and morphology of Kazumura Cave, Hawaii. *Journal Cave Karst Studies*, **59**, 67–80.

Andrejchuk, V. (2002) Collapse above the world's largest potash mine, Ural, Russia. *International Journal Speleology*, **31**, 137–158.

Arrington, D.V. and Lindquist, R.C. (1987) Thickly mantled karst of the Interlachen, Florida, area. In: Beck, B.F. and Wilson, W.L. (eds), *Karst Hydrogeology: Engineering and Environmental Applications*, pp. 31–39. Balkema: Rotterdam.

Arthur, J.C.R., Haas, C., Shilston, D.T. and Waltham, A.C. (2004). The Sivas karst, from terrain evaluation to ground truth. *Proceedings International Conference on Terrain and Geohazard Challenges Facing Onshore Oil and Gas Pipelines*. Thomas Telford: London.

Asef, M.R., Reddish, D.J. and Lloyd, P.W. (2000) Rock-support interaction analysis based on numerical modelling. *Geotechnical Geological Engineering*, **18**, 23–37.

A.S.T.M. (1998) *Standard test method for mechanical cone penetration tests of soil*. Designation: D3441, Volume 04.08, Soil and Rock. American Society for Testing and Materials: Philadelphia, 5pp.

A.S.T.M. (1999a) *Standard guide for selecting surface geophysical methods*. Designation D6429. American Society for Testing and Materials: Philadelphia, 11pp.

A.S.T.M. (1999b) *Standard test method for penetration test and split-barrel sampling of soils.* Designation: D1586, Volume 04.08, Soil and Rock. American Society for Testing and Materials: Philadelphia, 5pp.

Atapour, H. and Aftabi, A. (2002) Geomorphological, geochemical and geo-environmental aspects of karstification in the urban areas of Kerman city, southeastern Iran. *Environmental Geology*, **42**, 783–792.

Atkinson, T.C. and Smith, D.I. (1974) Rapid groundwater flow in fissures in chalk. *Quarterly Journal Engineering Geology*, **7**, 197–205.

Bahtijarevic, A. (1989) Sinkhole density of the Forest City Quadrangle. In Beck, B.F. (ed.) *Engineering and Environmental Impacts of Sinkholes and Karst*, pp. 75–82. Balkema: Rotterdam.

Balázs, D. (1973) Comparative investigation of karst waters in the Pacific. *Proceedings 6th International Congress Speleology*, **3**, 23–32. Academia: Prague.

Barner, W.L. (1999) Comparison of stormwater management in a karst terrain in Springfield, Missouri – case histories. *Engineering Geology*, **52**, 105–112.

Barton, N., Lien, R. and Lunde, J. (1974) Engineering classification of rock masses for tunnel design. *Rock Mechanics*, **6**, 189–236.

Baumgardner, R.W. Jr., Hoadley, A.D. and Goldstein, A.G. (1982) Formation of the Wink Sink, a salt dissolution and collapse feature, Winkler County, Texas. *University of Texas at Austin, Bureau of Economic Geology, Report of Investigations* 114, 38pp.

Beck, B.F. (1986) A generalized genetic framework for the development of sinkholes and karst in Florida, USA. *Environmental Geology Water Science*, **8**, 5–18.

Beck, B.F. (1999) There is no such thing as karstish – an introduction. *Engineering Geology*, **52**, 1–2.

Beck, B.F. (2000) Report on Italian sinkhole conference. *KWI Conduit*, **8**(1): www.karstwaters.org

Beck, B.F. and Sinclair, W.C. (1986) *Sinkholes in Florida: An introduction.* Florida Sinkhole Research Institute Report 85-86-4, 16pp.

Beddows, P.A. (2004) Yucatan phreas, Mexico. In: Gunn. J. (ed.), *Encyclopedia of Caves and Karst Science*, pp. 786–788. Fitzroy Dearborn: New York.

Bednar, D. and Aley, T. (2001) Groundwater dye tracing: An effective tool to use during the highway development process to avoid or minimize impacts to karst groundwater resources. In: Beck, B.F. and Herring, J.G. (eds), *Geotechnical and Environmental Applications of Karst Geology and Hydrology*, pp. 201–207. Balkema: Lisse.

Beese, A.P. and Creed, M.J. (1995) A database for subsidence sinkholes near Cork, Ireland. *Danish Geotechnical Society Bulletin*, **11**, 4.19–4.24.

Beggs, T.F. and Ruth, B.E. (1984) Factors affecting the collapse of cavities. In: Beck, B.F. (ed.), *Sinkholes: Their Geology, Engineering and Environmental Impact*, pp. 183–188. Balkema: Rotterdam.

Bekey, T.J. and Rinne, R.W. (1975) One hundred foot wide lava tube complicates hospital foundation design. *Proceedings Annual Engineering Geology and Soils Engineering Symposium, Boise, Idaho*, **13**, 75–83.

Bell, F.G. (1975) Salt and subsidence in Cheshire, England. *Engineering Geology*, **9**, 237–247.

Bell, F.G. (1992) Salt mining and associated subsidence in mid-Cheshire, England, and its influence on planning. *Bulletin Association Engineering Geologists*, **29**, 371–386.

Bell, F.G. (1993) *Engineering Treatment of Soils*, 346pp. Spon: London.

Bell, F.G. and Coulthard, J.M. (1990) Stabilization of clay soils with lime. *Municipal Engineer*, **7**, 125–140.

Bell, F.G., Cripps, J.C., Culshaw, M.G. and O'Hara, M. (1987) Aspects of geology in planning. *Geological Society Engineering Geology Special Publication*, **4**, 1–38.

Bell, F.G., Culshaw, M.G., Moorlock, B.S.P. and Cripps, J.C. (1992) Subsidence and ground movements in chalk. *Bulletin International Association Engineering Geology*, **45**, 75–82.

Bell, F.G., Culshaw, M.G. and Cripps, J.C. (1999) A review of selected engineering geological characteristics of English chalk. *Engineering Geology*, **54**, 237–269.

Bengtsson, T.O. (1987) The hydrologic effects from intense ground-water pumpage in east-central Hillsborough County, Florida. In: Beck, B.F. and Wilson, W.L. (eds), *Karst Hydrogeology: Engineering and Environmental Applications*, pp. 109–114. Balkema: Rotterdam.

Bennett, D. (1997) Finding a foothold. *New Civil Engineer*, December 4, 24–25.

Benson, R.C. and Kaufmann, R.D. (2001) Characterization of a highway sinkhole within the gypsum karst of Michigan. In: Beck, B.F. and Herring, J.G. (eds), *Geotechnical and Environmental Applications of Karst Geology and Hydrology*, pp. 103–112. Balkema: Lisse.

Benson R.C. and La Fountain, L.J. (1984) Evaluation of subsidence or collapse potential due to subsurface cavities. In: Beck, B.F. (ed.), *Sinkholes: Their Geology, Engineering and Environmental Impact*, pp. 210–215. Balkema: Rotterdam.

Benson, R.C. and Yuhr, L.B. (1987) Assessment and long term monitoring of localized subsidence using ground penetrating radar. In: Beck, B.F. and Wilson, W.L. (eds), *Karst Hydrogeology: Engineering and Environmental Applications*, pp. 161–169. Balkema: Rotterdam.

Benson, R.C., Yuhr, L. and Kaufmann, R.D. (2003) Assessing the risk of karst subsidence and collapse. *American Society Civil Engineers Geotechnical Special Publication* 122, 31–39.

Bergado, D.T. and Selvanayagam, A.N. (1987) Pile foundation problems in Kuala Lumpur Limestone, Malaysia. *Quarterly Journal Engineering Geology*, **20**, 159–175.

Bieniawski, Z.T. (1973) Engineering classification of jointed rock masses. *Transactions South African Institute Civil Engineers*, **15**, 335–343.

Bing, Z. and Greenhalgh, S.A. (2000) Cross-hole resistivity tomography using different electrode configurations. *Geophysical Prospecting*, **48**, 887–912.

Bishop, I., Styles, P., Emsley, S.J. and Ferguson, N.S. (1997) The detection of cavities using the microgravity technique: Case histories from mining and karstic environments. *Geological Society Special Publication*, **12**, 155–168.

Bloomberg, D., Upchurch, S.B., Hayden, M.L. and Williams, R.C. (1988) Cone penetrometer exploration of sinkholes: stratigraphy and soil properties. *Environmental Geology Water Science*, **12**, 99–105.

Bögli, A. (1960) Kalkosung und Karrenbildung. *Zeitschrift Geomorphologie*, Supplement 2, 4–21.

Bögli, A. (1964) Mischungskorrosion: Ein Beitrag zum Verkarstungsproblem. *Erdkunde*, **18**(2), 83–92.

Bögli, A. (1980) *Karst Hydrology and Physical Speleology*, 284pp. Springer-Verlag: Berlin.

Bonaparte, R. and Berg, R.R. (1987) The use of geosynthetics to support roadways over sinkhole prone areas. In: Beck, B.F. and Wilson, W.L. (eds), *Karst Hydrogeology: Engineering and Environmental Applications*, pp. 437–445. Balkema: Rotterdam.

Bosak, P., Knez, M., Otrubova, D., Pruner, P., Slabe, T. and Vendohova, D. (2000). Paleomagnetic research of fossil cave in the highway construction at Kozina (Slovenia). *Acta Carsologica*, **29**(2), 15–33.

Božičević, A., Budanur, H, Nonveiller, E. and Pavlin, B. (1981) The Keban dam foundation on karstified limestone – a case history. *Bulletin International Association Engineering Geology* 24, 45–51.

Branston, M.W. and Styles, P. (2003) The application of time-lapse microgravity for the investigation and monitoring of mining subsidence. *Quarterly Journal Engineering Geology Hydrogeology*, **36**, 231–244.

Brink, A.B.A. (1979) *Engineering Geology of Southern Africa* (Vol. 1), 319pp. Building Publications: Pretoria.

British Standards (1986) *BS8004, British standard code of practice for foundations*, 160pp. British Standards Institution: London.

British Standards (1995) *BS8006, Code of practice for strengthened/reinforced soils and other fills*, 171pp. British Standards Institution: London.

British Standards (1999) *BS5930, British standard code of practice for site investigations*, 204pp. British Standards Institution, London.

Brook, G.A. and Allison, T.L. (1986) Fracture mapping and ground subsidence susceptibility modelling in covered karst terrain: The example of Dougherty County, Georgia. *International Association Hydrological Sciences*, **151**, 595–606.

Bruce, D.A. (1994) Trends and developments in American grouting practice. In: Bell, F.G. (ed.), *Grouting in the Ground*, pp. 573–589. Thomas Telford: London.

Bruce, D.A. (2003) Sealing of massive water inflows through karst by grouting: Principles and practice. *American Society Civil Engineers Geotechnical Special Publication* 122, 615–626.

Brune, G. (1965) Anhydrite and gypsum problems in engineering geology. *Bulletin Association Engineering Geologists*, **3**, 26–38.

Budetta, G. and Del Negro, C. (1995) Magnetic field changes on lava flow to detect lava tubes. *Journal Volcanology Geothermal Research*, **65**, 237–248.

Bull, P.A. (1980) The antiquity of caves and dolines in the British Isles. *Zeitschrift Geomorphologie*, Supplement 36, 217–232.

Burland, J.B., Hancock, R.J. and May, J. (1983) A case history of a foundation problem on soft chalk. *Géotechnique*, **33**, 385–395.

Burnett, A.D. and Epps, R.J. (1979) The engineering geological description of the carbonate suite rocks and soils. *Ground Engineering*, **12**(2), 41–48.

Buttrick, D.B. and van Schalkwyk, A. (1995) The method of scenario supposition for stability evaluation of sites of dolomitic land in South Africa. *Journal South African Institution Civil Engineers*, **37**, 9–14.

Buttrick, D.B., van Schalkwyk, A., Kleywegt, R.J. and Watermeyer, R.B. (2001) Proposed method for dolomite land hazard and risk assessment in South Africa. *Journal South African Institution Civil Engineering*, **43**(2), 27–36.

Calembert, L. (1975) Engineering geological problems in karstic regions. *Bulletin International Association Engineering Geology*, **12**, 39–82.

Canace, R. and Dalton, R. (1984) A geological survey's cooperative approach to analysing and remedying a sinkhole related disaster in an urban environment. In: Beck, B.F. (ed.), *Sinkholes: Their Geology, Engineering and Environmental Impact*, pp. 343–348. Balkema: Rotterdam.

Cai, G. (1991) Karst environmental problems in China. In: Yuan, D. (ed.), *Karst of China*, pp. 136–157. Geological Publishing House: Beijing.

CH2MHill (2003) *Report to the Federal Energy Regulatory Commission on failure of the Swift No. 2 power canal embankment*, 520pp (on CD). Cowlitz County Public Utility District: Longview.

Chabert, C. (1981) *Les Grandes Cavités Françaises*, 154pp. Fédération Française de Spéléologie: Paris.

Chen G. (1994) Prediction of covered karst halls. *Proceedings 7th International Association Engineering Geology Congress*, pp. 1841–1845. Balkema: Rotterdam.

Chen J. (1988) Karst collapses in cities and mining areas, China. *Environmental Geology Water Science*, **12**, 29–35.

Chen J. and Beck, B.F. (1989) Qualitative modelling of the cover-collapse process. In: Beck, B.F. (ed.), *Engineering and Environmental Impacts of Sinkholes and Karst*, pp. 98–95. Balkema: Rotterdam.

Christiansen, E.A. (1971) Geology of the Crater Lake collapse structure in southeastern Saskatchewan. *Canadian Journal Earth Sciences*, **8**, 1505–1513.

Clark, R.G., Gutmanis, J.C., Furley, A.E. and Jordan, P.G. (1981) Engineering geology for a major industrial complex at Aughinish Island, Co. Limerick, Ireland. *Quarterly Journal Engineering Geology*, **14**, 231–239.

Clayton, C.R.I. (1983) The influence of diagenesis on some index properties of chalk in England. *Géotechnique*, **33**, 225–241.

Conrad, G., Geze, B. and Paloc, H. (1967) Observations sur des phenomenes karstique at pseudo-karstique du Sahara. *Revue Géographie Physique Géologie Dynamique* (Series 2), **9**(5), 357–370.

Cooley, T. (2001) Geological and geotechnical context of cover collapse and subsidence in mid-continent US clay-mantled karst. In: Beck, B.F. and Herring, J.G. (eds), *Geotechnical and Environmental Applications of Karst Geology and Hydrology*, pp. 19–24. Balkema: Lisse.

Cooley, T. (2002) Engineering approaches to conditions created by a combination of karst and faulting at a hospital in Birmingham, Alabama. *Engineering Geology*, **65**, 197–204.

Cooper, A.H. (1986) Foundered strata and subsidence resulting from the dissolution of Permian gypsum in the Ripon and Bedale areas, North Yorkshire. *Geological Society Special Publication* 22, 127–139.

Cooper, A.H. (1995) Subsidence hazards due to the dissolution of Permian gypsum in England: investigation and remediation. In: Beck, B.F. (ed.), *Karst Geohazards*, pp. 23–29. Balkema: Rotterdam.

Cooper, A.H. (1998) Subsidence hazards caused by the dissolution of Permian gypsum in England: geology, investigation and remediation. *Geological Society Engineering Group Special Publication* 15, 265–275.

Cooper, A.H. (1999). Description, interpretation and assessment of subsidence features at Ure Bank Terrace, Ripon, N. Yorkshire. *British Geological Survey Technical Report WA/99/09R.*

Cooper, A.H. and Saunders, J.M. (2002) Road and bridge construction across gypsum karst in England. *Engineering Geology*, **65**, 217–223.

Cooper, A.H. and Waltham, A.C. (1999) Subsidence caused by gypsum dissolution at Ripon, North Yorkshire. *Quarterly Journal Engineering Geology*, **32**, 305–310.

Cooper, A.H., Farrant, A.R., Adlam, K.A.M. and Walsby, J.C. (2001) The development of a national Geographic Information System (GIS) for British karst geohazards and risk assessment. In: Beck, B.F. and Herring, J.G. (eds), *Geotechnical and Environmental Applications of Karst Geology and Hydrology*, pp. 125–130. Balkema: Lisse.

Coussens, R.R.M. and Garrett, W.S. (1970) The flooding at West Dreifontein mine, South Africa. *Proceedings 9th Commonwealth Mining Metallurgical Congress*. Institution Mining Metallurgy: London, **1**, 931–972.

Crawford, N.C. (1984) Sinkhole flooding associated with urban development upon karst terrain, Bowling Green, Kentucky. In: Beck, B.F. (ed.), *Sinkholes: Their Geology, Engineering and Environmental Impact*, pp. 283–292. Balkema: Rotterdam.

Crawford, N.C. (2001) Environmental problems associated with urban development upon karst, Bowling Green, Kentucky. In: Beck, B.F. and Herring, J.G. (eds), *Geotechnical and Environmental Applications of Karst Geology and Hydrology*, pp. 397–424. Balkema: Lisse.

Crawford, N.C. (2003) Karst hydrogeological investigation for proposed Kentucky trimodal transpark. *American Society Civil Engineers Geotechnical Special Publication* 122, 385–403.

Crawford, N.C. and Groves, C.G. (1995) Sinkhole collapse and ground water contamination problems resulting from storm water drainage wells on karst terrain. In: Beck, B.F. (ed.), *Karst Geohazards*, pp. 257–264. Balkema: Rotterdam.

Crawford, N.C., and Ulmer, C.S. (1993) Groundwater investigations to determine contaminant movement in the karst aquifers of the Lewisburg, Tennessee area. In: Beck, B.F. (ed.), *Applied Karst Geology*, pp. 79–87. Balkema: Rotterdam.

Crawford, N.C., Lewis, M.A., Winter, S.A. and Webster, J.A. (1999) Microgravity techniques for subsurface investigations of sinkhole collapses and for detection of groundwater flow paths through karst aquifers. In: Beck, B.F., Pettit, A.J. and Herring, J.G. (eds), *Hydrogeology and Engineering Geology of Sinkholes and Karst*, pp. 203–218. Balkema: Rotterdam.

Culshaw, M. G. and Kelk, B. (1994) A national geo-hazard information system for the UK insurance industry – the development of a commercial product in a geological survey environment. In: *Proceedings 1st European Congress on Regional Geological Cartography and Information Systems, Bologna, Italy, 4, Paper 111, 3pp.*

Culshaw, M.G. and Waltham, A.C. (1987) Natural and artificial cavities as ground engineering hazards. *Quarterly Journal Engineering Geology*, **20**, 139–150.

Currin, W.E. and Barfus, B.L. (1989) Sinkhole distribution and characteristics in Pasco County, Florida. In: Beck, B.F. (ed.), *Engineering and Environmental Impacts of Sinkholes and Karst*, pp. 97–106. Balkema: Rotterdam.

Cuss, R.J. and Beamish, D. (2002) Ground penetrating radar and ground conductivity investigation of fissuring of the A690 road in Houghton-le-Spring. *British Geological Survey Report IR/02/142.*

Cvijic, J. (1893) Das Karstphaenomen. *Geographische Abhandlungen Wien*, **5**(3), 218–329.

Daily, W. and Ramirez, A.L. (1999) Electrical imaging of engineered hydraulic barriers. *Geophysics*, **65**, 83–94.

Daniel, A.J. and Styles, P. (1997) Topographic accessibility and the tectonic interpretation of gravity data. *Geophysical Prospecting*, **45**(6), 1013–1026.

Davies, W.E. (1951) Mechanics of cavern breakdown. *Bulletin National Speleological Society*, **13**, 36–43.

Davis, S.T. (1997) Interstate assessment of governmental regulations on landfills in karst areas. In: Beck, B.F. and Stephenson, J.B. (eds), *Engineering Geology and Hydrogeology of Karst Terrains*, pp. 433–438. Balkema: Rotterdam.

Day, M.J. (2003) An assessment of karstic collapse hazards at Mount Rooser, Ewarton, Jamaica. *American Society Civil Engineers Geotechnical Special Publication* 122, 40–49.

De Bruyn, I.A. and Bell, F.G. (2001) The occurrence of sinkholes and subsidence depressions in the Far West Rand and Gauteng Province, South Africa, and their engineering implications. *Environmental Engineering Geoscience*, **7**, 281–295.

Derbyshire, E., Meng X. and Dijkstra, T.A. (eds) (2000) *Landslides In The Thick Loess Terrain of North-west China*, 288pp. Wiley: New York.

Destephen, R.A. and Wargo, R.H. (1992) Foundation design in karst terrain. *Bulletin Association Engineering Geologists*, **29**, 165–175.

Devilbiss, T.S. (1995) A local government approach to mitigating impacts of karst. In: Beck, B.F. (ed.), *Karst Geohazards*, pp. 499–504. Balkema: Rotterdam.

Dogan, U. and Cicek, I. (2002) Occurrence of cover-collapse sinkholes in the May Dam reservoir area (Konya, Turkey). *Cave Karst Science*, **29**, 111–116.

Dotson, D.W. and Tarquinio, F.S. (2003) A creative solution to problems with foundation construction in karst. *American Society Civil Engineers Geotechnical Special Publication* 122, 627–634.

Dougherty, P.H. (1991). Spatial-temporal characteristics of karst subsidence in the Lehigh Valley of Pennsylvania. In Kastning, E.H. and Kastning, K.M. (eds), *Appalachian Karst*, pp. 138–145. National Speleological Society: Huntsville.

Dougherty, P.H. and Perlow, M. (1988) The Macungie sinkhole, Lehigh Valley, Pennsylvania: cause and repair. *Environmental Geology Water Science*, **12**, 89–98.

Dragoni, M., Piombo, A. and Tallarico, A. (1995) A model for the formation of lava tubes by roofing over a channel. *Journal Geophysical Research*, **100**, 8435–8447.

Drew, D.P. (1972) Geomorphology of the Big Muddy Valley area, southern Saskatchewan, with special reference to the occurrence of piping. *Proceedings Southern Prairies Symposium, International Geographical Union Conference, Montreal, 197–212.*

Drew, D. (1985) *Karst Processes and Landforms*, 240pp. Macmillan: London.

Dreybrodt, W. (2000) Equilibrium chemistry of karst water in limestone terranes. In: Klimchouk, A.B., Ford, D.C., Palmer, A.N. and Dreybrodt, W. (eds), *Speleogenesis: Evolution of Karst Aquifers*, pp. 126–135. National Speleological Society: Huntsville.

Dreybrodt, W. and Eisenlohr, L. (2000) Limestone dissolution rates in karst environments. In: Klimchouk, A.B., Ford, D.C., Palmer, A.N. and Dreybrodt, W. (eds), *Speleogenesis: Evolution of Karst Aquifers*, pp. 136–148. National Speleological Society: Huntsville.

Dreybrodt, W., Romanov, D. and Gabrovsek, F. (2002) Karstification below dam sites: A model of increasing leakage from reservoirs. *Environmental Geology*, **42**, 519–524.

Drumm, E.C., Kane, W.F. and Yoon, C.J. (1990) Application of limit plasticity to the stability of sinkholes. *Engineering Geology*, **29**, 213–225.

Dunham, R.J. (1962) Classification of carbonate rocks. *American Association Petroleum Geologists Memoir* 1, 108–121.

Eck, W. and Redfield, R.C. (1965) Engineering geology problems at Sanford Dam, Borger, Texas. *Bulletin Association Engineering Geologists*, **3**, 15–25.

Edmonds, C.N., Green, C.P. and Higginbottom, I.E. (1987) Subsidence hazard prediction for limestone terrains, as applied to the English Cretaceous chalk. *Geological Society Engineering Geology Special Publication* 4, 283–293.

Edmonds, C.N. (1988) Induced subsurface movements associated with the presence of natural and artificial underground openings in areas underlain by Cretaceous Chalk. *Geological Society Engineering Geology Special Publication* 5, 205–214.

Edmonds, C.N. (2001) Predicting natural cavities in chalk. *Geological Society Engineering Geology Special Publication* 18, 29–38.

Einstein, H.H. (1997) Landslide risk systematic approaches to assessment and management. In: Cruden, D. M. and Fell, R. (eds), *Proceedings International Workshop Landslide Risk Assessment, Honolulu*, pp. 25–50. Balkema: Rotterdam.

Erwin, J.W. and Brown, R.A. (1988) Karstic foundation problems Sunny Point Railroad. *American Society of Civil Engineers Geotechnical Special Publication* 14, 74–85.

Fell, R. and Hartford, D. (1997) Landslide risk management. In: Cruden, D. M. and Fell, R. (eds), *Proceedings International Workshop Landslide Risk Assessment, Honolulu*, pp. 51–109. Balkema: Rotterdam.

Field, M.S. (2002) A lexicon of cave and karst terminology with special reference to environmental karst hydrology. `http://www.speleogenesis.info/glossary/`

Fischer, J.A. (1999) Limestone ordinances of New Jersey and Pennsylvania: A practioner's experiences. *Engineering Geology*, **52**, 61–66.

Fischer, J.A. and Lechner, H. (1989) A karst ordinance – Clinton Township, New Jersey. In: Beck, B.F. (ed.), *Engineering and Environmental Impacts of Sinkholes and Karst*, pp. 357–361. Balkema: Rotterdam.

Fischer, J.A. and Fischer, J.J. (1995) Karst site remediation grouting. In: Beck, B.F. (ed.), *Karst Geohazards*, pp. 325–330. Balkema: Rotterdam.

Folk, R.L. (1962) Spectral subdivision of limestone types. *American Association Petroleum Geologists Memoir* 1, 62–84.

Folk, R.L. (1973) Carbonate petrology in the post-Sorbian age. In: Ginsburg, R.N. (ed.), *Evolving Concepts in Sedimentology*. Johns Hopkins University Press: Baltimore.

Fookes, P.G. (1997) Geology for engineers: The geological model, prediction and performance. *Quarterly Journal Engineering Geology*, **30**, 293–424.

Foose, R.M. (1953) Ground-water behaviour in the Hershey Valley, Pennsylvania. *Bulletin Geological Society America*, **64**, 623–645.

Foose, R.M. (1968) Surface subsidence and collapse caused by ground water withdrawal in carbonate rock areas. *Proceedings 23rd International Geological Congress*, **12**, 155–166.

Foose, R.M. and Humphreville, J.A. (1979) Engineering geological approaches to foundations in the karst terrain of the Hershey Valley. *Bulletin Association Engineering Geologists*, **16**, 355–381.

Ford, D.C. and Williams, P.F. (1989) *Karst Geomorphology and Hydrology*, 601pp. Unwin Hyman: London.

Ford, D.C. and Williams, P.F. (2005) *Karst Hydrogeology and Geomorphology*. Wiley: New York.

Ford, T.D. (1984). Paleokarsts in Britain. *Cave Science*, **11**, 246–264.

Franklin, J.A. (1989) Stability of shallow caverns. In: Malone, A.W. and Whiteside, P.G.D. (eds), *Rock Cavern – Hong Kong*, pp. 203–212. Institution Mining Metallurgy: London.

Frumkin, A. and Raz, E. (2001) Collapse and subsidence associated with salt karstification along the Dead Sea. *Carbonates and Evaporites*, **16**, 117–130.

Frye, J.C. and Schoff, S.L. (1942) Deep-seated solution in the Meade basin and vicinity, Kansas and Oklahoma. *American Geophysical Union Transactions*, **23**, 35–39.

Gao Y., Alexander, E.C. and Lei M. (2001) A comparison of human-induced sinkholes between China and the United States. In: Beck, B.F. and Herring, J.G. (eds), *Geotechnical and Environmental Applications of Karst Geology and Hydrology*, pp. 25–30. Balkema: Lisse.

Garlanger, J.E. (1991) Foundation design in Florida karst. *Concrete International*, **13**(4), 56–62.

Gascoyne, M., Ford, D.C. and Schwartz, H.P. (1983) Rates of cave and landform development in the Yorkshire Dales from speleothem age data. *Earth Surface Processes Landforms*, **8**, 557–568.

Gilli, E. (1986) Les grandes cavités souterraines: Études et applications. *Karstologia*, **7**, 3–11.

Gilli, E. (1993) Les grands volumes souterrains du massif de Mulu. *Karstologia*, **22**, 1–13.

Grainger, P., McCann, D.M. and Gallois, R.W. (1973) The application of the seismic refraction technique to the study of fracturing of the Middle Chalk at Mundford, Norfolk. *Géotechnique*, **28**, 219–232.

Grosch, J.J., Touma, F.T. and Richards, D.P. (1987) Solution cavities in the limestone of eastern Saudi Arabia. In: Beck, B.F. and Wilson, W.L. (eds), *Karst Hydrogeology: Engineering and Environmental Applications*, pp. 73–78. Balkema: Rotterdam.

Grouting Committee of Geo-Institute (1998) Grouts and grouting. *American Society Civil Engineers Geotechnical Special Publication* 80, 195pp.

Guo X. (1991) *Geological Hazards of China and their Prevention and Control*, 260pp. Geological Publishing House: Beijing.

Guyot, C.A. (1984) Collapse and compaction of sinkholes by dynamic compaction. In: Beck, B.F. (ed.), *Sinkholes: Their Geology, Engineering and Environmental Impact*, pp. 419–423. Balkema: Rotterdam.

Gutiérrez, F. and Cooper, A.H. (2002) Evaporite dissolution subsidence in the historical city of Calatayud, Spain: Damage appraisal and prevention. *Natural Hazards*, **25**, 259–288.

Gutiérrez, F., García-Hermoso, F. and Cooper, A.H. (2000) Spatial assessment, mitigation and prevention of evaporate dissolution subsidence damage in the historical city of Calatayud, Spain. In: Carbognin, L., Gambolati, G. and Johnson, A.I. (eds), *Proceedings Sixth International Symposium Land Subsidence, Ravenna, Italy*, **1**, 237–248.

Habič, P., Knez, M., Kogovšek, J., Kranjc, A., Mihevc, A., Slabe, T., Šebela, S. and Zupan, N. (1989) Škocjanske Jame speleological review. *International Journal Speleology*, **18**(1–2), 1–42.

Hagerty, D.J. and Pavoni, J.L. (1973) Geologic aspects of landfill refuse disposal. *Engineering Geology*, **7**, 219–230.

Hall, P.J., McCusker, A.J. and Bingham, C.T. (1995) Site investigation, monitoring and successful landfill closure within a karst terrain. In: Beck, B.F. (ed.), *Karst Geohazards*, pp. 189–194. Balkema: Rotterdam.

Halliday, W.R. (2004a). Piping caves and badlands pseudokarst. In: Gunn, J. (ed.), *Encyclopedia of Caves and Karst Science*, pp. 589–593. Fitzroy Dearborn: New York.

Halliday, W.R. (2004b). Pseudokarst. In: Gunn, J. (ed.), *Encyclopedia of Caves and Karst Science*, pp. 604–608. Fitzroy Dearborn: New York.

Handfelt, L.D. and Attwooll, W.J. (1988) Exploration of karst conditions in central Florida. *American Society Civil Engineers Geotechnical Special Publication* 14, 40–52.

Harrison, H.D. and Hiltunen, D.R. (2003) Characterization of karst terrane via SASW seismic wave method. *American Society Civil Engineers Geotechnical Special Publication* 122, 519–528.

He K., Liu C. and Wang S. (2003) Karst collapse related to over-pumping and a criterion for its stability. *Environmental Geology*, **43**, 720–724.

Heath, W.E. (1995) Drilled pile foundations in porous, pinnacled carbonate rock. In: Beck, B.F. (ed.), *Karst Geohazards*, pp. 371–374. Balkema: Rotterdam.

Henry, J.F. (1987) The application of compaction grouting to karstic limestone problems. In: Beck, B.F. and Wilson, W.L. (eds), *Karst Hydrogeology: Engineering and Environmental Applications*, pp. 447–450. Balkema: Rotterdam.

Herbert, S.M., Roche, D.P. and Card, G.B. (1987) The value of engineering geological desk study appraisals in scheme planning. *Geological Society Engineering Geology Special Publication* 4, 151–154.

Higginbottom, I.E. (1966) The engineering geology of chalk. *Proceedings Symposium on Chalk in Earthworks and Foundations*, pp. 1–13. Institute Civil Engineers: London.

Hill, C.A. (2000) Sulfuric acid, hypogene karst in the Guadalupe Mountains of New Mexico and West Texas, USA. In: Klimchouk, A.B., Ford, D.C., Palmer, A.N. and Dreybrodt, W. (eds), *Speleogenesis: Evolution of Karst Aquifers*, pp. 309–316. National Speleological Society: Huntsville.

Hobson, D.M. (1992) Site investigation – making the most of your money. In: Forde, M.C. (ed.), *Proceedings 2nd International Conference on Construction on Polluted and Marginal Land*, pp. 151–158. Engineering Technics Press: Edinburgh.

Hoenstine, R.W., Lane, E., Spencer, S.M. and O'Carroll, T. (1987) A landfill site in a karst environment, Madison County, Florida – a case study. In: Beck, B.F. and Wilson, W.L. (eds), *Karst Hydrogeology: Engineering and Environmental Applications*, pp. 253–258. Balkema: Rotterdam.

Holliday, D.W. (1970) The petrology of secondary gypsum rocks: A review. *Journal Sedimentary Petrology*, **40**, 734–744.

Hon, K., Kauahikaua, J., Denlinger, R. and McKay, K. (1994). Emplacement and inflation of pahoehoe sheet flows – observations and measurements of active lava flows on Kilauea Volcano, Hawaii. *Bulletin Geological Society America*, **106**, 351–370.

Hoover, R.A. (2003) Geophysical choices for karst investigations. *American Society Civil Engineers Geotechnical Special Publication* 122, 529–538.

Horwitz, H.H. and Smith, T.J. (2003) Characteristics of stable and reactivated in-filled paleokarst features in west-central Florida. *American Society Civil Engineers Geotechnical Special Publication* 122, 50–60.

Hose, L.D. and Strong, T.R. (1981) The genetic relationship between breccia pipes and caves in non-karstic terranes in northern Arizona. *Proceedings 8th International Congress Speleology*, pp. 136–138. National Speleological Society: Huntsville.

Hyatt, J.A., Wilkes, H.P. and Jacobs, P.M. (1999) Spatial relationships between new and old sinkholes in covered karst, Albany, Georgia, USA. In: Beck, B.F., Pettit, A.J. and Herring, J.G. (eds), *Hydrogeology and Engineering Geology of Sinkholes and Karst*, pp. 37–44. Balkema: Rotterdam.

Hyatt, J.A., Wilson, R., Givens, J.S. and Jacobs, P.M. (2001) Topographic, geologic and hydrogeologic controls on dimensions and locations of sinkholes in thick covered karst, Lowndes County, Georgia. In: Beck, B.F. and Herring, J.G. (eds), *Geotechnical and Environmental Applications of Karst Geology and Hydrology*, pp. 37–45. Balkema: Lisse.

I.C.E. (1977) *Ground Subsidence*, 99pp. Institution Civil Engineers: London.

Jackson, P.D. and McCann, D.M. (1997) Cross-hole seismic tomography for engineering site investigation. *Geological Society Engineering Geology Special Publication* 12, 247–264.

Jackson, P.D., Gunn, D.A., Flint, R.C., McCann, D.M., Bent, M. and Howes, D. (2001) Cross-hole seismic measurements for detection of disturbed ground beneath existing structures. *Non-Destructive Testing Exploration International*, **34**, 155–162.

Jammal, S.E. (1984) Maturation of the Winter Park sinkhole. In: Beck, B.F. (ed.), *Sinkholes: Their Geology, Engineering and Environmental Impact*, pp. 363–369. Balkema: Rotterdam.

James, A.N. and Kirkpatrick, I.M. (1980) Design of foundations of dams containing soluble rocks and soils. *Quarterly Journal Engineering Geology*, **13**, 189–198.

James, A.N. and Lupton, A.R.R. (1978) Gypsum and anhydrite in foundations of hydraulic structures. *Géotechnique*, **28**, 249–272.

James, N.P. and Choquette, P.W. (1988) *Paleokarst*, 416pp. Springer: New York.

Jancin, M. and Clark, D.D. (1993) Subsidence sinkhole development in light of mud infiltrate structures within interstratal karst of the coastal plain, southeast United States. *Environmental Geology*, **22**, 330–336.

Jansen, J., Anklam, J., Goodwin, C. and Roof, A. (1993) Electromagnetic induction and seismic refraction surveys to detect bedrock pinnacles. In: Beck, B.F. (ed.), *Applied Karst Geology*, pp. 115–122. Balkema: Rotterdam.

Jappelli, R. and Liguori, V. (1979) An unusually complex underground cavity. *Proceedings International Symposium Geotechnics of Structurally Complex Formations, Associazione Geotecnica Italiana, Rome*, **2**, 79–90.

Jennings, J.E. (1966) Building on dolomites in the Transvaal. *The Civil Engineer in South Africa*, **8**, 41–62.

Jennings, J.N. (1975) Doline morphometry as a morphogenetic tool: New Zealand examples. *New Zealand Geographer*, **31**, 6–28.

Jennings, J.N. (1983) The disregarded karst of the arid and semiarid domain. *Karstologia*, **1**, 61–73.

Jennings, J.N. (1985) *Karst Geomorphology*, 293pp. Blackwell: Oxford.

Johnson, K.S. (1989a) Salt dissolution, interstratal karst and ground subsidence in the northern part of the Texas Panhandle. In: Beck, B.F. (ed.), *Engineering and Environmental Impacts of Sinkholes and Karst*, pp. 115–121. Balkema: Rotterdam.

Johnson, K.S. (1989b) Development of the Wink Sink in west Texas, USA, due to salt dissolution and collapse. *Environmental Geology Water Science*, **14**, 81–92.

Johnson, K.S. (1997) Evaporite karst in the United States. *Carbonates and Evaporites*, **12**, 2–14.

Johnson, K.S. (1998) Land subsidence above man-made salt-dissolution cavities. *Association Engineering Geologists Special Publication*, **8**, 385–392.

Johnson, K.S. and Neal, J.T. (2003) Evaporite karst and engineering/environmental problems in the United States. *Oklahoma Geological Survey Circular*, **109**, 353pp.

Johnson, K.S., Collins, E.W. and Seni, S.J. (2003) Sinkholes and land subsidence due to salt dissolution near Wink, West Texas, and other sites in western Texas and New Mexico. *Oklahoma Geological Survey Circular* 109, 183–195.

Kaderabek, T.J. and Reynolds, R.T. (1981) Miami Limestone foundation design and construction. *Proceedings American Society Civil Engineers*, **107**(GT7), 859–872.

Kambesis, P., Crawford, N., Croft, L.A., Moore, R. and Pfaff, R. (2003) Dishman Lane collapse, Bowling Green, Kentucky. *American Society Civil Engineers Geotechnical Special Publication* 122, 404–414.

Kannan, R.C. (1999) Designing foundations around sinkholes. *Engineering Geology*, **52**, 75–82.

Kauahikaua, J., Cashman, K.V., Mattox, T.N., Heliker, C.C., Hon, K.A., Mangan, M.T. and Thornber, C.R. (1998) Observations on basaltic lava streams in tubes from Kilauea volcano, island of Hawaii. *Journal Geophysical Research*, **103**, 27303–27323.

Kaufmann, O. and Quinif, Y. (2001) An application of cone penetration tests and combined array 2D electrical resistivity tomography to delineate cover-collapse sinkholes prone areas. In: Beck, B.F. and Herring, J.G. (eds), *Geotechnical and Environmental Applications of Karst Geology and Hydrology*, pp. 359–364. Balkema: Lisse.

Kaufmann, O. and Quinif, Y. (2002) Geohazard map of cover-collapse sinkholes in the Tournaisis area, southern Belgium. *Engineering Geology*, **65**, 117–124.

Kemmerly, P.R. (1976) Definitive doline characteristics in the Clarkesville quadrangle, Tennessee. *Bulletin Geological Society America*, **87**, 42–46.

Kemmerly, P. (1981) The need for recognition and implementation of a sinkhole-floodplain hazard designation in urban karst terrains. *Environmental Geology*, **3**, 281–292.

Kesserü, Z. (1997) Assessing the risk of cave-collapse sinkholes using analogous information from mining. In: Beck, B.F. and Stephenson, J.B. (eds), *Engineering Geology and Hydrogeology of Karst Terrains*, pp. 55–60. Balkema: Rotterdam.

Kleywegt, R.J. and Enslin, J.F. (1973) The application of the gravity method to the problem of ground settlement and sinkhole formation in dolomite on the Far West Rand, South Africa. *Proceedings Hannover Symposium on Sinkholes and Subsidence*, T301–T315. International Association Engineering Geology.

Klimchouk, A.B. (2000) The formation of epikarst and its role in vadose speleogenesis. In: Klimchouk, A.B., Ford, D.C., Palmer, A.N. and Dreybrodt, W. (eds), *Speleogenesis: Evolution of Karst Aquifers*, pp. 91–99. National Speleological Society: Huntsville.

Klimchouk, A. and Andrejchuk, V. (1996) Breakdown development in cover beds and landscape features induced by interstratal gypsum karst. *International Journal Speleology*, **24**(3–4), 127–144.

Klimchouk, A. and Andrejchuk, V. (2003) Karst breakdown mechanisms from observations in the gypsum caves of the western Ukraine: Implications for subsidence hazard assessment. *Speleogenesis and Evolution of Karst Aquifers* (www.speleogenesis.info), **1**(1), 1–20.

Klimchouk, A., Lowe, D., Cooper, A. and Sauro, U. (1996) Gypsum karst of the world. *International Journal Speleology*, **25**(3–4), 1–307.

Klimchouk, A., Cucchi, F., Calaforra, J.M., Aksem, S., Finocchiaro, F. and Forti, P. (1996) Dissolution of gypsum from field obsevations. *International Journal Speleology*, **25**, 37–48.

Knez, M. and Slabe, T. (1999) Unroofed caves and recognising them in karst relief (discovered during expressway construction at Kozina, South Slovenia). *Acta Carsologica*, **28**(2), 103–112.

Knez, M. and Slabe, T. (2001) Karstology and expressway construction. *Proceedings 14th IRF Road World Congress, Paris* (on CD-Rom). International Road Federation: Geneva.

Knez, M. and Slabe, T. (2002a) Unroofed caves are an important feature of karst surfaces: Examples from the classical karst. *Zeitschrift Geomorphologie*, **46**, 181–191.

Knez, M. and Slabe, T. (2002b) Karstology and the opening of caves during motorway construction in the karst region of Slovenia. *International Journal Speleology*, **31**, 159–168.

Kochanov, W.E. and Kochanov, J.S. (1997) The Pennsylvania sinkhole inventory: design and application in a relational database. In: Beck, B.F. and Stephenson, J.B. (eds), *Engineering Geology and Hydrogeology of Karst Terrains*, pp. 49–54. Balkema: Rotterdam.

Kortnik, J. (2002) Stability appraisal of the Medvedova Konta pothole. *International Journal Speleology*, **31**, 129–135.

Kozary, M.T., Dunlap, J.C. and Humphrey, W.E. (1968) Incidence of saline deposits in geologic time. *Geological Society America Special Paper* 88, 43–57.

Knight, F.J. (1971) Geologic problems of urban growth in limestone terrains in Pennsylvania. *Bulletin Association Engineering Geologists*, **8**, 91–101.

Knott, D.L., Newman, F.B., Rojas-Gonzalez, L.F. and Gray, R.E. (1993) Foundation engineering practice for bridges in karst areas in Pennsylvania. In: Beck, B.F. (ed.), *Applied Karst Geology*, pp. 225–230. Balkema: Rotterdam.

Kortnik, J. and Šušteršič, F. (2000) Modelling the stability of a very large cave room, case study, Brezno pri Medvedovi Konti. *Acta Carsologica*, **29**, 149–160.

LaMoreaux, P.E. and Newton, J.G. (1986) Catastrophic subsidence: an environmental hazard, Shelby County, Alabama. *Environmental Geology Water Science*, **8**, 25–40.

LaValle, P. (1967) Some aspects of linear karst depression development in south-central Kentucky. *Association American Geographics Annals*, **57**, 49–71.

Leddin, M.M. (1994). Sinkhole. *BOCA, Building Official and Code Administrator Magazine*, **28**(3), 28–34.

Lei M., Jiang X. and Yu L. (2001) New advances of karst collapse research in China. In: Beck, B.F. and Herring, J.G. (eds), *Geotechnical and Environmental Applications of Karst Geology and Hydrology*, pp. 145–151. Balkema: Lisse.

Li G. and Zhou W. (1999) Sinkholes in karst mining areas of China and some methods of prevention. *Engineering Geology*, **52**, 45–50.

Li J., Niu J., Liu Q. and Li G. (1983) Subsidence and its treatment in karst coal mines in China. *Proceedings 11th International Congress Speleology, Beijing*, pp. 180–181.

Littlefield, J.R., Culbreth, M.A., Upchurch, S.B. and Stewart, M.T. (1984) Relationship of modern sinkhole development to large scale photolinear features. In: Beck, B.F. (ed.), *Sinkholes: Their Geology, Engineering and Environmental Impact*, pp. 189–195. Balkema: Rotterdam.

Loke, M.H. and Barker, R.D. (1996) Rapid least-squares inversion of apparent resistivity pseudosections by quasi-Newton method. *Geophysical Prospecting*, **40**, 131–152.

Lord, J.A., Clayton, C.R.I. and Mortimore, R.N. (2002) *Engineering in Chalk*. Publication C574, CIRIA: London.

Lowe, D.J. (1992) Chalk caves revisited. *Cave Science*, **19**, 55–58.

Lowe, D. and Waltham, T. (2002) Dictionary of karst and caves. *British Cave Research Association Cave Studies*, **10**, 40pp.

Lowry, D.C. and Jennings, J.N. (1974) The Nullarbor karst, Australia. *Zeitschrift Geomorphologie*, **18**, 35–81.

Lu Y. (1986) *Karst in China*, 288pp. Geological Publishing House: Beijing.

Lu Y. and Cooper, A.H. (1997) Gypsum karst geohazards in China. In: Beck, B.F. and Stephenson, J.B. (eds), *Engineering Geology and Hydrogeology of Karst Terrains*, pp. 117–126. Balkema: Rotterdam.

Luke, B.A. and Chase, D.S. (1997) Detecting caves using seismic surface waves: a feasibility study. In: Beck, B.F. and Stephenson, J.B. (eds), *Engineering Geology and Hydrogeology of Karst Terrains*, pp. 419–424. Balkema: Rotterdam.

Magdalene, S. and Alexander, E.C. (1995) Sinkhole distribution in Winona County, Minnesota, revisited. In: Beck, B.F. (ed.), *Karst Geohazards*, pp. 43–51. Balkema: Rotterdam.

Maire, R. (1981) Giant shafts and underground rivers of the Nakanai Mountains, New Britain. *Spelunca*, Supplement 3, 8–30.

Marinos, P.G. (2001) Tunnelling and mining in karstic terrane: an engineering challenge. In: Beck, B.F. and Herring, J.G. (eds), *Geotechnical and Environmental Applications of Karst Geology and Hydrology*, pp. 3–16. Balkema: Lisse.

Martinez, J.D., Johnson, K.S. and Neal, J.T. (1998) Sinkholes in evaporite rocks. *American Scientist*, **86**, 38–51.

Martini, J. (2004) Silicate karst. In: Gunn, J. (ed.), *Encyclopedia of Caves and Karst Science*, pp. 649–653. Fitzroy Dearborn: New York.

Matthews, M.C., Clayton, C.R.I. and Rigby-Jones, J. (2000) Locating dissolution features in the Chalk. *Quarterly Journal Engineering Geology Hydrogeology*, **33**, 125–140.

McCann, D.M., Jackson, P.D. and Culshaw, M.G. (1987) The use of geophysical surveying methods in the detection of natural cavities and mineshafts. *Quarterly Journal Engineering Geology*, **20**, 59–73.

McCann, D.M., Culshaw, M.G. and Fenning, P.J. (1997) Setting the standard for geophysical surveys in site investigation. *Geological Society Engineering Geology Special Publication* 12, 3–34.

McDowell, P.W. (1975) Detection of clay-filled sink-holes in the chalk by geophysical methods. *Quarterly Journal Engineering Geology*, **8**, 303–310.

McDowell, P.W. (1989) Ground subsidence associated with doline formation in chalk areas of southern England. In: Beck, B.F. (ed.), *Engineering and Environmental Impacts of Sinkholes and Karst*, pp. 129–134. Balkema: Rotterdam.

McDowell, P.W. and Hope, V. (1993) The location and delineation of karst and solution collapse features by acoustic tomography. In: Beck, B.F. (ed.), *Applied Karst Geology*, pp. 123–129. Balkema: Rotterdam.

McDowell, P.W. and Poulsom, A.J. (1996) Ground subsidence related to dissolution of chalk in southern England. *Ground Engineering*, **29**(2), 29–33.

McDowell, P.W., Barker, R.D., Butcher, A.P., Culshaw, M.G., Jackson, P.D., McCann, D.M., Skipp, B.O., Matthews, S.L. and Arthur, J.C.R. (2002) *Geophysics in Engineering Investigations*. Construction Industry Research and Information Association Report C592 (and Geological Society Engineering Geology Special Publication 19) CIRIA, London. 252pp.

Mendonca, A.F., Pires, A.C.B. and Barros, J.G.C. (1993) Pseudo-sinkholes in lateritic terrains, Brasilia, Brazil. In: Beck, B.F. (ed.), *Applied Karst Geology*, pp. 43–49. Balkema: Rotterdam.

Merrill, R.H. (1957) Roof-span studies in limestone. *US Bureau of Mines Report of Investigations* 5348, 38pp.

Metcalfe, S.J. and Hall, L.E. (1984) Sinkhole collapse induced by groundwater pupage for freeze protection irrigation near Dover, Florida, January 1977. In: Beck, B.F. (ed.), *Sinkholes: Their Geology, Engineering and Environmental Impact*, pp. 29–33. Balkema: Rotterdam.

Mihevc, A. (1998) Dolines, their morphology and origin. Case study: dolines from the Kras, west Slovenia (the Skocjan karst). *Geografia Fisica Dinamica Quaternaria Supplement 3 (Proceedings Fourth International Conference Geomorphology)*, **4**, 69–74.

Milanovic, P.T. (2000) *Geological Engineering in Karst*, 347pp. Zebra: Belgrade.

Milanovic, P.T. (2003) Prevention and remediation in karst engineering. *American Society Civil Engineers Geotechnical Special Publication* 122, 3–28.

Miller, T. (1987) Fluvial and collapse influences on cockpit karst of Belize and eastern Guatemala. In: Beck, B.F. and Wilson, W.L. (eds), *Karst Hydrogeology: Engineering and Environmental Applications*, pp. 53–57. Balkema: Rotterdam.

Mirabito, M.R., Dermatas, D. and Jancin, M. (2001) Characterization and grouting beneath an impending landfill cap. In: Beck, B.F. and Herring, J.G. (eds), *Geotechnical and Environmental Applications of Karst Geology and Hydrology*, pp. 297–305. Balkema: Lisse.

Mishu, L.P., Godfrey, J.D. and Mishu, J.R. (1997) Foundation remedies for residential construction over karst limestone in Nashville, Tennessee. In: Beck, B.F. and Stephenson, J.B. (eds), *Engineering Geology and Hydrogeology of Karst Terrains*, pp. 319–327. Balkema: Rotterdam.

Mitchell, J.K. (1986) Hazardous waste containment. *Geological Society Engineering Geology Special Publication* 3, 145–157.

Molecke, M.A. (ed.) (2000) Final status of the Strategic Petroleum Reserve (SPR) Weeks Island mine. *Sandia National Laboratories Report*, 56pp. SAND2000-0400: Albuquerque.

Moneymaker, B.C. (1968) Reservoir leakage in limestone terrains. *Bulletin Association Engineering Geologists* 6, 3–30.

Monroe, W.H. (1976) The karst landforms of Puerto Rico. *US Geological Survey Professional Paper* 899, 1–69.

Moore, H.L. (1984) Geotechnical considerations in the location, design and construction of highways in karst terrain – the Pellissippi Parkway extension, Knox-Blount counties, Tennessee. In: Beck, B.F. (ed.), *Sinkholes: Their Geology, Engineering and Environmental Impact*, pp. 385–389. Balkema: Rotterdam.

Moore, H. (1988) Treatment of karst along Tennessee highways. *American Society Civil Engineers Geotechnical Special Publication* 14, 133–148.

Moore, H. and Amari, D. (1987) Sinkholes and gabions – a solution to the solution problem. In: Beck, B.F. and Wilson, W.L. (eds), *Karst Hydrogeology: Engineering and Environmental Applications*, pp. 305–310. Balkema: Rotterdam.

Mossop, G.D. and Shearman, D.J. (1973) The origin of secondary gypsum. *Transactions Institution Mining Metallurgy*, **82**, B147–B154.

Myers, P.B. and Perlow, M. (1984) Development, occurrence and triggering mechanisms of sinkholes in the carbonate rocks of the Lehigh Valley, eastern Pennsylvania. In: Beck, B.F. (ed.), *Sinkholes: Their Geology, Engineering and Environmental Impact*, pp. 111–115. Balkema: Rotterdam.

Mylroie, J.E. and Carew, J.L. (1987) Field evidence of the minimum time for speleogenesis. *Bulletin National Speleological Society*, **49**, 67–72.

N.C.B. (1975) *Subsidence Engineers' Handbook*, 111pp. National Coal Board: London.

Neal. J.T. (1995) Supai salt karst features: Holbrook basin, Arizona. In: Beck, B.F. (ed.), *Karst Geohazards*, pp. 53–59. Balkema: Rotterdam.

Neal, J.T., Bauer, S.J. and Ehgartner, B.L. (1998) Mine-induced sinkholes over the U.S. Strategic Petroleum Reserve (SPR) storage facility at Weeks Island, Louisiana; geologic causes and effects. *Association Engineering Geologists Special Publication* 8, 327–336.

Neal, J.T. and Myers, R.E. (1995) Origin, diagnostics, and mitigation of a salt dissolution sinkhole at the US Strategic Petroleum Reserve storage site, Weeks Island, Louisiana. *International Association Hydrological Sciences Publication* 234, 187–195.

Newton, J.G. (1981) Induced sinkholes: an engineering problem. *Journal Irrigation Drainage, American Society Civil Engineers*, **107**(IR2), 175–185.

Newton, J.G. (1987) Development of sinkholes resulting from man's activities in the eastern United States. *US Geological Survey Circular* 968, 54pp.

Newton, J.G. and Hyde, L.W. (1971) Sinkhole problem in and near Roberts industrial subdivision, Birmingham, Alabama. *Geological Survey Alabama Circular* 68, 42pp.

Nichol, D. (1998) Sinkholes at Glan Llyn on the A55 North Wales Coast Road, UK. *Engineering Geology*, **50**, 101–109.

Ogden, A.E., Curry, W.A. and Cummings, J.L. (1989) Morphometric analysis of sinkholes and caves in Tennessee comparing the Eastern Highland Rim and Valley and Ridge physiographic provinces. In: Beck, B.F. (ed.), *Engineering and Environmental Impacts of Sinkholes and Karst*, pp. 135–142. Balkema: Rotterdam.

Owen, T.E. (1983) Detection and mapping of tunnels and caves. *Developments in Geophysical Exploration Methods*, **5**, 161–258.

Padgett, D.A. (1993) Remote sensing application for identifying areas of vulnerable hydrogeology and potential sinkhole collapse within highway transportation corridors. In: Beck, B.F. (ed.), *Applied Karst Geology*, pp. 285–290. Balkema: Rotterdam.

Palmer, A.N. (1991) Origin and morphology of limestone caves. *Bulletin Geological Society America*, **103**, 1–21.

Panno, S.V., Weibel, C.P., Krapac, I.G. and Storment, E.C. (1997) Bacterial contamination of groundwater from private septic systems in Illinois' sinkhole plain: Regulatory considerations. In: Beck, B.F. and Stephenson, J.B. (eds), *Engineering Geology and Hydrogeology of Karst Terrains*, pp. 443–447. Balkema: Rotterdam.

Parry-Davies, R. (1992) Consolidation of old mine workings. *Proceedings Symposium on Construction over Mined Areas, Pretoria, South African Institution of Civil Engineers*, pp. 223–227.

Partridge, T.C., Harris, G.M. and Diesel, V.A. (1981) Construction upon dolomites of the south-western Transvaal. *Bulletin International Association Engineering Geology*, **24**, 125–135.

Patterson, D.A, Davey, J.C., Cooper, A.H. and Ferris, J.K. (1995) The investigation of dissolution subsidence incorporating microgravity geophysics at Ripon, Yorkshire. *Quarterly Journal Engineering Geology*, **28**, 83–94.

Paukštys, B. (1999). Is human activity accelerating karst evolution? In: Beck, B.F., Pettit, A.J. and Herring, J.G. (eds), *Hydrogeology and Engineering Geology of Sinkholes and Karst*, pp. 103–106. Balkema: Rotterdam.

Paukštys, B., Cooper, A.H. and Arustien, J. (1997) Planning for gypsum geohazards in Lithuania and England. In: Beck, B.F. and Stephenson, J.B. (eds), *Engineering Geology and Hydrogeology of Karst Terrains*, pp. 127–135. Balkema: Rotterdam.

Paukštys, B. and Karise, V. (1998) Karst in the Baltic Republics. In: Yuan D. and Liu Z. (eds), *Global Karst Correlation*, pp. 91–108. Science Press: New York.

Paukštys, B., Cooper, A.H. and Arustiene, J. (1999) Planning for gypsum geohazards in Lithuania and England. *Engineering Geology*, **52**, 93–103.

Pazuniak, B.L. (1989) Subsurface investigation response to sinkhole activity at an eastern Pennsylvania site. In: Beck, B.F. (ed.), *Engineering and Environmental Impacts of Sinkholes and Karst*, pp. 263–269. Balkema: Rotterdam.

Perlow, M. (2003) An overview of recent Lehigh Valley, Pennsylvania, sinkhole problems. *American Society Civil Engineers Geotechnical Special Publication* 122, 644–651.

Petersen, W.K., Meyers, J.R. and Mackey, R.R. (2003) Sinkhole remediation measures for a highway in King of Prussia, Pennsylvania. *American Society Civil Engineers Geotechnical Special Publication* 122, 569–579.

Peterson, D.W., Holcomb, R.T., Tilling, R.I. and Christiansen, R.L. (1994). Development of lava tubes in the light of observations at Mauna Ulu, Kilauea volcano, Hawaii. *Bulletin Volcanology*, **56**, 343–360.

Pitman, J.I. (1983) Chemical weathering of the East Yorkshire chalk. In: Paterson, K. and Sweeting, M.M. (eds), *New Directions in Karst*, pp. 77–113. Geo Books: Norwich.

Pope, S.R. (2001) Karst void mitigation for water quality and quantity protection in Austin, Texas. In: Beck, B.F. and Herring, J.G. (eds), *Geotechnical and Environmental Applications of Karst Geology and Hydrology*, pp. 307–312. Balkema: Lisse.

Prandini, F.L., Nakazawa, V.A., Dantas, A.M.A. and Holzer, T.L. (1990) Karst and urbanization: Investigation and monitoring in Cajamar, Sao Paulo State, Brazil. *International Association of Hydrogeologists, Selected Papers from 28th International Geological Congress*, **1**, 53–65.

Qubain, B.S., Seksinsky, E.J. and Aldin, E.G. (1995) Techniques to investigate and remedy sinkholes. In: Beck, B.F. (ed.), *Karst Geohazards*, pp. 341–347. Balkema: Rotterdam.

Quinlan, J.F. and Ewers, R.O. (1989) Subsurface drainage in the Mammoth Cave area. In: White W.B. and White, E.L. (eds), *Karst Hydrology, Concepts from the Mammoth Cave Area*, pp. 65–103. Van Norstrand Reinhold: New York.

Quinlan, J.F., Smith, A.R. and Johnson, K.S. (1986) Gypsum karst and salt karst of the United States of America. *Le Grotte d'Italia*, **4**(13), 73–92.

Raghu, D. (1987) Determination of pile lengths and proofing of the bearing stratum of piles in cavernous carbonate formations. In: Beck, B.F. and Wilson, W.L. (eds), *Karst Hydrology: Engineering and Environmental Applications*, pp. 307–402. Balkema: Rotterdam.

Rahn, P.H. and Davis, A.D. (1996) Gypsum foundation problems in the Black Hills area, South Dakota. *Environmental Engineering Geoscience*, **2**, 213–223.

Reynolds, J.M. (2005) *An Introduction to Applied and Environmental Geophysics*, 544pp. Wiley: Chichester.

Rhodes, S.J. and Marychurch, I.M. (1998) Chalk solution features at three sites in southeast England: their formation and treatment. *Geological Society Engineering Geology Special Publication* 15, 277–289.

Rigby-Jones, J., Clayton, C.R.I. and Matthews, M.C. (1993) Dissolution features in the chalk: From hazard to risk. In: Skipp, B.O. (ed.), *Risk and Reliability in Ground Engineering*, pp. 87–99. Thomas Telford: London.

Rigby-Jones, J., Clayton, C.R.I. and Matthews, M.C. (1994) Dissolution features in the chalk: From hazard to risk. *Geotechnical Engineering Bulletin*, **3**(3), 187–201.

Rigby-Jones, J., Matthews, M.C. and McDowell, P.W. (1997) Electrical resistivity imaging systems for ground investigations, with particular reference to dissolution features in chalk areas. *Geological Society Engineering Geology Special Publication* 12, 235–246.

Rogers, C.D.F., Glendinning, S. and Dixon, N. (1996) *Lime Stabilisation*, 182pp. Thomas Telford: London.

Roth, M.J.S., Mackey, J.R., Mackey, C. and Nyquist, J.E. (2002) A case study of the reliability of multi-electrode earth resistivity testing for geotechnical investigations in karst terrains. *Engineering Geology*, **65**, 225–232.

Ryan, M. and Meiman, J. (1996) An examination of short-term variation in water quality at a karst spring in Kentucky. *Ground Water*, **34**, 23–30.

Sauro, U. (1996) Geomorphological aspects of gypsum karst areas with special emphasis on exposed karst. *International Journal Speleology*, **25**(3–4), 105–114.

Schmidt, L.A. (1943) Flowing water in underground channels, Hales Bar dam, Tennessee. *Proceedings American Society Civil Engineers*, **69**, 1417–1446.

Šebela, S. (1996) The dolines above the collapse chambers of Postojna Jama. *Acta Carsologica*, **25**, 241–250.

Self, S., Thordarson, Th., Keszthelyi, L., Walker, G.P.L., Hon, K., Murphy, M.T., Long, P. and Finnemore, S. (1996) A new model for the emplacement of Columbia River basalts as large inflated pahoehoe lava flow fields. *Geophysical Research Letters*, **23**, 2689–2692.

Senior, K. (1995) The Yangtze Gorges expedition: China Caves Project 1994. *Cave Karst Science*, **22**, 53–89.

Shugg, A. (1998) Whey disposal, Allansford, western Victoria. *Proceedings Conference International Association Hydrogeologists, Melbourne*, 563–568.

Siegel, T.C., Cargill, P.E. and McCracken, D.W. (2001) Geosynthetic reinforcement above sinkholes to protect landfill liners. In: Beck, B.F. and Herring, J.G. (eds), *Geotechnical and Environmental Applications of Karst Geology and Hydrology*, pp. 173–176. Balkema: Lisse.

Siegel, T.C. and McCracken, D.W. (2001) Geotechnical characterization and modelling of a shallow karst bedrock site. In: Beck, B.F. and Herring, J.G. (eds), *Geotechnical and Environmental Applications of Karst Geology and Hydrology*, pp. 169–172. Balkema: Lisse.

Siegel, T.C., Belgeri, J.J., McCracklin, D.W. and Cargill, P.E. (2003) Seismic stability analysis of shallow caves. *American Society Civil Engineers Geotechnical Special Publication* 122, 457–465.

Simpson, D. (2001) It's a vision thing. *Ground Engineering*, **34**, 22–23.

Sinclair, W.C. (1982) Sinkhole development resulting from ground-water withdrawal in the Tampa area, Florida. *US Geological Survey Water Resources Investigation* 81-50, 19pp.

Sinclair, W.C. and Stewart, J.W. (1985) Sinkhole type, development and distribution in Florida. *US Geological Survey Map Series*, 110.

Slabe, T. (1997) Karst features discovered during motorway construction in Slovenia. *Environmental Geology*, **32**, 186–190.

Slifer, D.W. and Erchul, R.A. (1989) Sinkhole dumps and the risk to groundwater in Virginia's karst areas. In: Beck, B.F. (ed.), *Engineering and Environmental Impacts of Sinkholes and Karst*, pp. 207–212. Balkema: Rotterdam.

Smart, P.L. and Waltham, A.C. (1987) Cave dams of the Guanyan system, Guangxi, China. *Quarterly Journal Engineering Geology*, **20**, 239–243.

Smith, D.I. (1972) The solution of limestone in an Arctic environment. *Transactions Institute British Geographers Special Publication* 4, 187–200.

Smith, D.I. and Atkinson, T.C. (1976) Process, landform and climate in limestone regions. In: Derbyshire, E. (ed.), *Geomorphology and Climate*, pp. 367–409. Wiley: New York.

Sofianos, A.I. (1996) Analysis and design of an underground hard rock voussoir beam roof. *International Journal Rock Mechanics, Mining Science, Geomechanics Abstracts*, **33**, 153–166.

Soriano, M.A. and Simon, J.L. (2001) Subsidence rates of alluvial dolines in the central Ebro basin, Northeastern Spain. In: Beck, B.F. and Herring, J.G. (eds), *Geotechnical and Environmental Applications of Karst Geology and Hydrology*, pp. 47–52. Balkema: Lisse.

Sotiropoulos, E. and Cavounidis, S. (1979) Cast in situ piles in karstic limestone. *Proceedings Conference on Recent Developments in the Design and Construction of Piles, Institution Civil Engineers, London*, pp. 59–66.

Sowers, G.F. (1975) Failures in limestones in humid subtropics. *Proceedings American Society Civil Engineers*, **101**(GT8), 771–787.

Sowers, G.F. (1986) Correction and protection in limestone terrane. *Environmental Geology Water Science*, **8**, 77–82.

Sowers, G.F. (1996) *Building on Sinkholes*, 202pp. ASCE Press: New York.

Sperling, C.H.B., Goudie, A.S., Stoddart, D.R. and Poole, G.G. (1977) Dolines of the Dorset chalklands and other areas in southern Britain. *Transactions Institute British Geographers*, **2**, 205–223.

Spooner, J. (1971) Mufulira interim report. *Mining Journal*, **276**, 122.

Stangland, H.G. and Kuo, S.S. (1987). Use of ground penetrating radar techniques to aid in site selection for land application sites. In: Beck, B.F. and Wilson, W.L. (eds), *Karst Hydrogeology: Engineering and Environmental Applications*, pp. 171–177. Balkema: Rotterdam.

Stapleton, D.C., Corso, D. and Blakita, P. (1995) A case history of compaction grouting to improve soft soils over karstic limestone. In: Beck, B.F. (ed.), *Karst Geohazards*, pp. 383–387. Balkema: Rotterdam.

Statham, I. and Baker, M. (1986) Foundation problems on limestone: A case history from the Carboniferous Limestone at Chepstow, Gwent. *Quarterly Journal Engineering Geology*, **19**, 191–201.

Stelmack, M.J., Sheridan, T. and Beck, B.F. (1995) Cap reinforcement over a sinkhole as part of a landfill closure. In: Beck, B.F. (ed.), *Karst Geohazards*, pp. 389–395. Balkema: Rotterdam.

Stephenson, J.B., Kaufmann, R.D., Archer, J.A., Bertz, C. and Bradley, P. (2003) Environmental investigation of a diesel spill in a karst area of Kentucky. *American Society Civil Engineers Geotechnical Special Publication* 122, 415–428.

Styles, P. and Thomas, E. (2001) The use of microgravity for the characterization of karstic cavities on Grand Bahama, Bahamas. In: Beck, B.F. and Herring, J.G. (eds), *Geotechnical and Environmental Applications of Karst Geology and Hydrology*, pp. 389–394. Balkema: Lisse.

Šušteršič, F. (1998) Rakovska kukava – collapse or tumour doline. *Acta Carsologica*, **27**, 231–259.

Swart, C.J.U., Stoch, E.J., van Jaarsveld, C.F. and Brink, A.B.A. (2003) The Lower Wonderfontein Spruit: An exposé. *Environmental Geology*, **43**, 635–653.

Sweeting, M.M. (1972) *Karst Landforms*, 362pp. Macmillan: London.

Swiger, W.F. and Estes, H.M. (1959) Major power station foundation in broken limestone. *Proceedings American Society Civil Engineers*, **85**(SM5), 77–86.

Tan, B.K. (1987) Some geotechnical aspects of urban development over limestone terrain in Malaysia. *Bulletin International Association Engineering Geology*, **35**, 57–63.

Tan, B.K. and Batchelor, B. (1981) Foundation problems in limestone areas: A case study in Kuala Lumpur, Malaysia. *Proceedings International Conference on Weak Rock, Tokyo*, pp. 1461–1463. Balkema: Rotterdam.

Tan, B.K. and Komoo, I. (1990) Urban geology case study of Kuala Lumpur, Malaysia. *Engineering Geology*, **28**, 71–94.

Tarquinio, F.S. and Pearlman, S.L. (2001) Pin piles in karst topography. In: Beck, B.F. and Herring, J.G. (eds), *Geotechnical and Environmental Applications of Karst Geology and Hydrology*, pp. 177–181. Balkema: Lisse.

Telford, W.M., Geldart, L.P., Sherriff, R.E. and Keys, D.A. (1990) *Applied Geophysics*, 769pp. Cambridge University Press: Cambridge.

Terashi, M. and Juran, I. (2000) Ground improvement – state of the art. *Proceedings GeoEng2000: Melbourne* (on CD-Rom).

Tharp, T.M. (1995) Design against collapse of karst caverns. In: Beck, B.F. (ed.), *Karst Geohazards*, pp. 397–406. Balkema: Rotterdam.

Tharp, T.M. (1999) Mechanics of upward propagation of cover-collapse sinkholes. *Engineering Geology*, **52**, 23–33.

Tharp, T.M. (2001) Cover-collapse sinkhole formation and piezometric surface drawdown. In: Beck, B.F. and Herring, J.G. (eds), *Geotechnical and Environmental Applications of Karst Geology and Hydrology*, pp. 53–58. Balkema: Lisse.

Tharp, T.M. (2003) Cover-collapse sinkhole formation and soil plasticity. *American Society Civil Engineers Geotechnical Special Publication* 122, 110–123.

Thomas, T.M. (1963) Solution subsidence in southeast Carmarthenshire and southwest Breconshire. *Transactions Institute British Geographers* 33, 45–60.

Thomas, T.M. (1974) The South Wales interstratal karst. *Transactions British Cave Research Association*, **1**, 131–152.

Thompson, A., Hine, P.D., Greig, J.R. and Peach, D.W. (1996) *Assessemnt of subsidence arising from gypsum dissolution: Technical report for the Department of the Environment*, 288pp. Symonds Group: East Grinstead.

Tinjun, J.M., Krantz, B.J., Ditto, W.L. and Aldern, J.L. (2003) Investigation and remediation of karst features at foundry landfill site. *American Society Civil Engineers Geotechnical Special Publication* 122, 349–358.

Toulemont, M. (1987) Les risques d'instabilité liés au karst gypseux lutétien de la région parisienne. *Bulletin liaison Laboratoires Ponts et Chaussées* 150/151, 109–116.

Trudgill, S.T. (1976) The marine erosion of limestones on Aldabra Atoll, Indian Ocean. *Zeitschrift Geomorphologie*, Supplement 26, 164–200.

Upchurch, S.B. and Littlefield, J.R. (1987) Evaluation of data for sinkhole-development risk models. In: Beck, B.F. and Wilson, W.L. (eds), *Karst Hydrogeology: Engineering and Environmental Applications*, pp. 359–364. Balkema: Rotterdam.

Vandevelde, G.T. and Schmitt, N.G. (1988) Geotechnical exploration and site preparation techniques for a large mall in karst terrain. *American Society Civil Engineers Geotechnical Special Publication* 14, 86–96.

Van Rooy, J.L. (1989) A new proposed classification system for dolomitic areas south of Pretoria. *Contributions to Engineering Geology*, **1**, pp. 57–65 (Pretoria).

Van Schalkwyk, A. (1998) Legal aspects of development on dolomite land in South Africa. *Environmental Geology*, **36**, 167–169.

Varnes, D. J. (1984) Landslide hazard zonation: A review of principles and practice. *Natural Hazards 3*, 63pp. UNESCO: Paris.

Vasquez Castillo, L.V. and Rodriguez Molina, C.R. (1999) Geotechnical engineering and geology for a highway through cone karst in Puerto Rica. In: Beck, B.F., Pettit, A.J. and Herring, J.G. (eds), *Hydrogeology and Engineering Geology of Sinkholes and Karst*, pp. 431–445. Balkema: Rotterdam.

Venter, I.S. and Gregory, B.J. (1987) A risk assessment in dolomitic terrain – a case history. *Geological Society Engineering Geology Special Publication* 4, 329–334.

Villard, P., Gourc, J.P. and Giraud, H. (2000) A geosynthetic reinforcement solution to prevent the formation of localized sinkholes. *Canadian Geotechnical Journal*, **37**, 987–999.

Wagener, F.v.M. (1985) Dolomites; problems of soils in South Africa. *The Civil Engineer in South Africa*, **27**, 395–407.

Wagener, F.v.M. and Day, P.W. (1986) Construction on dolomite in South Africa. *Environmental Geology Water Science*, **8**, 83–89.

Walsh, P.T., Boulter, M.C., Itjaba, M. and Urbani, D.M. (1972) The preservation of the Neogene Brassington Formation of the southern Pennines and its bearing on the evolution of upland Britain. *Quarterly Journal Geological Society*, **128**, 519–559.

Walters, R.F. (1977) Land subsidence in central Kansas related to salt dissolution. *Bulletin Kansas Geological Survey* 214, 82pp.

Waltham, A.C. (1989) *Ground Subsidence*, 202pp. Blackie: Glasgow.

Waltham, A.C. (1996) Limestone karst morphology in the Himalayas of Nepal and Tibet. *Zeitschrift Geomorphologie*, **40**, 1–22.

Waltham, T. (1997) Mulu – the ultimate in cavernous karst. *Geology Today*, **13**, 216–222.

Waltham, T. (2002a) Gypsum karst near Sivas, Turkey. *Cave Karst Science*, **29**, 39–44.

Waltham, T. (2002b) The engineering classification of karst with respect to the role and influence of caves. *International Journal Speleology*, **31**, 19–35.

Waltham, T. and Cooper, A. (1998) Features of gypsum caves and karst at Pinega (Russia) and Ripon (England). *Cave Karst Science*, **25**, 131–140.

Waltham, A.C. and Fookes, P.G. (2003) Engineering classification of karst ground conditions. *Quarterly Journal Engineering Geology Hydrogeology*, **36**, 101–118.

Waltham, T. and Hatherley, P. (1983) The caves of Leck Fell. *Cave Science*, **10**, 245–251.

Waltham, T. and Middleton, J. (2000) The Khammouan karst of Laos. *Cave Karst Science*, **27**, 113–120.

Waltham, A.C. and Park, H.D. (2002) Roads over lava tubes in Cheju Island, South Korea. *Engineering Geology*, **66**, 53–64.

Waltham, A.C. and Smart, P.L. (1988) Civil engineering difficulties in the karst of China. *Quarterly Journal Engineering Geology*, **21**, 2–6.

Waltham, A.C. and Swift, G.M (2004) Bearing capacity of rock over mined cavities in Nottingham. *Engineering Geology*, **75**, 15–31.

Waltham, A.C., Smart, P.L., Friederich, H., Eavis, A.J. and Atkinson, T.C. (1983) The caves of Gunung Sewu, Java. *Cave Science*, **10**, 55–96.

Waltham, A.C., Vandenven, G. and Ek, C.M. (1986) Site investigations on cavernous limestone for the Remouchamps viaduct, Belgium. *Ground Engineering*, **19**(8), 16–18.

Waltham, T., Brook, D. and Bottrell, S. (1993) The caves and karst of Xingwen, China. *Cave Science*, **20**, 75–86.

Waltham, A.C., Simms, M.J., Farrant, A.S. and Goudie, H.S. (1997) *Karst and Caves of Great Britain* (Geology Conservation Review Series, 12), 358pp. Chapman and Hall: London.

Warren, C.D. and Mortimore, R.N. (2003) Chalk engineering geology – Channel Tunnel Rail Link and North Downs Tunnel. *Quarterly Journal Engineering Geology and Hydrogeology*, **36**, 17–34.

Wassmann, T.H. (1979) Mining subsidence in the east Netherlands. In: Saxena, S.K. (ed.), *Evaluation and Prediction of Subsidence*, pp. 283–302. American Society Civil Engineers: New York.

Welsh, J.P. (1988) Sinkhole rectification by compaction grouting. *American Society Civil Engineers Geotechnical Special Publication* 14, 115–132.

Wenrich, K.J. and Sutphin, H.B. (1994) Grand Canyon caves, breccia pipes and mineral deposits. *Geology Today*, **10**, 97–104.

West, G. and Dumbleton, A.J., (1972) Some observations on swallow holes and mines in the Chalk. *Quarterly Journal of Engineering Geology*, **5**, 171–177.

Weyl, P.K. (1959) The solution kinetics of calcite. *Journal Geology*, **66**, 163–176.

Whittaker, B.N. and Reddish, D.J. (1989) *Subsidence; Occurrence, Prediction and Control*, 528pp. Elsevier: Amsterdam.

White, E.L., Aron, G. and White, W.B. (1986) The influence of urbanization on sinkhole development in central Pennsylvania. *Environmental Geology Water Science*, **8**, 91–97.

White, E.L. and White, W.B. (1969) Processes of cave breakdown. *Bulletin National Speleological Society*, **31**, 83–96.

White, W.B. (1988) *Geomorphology and Hydrology of Karst Terrains*, 464pp. Oxford University Press: New York.

White, W.B. and White, E.L. (2003) Gypsum wedging and cavern breakdown: studies in the Mammoth Cave System, Kentucky. *Journal Cave Karst Studies*, **65**, 43–52.

Williams, D.M. and McNamara, K. (1992) Limestone to dolomite to dedolomite conversion and its effect on rock strength: a case study. *Quarterly Journal Engineering Geology*, **24**, 131–135.

Williams, P.W. (1983) The role of the subcutaneous zone in karst hydrology. *Journal Hydrology*, **61**, 45–67.

Williams, P.W. (1985) Subcutaneous hydrology and the development of doline and cockpit karst. *Zeitschrift Geomorphologie*, **29**, 463–482.

Williams, P.W. (2004) Dolines. In: Gunn, J. (ed.), *Encyclopedia of Caves and Karst Science*, pp. 304–310. Fitzroy Dearborn: New York.

Wilson. W.L. (1995) Sinkhole and buried sinkhole densities and new sinkhole frequencies in karsts of northwest peninsular Florida. In: Beck, B.F. (ed.), *Karst Geohazards*, pp. 79–91. Balkema: Rotterdam.

Wilson, W.L. (2000) *Reported sinkhole density map of Florida*. Subsurface Evaluations Inc.: Tampa, Florida.

Wilson, W.L. and Beck, B.F. (1988) Evaluating sinkhole hazards in mantled karst terrane. *American Society Civil Engineers Geotechnical Special Publication* 14, 1–24.

Wilson, W.L., Mylroie, J.E. and Carew, J.L. (1995) Caves as a geologic hazard: a quantitative analysis from San Salvador Island, Bahamas. In: Beck, B.F. (ed.), *Karst Geohazards*, pp. 487–495. Balkema: Rotterdam.

Wood, C. (1981) Exploration and geology of some lava tube caves on the Hawaiian volcanoes. *Transactions British Cave Research Association*, **8**, 111–129.

Wood, C., Watts, R., Waters, E. and Cheetham, P. (2002) *Laki Underground 2001*, 32pp. Bournemouth University: Bournemouth.

Wyllie, D.C. (1999) *Foundations on Rock*, 401pp. Spon: New York.

Xiang S., Chen J., Wilson, W.L. and Beck, B.F. (1988) Sinkholes as a consequence of groundwater development in karst regions. *American Institute of Hydrology Journal*, **4**, 160–173.

Xu W. and Zhao G. (1988) Mechanism and prevention of karst collapse near mine areas in China. *Environmental Geology Water Science*, **12**, 37–42.

Yechieli, Y., Abelson, M., Wachs, D., Shtivelman, V., Crouvi, O. and Baer, G. (2003) Formation of sinkholes along the shore of the Dead Sea – preliminary investigation. *American Society Civil Engineers Geotechnical Special Publication* 122, 184–192.

Yuan D. (1987) Environmental and engineering problems of karst geology in China. In: Beck, B.F. and Wilson, W.L. (eds), *Karst Hydrogeology: Engineering and Environmental Applications*, pp. 1–11. Balkema: Rotterdam.

Zenger, D.H., Dunham, J.B. and Etherington, R.L. (1980) Concepts and models of dolomitisation. *Society Economic Palaeontology Mineralogy Special Publication* 28, 320pp.

Zhou W., Beck, B.F. and Adams, A.L. (2003) Sinkhole risk assessment along highway 1-70 near Frederick, Maryland. *American Society Civil Engineers Geotechnical Special Publication* 122, 591–601.

Zhu X. (2001) China's karst tiankeng and its value for science. *Science and Technology Review Beijing*, **160**(10), 60–63.

Zhu X. and Zhang Y. (1995) The world's largest dolines and Great Crack Gorge in southern Sichuan. *Carsologica Sinica*, **14**, Supplement, 1–11.

Zhu X., Huang B., Zhu D. and Chen W. (2003) *Dashiwei Tiankeng Group, Leye, Guangxi*, 184pp. CIP: Beijing.

Zisman, E.D. (2001) A standard method for sinkhole detection in the Tampa, Florida, area. *Environmental and Engineering Geoscience*, **7**, 31–50.

Zisman, E.D. (2003) Guilty until proven innocent – sinkhole definition and identifying features. *American Society Civil Engineers Geotechnical Special Publication* 122, 124–130.

A major data resource is the series of volumes containing the papers from the multisdisciplinary conferences on sinkholes and their various aspects of geology, geotechnics, hydrogeology and engineering, that have been held in America since 1984. The volume references are as follows:

First, 1984: Beck, B.F. (ed.), *Sinkholes: their Geology, Engineering and Environmental Impact*, 429pp. Balkema: Rotterdam.

Second, 1987: Beck, B.F. and Wilson, W.L. (eds), *Karst Hydrogeology: Engineering and Environmental Applications*, 467pp. Balkema: Rotterdam.

Third, 1989: Beck, B.F. (ed.), *Engineering and Environmental Impacts of Sinkholes and Karst*, 384pp. Balkema: Rotterdam.

Fourth, 1993: Beck, B.F. (ed.), *Applied Karst Geology*, 295pp. Balkema: Rotterdam.

Fifth, 1995: Beck, B.F. (ed.), *Karst Geohazards*, 582pp. Balkema: Rotterdam.

Sixth, 1997: Beck, B.F. and Stephenson, J.B. (eds), *Engineering Geology and Hydrogeology of Karst Terrains*, 528pp. Balkema: Rotterdam.

Seventh, 1999: Beck, B.F., Pettit, A.J. and Herring, J.G. (eds), *Hydrogeology and Engineering Geology of Sinkholes and Karst*, 478pp. Balkema: Rotterdam.

Eighth, 2001: Beck, B.F. and Herring, J.G. (eds), *Geotechnical and Environmental Applications of Karst Geology and Hydrology*, 437pp. Balkema: Lisse.

Ninth, 2003: Beck, B.F. (ed.), Sinkholes and the engineering and environmental impacts of karst. *American Society of Civil Engineers Geotechnical Special Publication* 122, 737pp.

Location index

Subject index

Printing: Mercedes-Druck, Berlin
Binding: Stein+Lehmann, Berlin

The manufacturer's authorised representative in the EU is Springer Nature Customer Service Centre GmbH, Europaplatz 3, 69115 Heidelberg, Germany. If you have any concerns regarding our products, please contact ProductSafety@springernature.com

Printed and bound by CPI Group (UK) Ltd, Croydon, CR0 4YY
05/10/2026
02228131-0014